Computational Methods in Applied Sciences

Volume 52

This series publishes monographs and carefully edited books inspired by the thematic conferences of ECCOMAS, the European Committee on Computational Methods in Applied Sciences. As a consequence, these volumes cover the fields of Mathematical and Computational Methods and Modelling and their applications to major areas such as Fluid Dynamics, Structural Mechanics, Semiconductor Modelling, Electromagnetics and CAD/CAM. Multidisciplinary applications of these fields to critical societal and technological problems encountered in sectors like Aerospace, Car and Ship Industry, Electronics, Energy, Finance, Chemistry, Medicine, Biosciences, Environmental sciences are of particular interest. The intent is to exchange information and to promote the transfer between the research community and industry consistent with the development and applications of computational methods in science and technology.

Book proposals are welcome at
Eugenio Oñate
International Center for Numerical Methods in Engineering (CIMNE)
Technical University of Catalunya (UPC)
Edificio C-1, Campus Norte UPC Gran Capitán
s/n08034 Barcelona, Spain
onate@cimne.upc.eduwww.cimne.com
or contact the publisher, Dr. Mayra Castro, mayra.castro@springer.com

Indexed in SCOPUS, Google Scholar and SpringerLink.

More information about this series at http://www.springer.com/series/6899

Ning Qin · Jacques Periaux · Gabriel Bugeda
Editors

Advances in Effective Flow Separation Control for Aircraft Drag Reduction

Modeling, Simulations and Experimentations

Springer

Editors
Ning Qin
Department of Mechanical Engineering
University of Sheffield
Sheffield, UK

Jacques Periaux
Universitat Politècnica de Catalunya
International Center for Numerical Methods
in Engineering
Barcelona, Spain

Gabriel Bugeda
Universitat Politècnica de Catalunya
International Center for Numerical Methods
in Engineering
Barcelona, Spain

ISSN 1871-3033
Computational Methods in Applied Sciences
ISBN 978-3-030-29687-2   ISBN 978-3-030-29688-9 (eBook)
https://doi.org/10.1007/978-3-030-29688-9

This Springer imprint is published by the registered company Springer Nature Switzerland AG
The registered company address is: Gewerbestrasse 11, 6330 Cham, Switzerland

Preface

This book includes a collection of chosen research papers from the partners of a research project (the Seventh Framework Programme Grant Agreement (Grant Number: 266326)) named Manipulation of Reynolds Stress for Separation Control and Drag Reduction (MARS) conducted between 2010 and 2014. The project is jointly funded by the European Commission (EC) and the Ministry of Industrial and Information Technology (MIIT) of China. As you will notice, many of the papers reflect the bilateral EU-China collaborations as a result of the project. The MARS consortium was led by International Centre for Numerical Methods in Engineering (CIMNE) in Spain and the University of Sheffield in UK on the European side, Chinese Aeronautical Establishment (CAE) and AVIC Aeronautical Research Institute (ARI) on the Chinese side.

The consortium (details can be found at https://www.cimne.com/mars) was composed of 12 European partners and 10 Chinese partners, who have contributed to the volume.

Based on the MARS project, the contributed papers included in the volume are devoted to the analysis and understanding of the phenomena involved in the birth of the aerodynamic drag, focusing on the turbulence phenomena, in particular, the effects of Reynolds stress on flow separation and therefore the drag force. A review is included in Chap. 1 on the flow physics, experimental and computational tools in relation to active flow control, introducing the following contributions in the context of the state of the art. In the following contents, various active flow control devices have been investigated focusing on controlling flow separation for drag reduction. A couple of representative common baseline test cases were specified beforehand, the details of which can be found in Chap. 2. These include the backward-facing step (BFS) and the NACA 0015 airfoil. The BFS case presents a case with fixed separation point from the step corner, while the airfoil case presents a case with separation from a smooth curved surface. The former test case has a very sensitive reattachment, while the latter has a sensitive separation location depending on the flow control mechanism. As can be observed from the papers in this volume, the partners have investigated a comprehensive range of flow control actuators, including plasma, synthetic jets, pulsed jets, micro-blowing and spanwise vortex generators. Both experimental and simulation tools have been employed for the study of various active flow control devices. On the experimental side, advanced techniques such as time-resolved tomographic measurement of the Reynolds stress and full surface skin friction technique have been used to extract the detailed flow field and surface data. On the simulation side, the main tool used by the authors is the hybrid RANS-LES simulation to resolve the detailed turbulent flow structure around and downstream of the flow control actuators. The effect of flow control on the coherent turbulence structures and therefore Reynolds stress have been revealed to gain further insight into the control mechanism. Some parametric studies can be found in the volume in relation to the key flow control design variables. Considering the huge computational demand for the hybrid RANS-LES simulations for the flow control study, direct coupling of experiment set-up with a genetic algorithm-based optimization has been carried out to optimize the plasma control design parameters. This is the first time that such approach has been used for flow control studies. In addition, some novel flow control devices are presented in the volume, including the spanwise vortex generator, providing an alternative mechanism for separation control. The authors believe that this volume should be a useful reference for those who are involved in flow control research or interested in applying flow control to improve the performance of engineering product. The potential scientific and technological impacts of the content of the book are:

- a basic understanding of the effects of various flow control devices;
- ways of manipulating the turbulent Reynolds stress for effective flow control;
- identification of practical issues in aeronautical applications;
- experimental and numerical flow control data for future validation.

Sheffield, UK
October 2019

Ning Qin
Jacques Periaux
Gabriel Bugeda

Acknowledgements

We would like to take this opportunity to thank the EC and MIIT for funding the research investigated in this volume and all the contributing authors for their effort in writing up the research despite its delayed publication.

Many thanks are due to Dr. Dietrich Knoerzer, EC Scientific Officer, and Mr. Benjian Li, MIIT Director of International Collaboration, for the support and continuous technical interest provided throughout the MARS project and their interest in the production of this volume. The other two coordinators, Ms Jian Sun (CAE) and Dr Jun Dong (AVIC-ARI), are also appreciated for their contribution to the challenge of the management of the MARS project, one of the first two such collaborative projects along with the University of Sheffield and CIMNE coordinators, which opened the door to excellent contributions from partners in China and Europe in the field of active flow control for drag reduction.

Finally, the editors sincerely acknowledge Nathalie Jacobs, Ram Prasad Chandrasekar and Muruga Prashanth Rajendran of Springer Nature for their patience in proofreading and accepting with patience the delays of the manuscripts and corrections. We also thank Prof. Eugenio Oñate (CIMNE) for his interest shown to this book to be published in the SPRINGER-ECCOMAS Series « Computational Methods in Applied Sciences.

Contents

Chapter 1
Active Flow Control Strategies and Tools for Turbulent Flows

Jean-Paul Bonnet and Ning Qin

Abstract In this chapter we present considerations on the impact of the turbulence characteristics of most flows of industrial interest on the flow control interpretations or strategies. In the second part we will give an overview of the CFD approaches, including actuator integration methods. The actuators characteristics will then be presented with the associated detection methods. Considerations on the spatial and time characteristics of flow control will be provided.

Keywords Flow control methods · Turbulence management · CFD methods for flow control · Actuators · Sensors

Increased attention has been devoted to the development of techniques capable of enhancing our ability to control unsteady flows in a wide variety of configurations and applications. The revolutionary work of Schubauer and Skramstad [1], Poisson-Quinton and Lepage [2] on steady state tools and mechanisms for flow management set the foundation for flow control and aeronautical applications. A comprehensive review on flow control technology was given by Gad-el-Hak and Bushnell [3], Gad-el-Hak et al. [4] and King [5]. While passive devices, such as micro vortex generators, have been incorporated into some production vehicles, these devices are primarily limited to a single function or are effective over small operational ranges. Although decades of research have been conducted using pseudo-active controls (e.g., laminar flow control by suction), questions of system complexity and/or cost versus benefit have prevented such controls from reaching marketable applications.

Active control can be divided into predetermined (open-loop) or reactive (open-/closed-loop) (Gad-el-Hak 2003). In a closed loop feedback flow-control process, the current flow state is measured and then utilized along with models of the state to devise a control input to alter the system state in the desired direction. An essential ingredient for success of a closed-loop control system is the availability of an appropriate

J.-P. Bonnet (✉)
Institut PPRIME, CNRS-Université de Poitiers ISAE/ENSMA TSA51124, 86073 Poitiers, France
e-mail: jean.paul.bonnet@univ-poitiers.fr

N. Qin
Department of Mechanical Engineering, University of Sheffield, Sheffield S1 3JD, UK

N. Qin et al. (eds.), *Advances in Effective Flow Separation Control for Aircraft Drag Reduction*, Computational Methods in Applied Sciences 52,
https://doi.org/10.1007/978-3-030-29688-9_1

model of the state. Another important issue in formulating closed-loop control is in accounting for the effect of transient growth of stable modes [6]. Therefore, actuators and sensors in the closed-loop flow-control system should be designed and utilized in an effective way. In general, there are several approaches to controller designs, i.e. physical model-based control, classic control, modern control, and so on.

For turbulent wall bounded flows, physical model-based control strategy is a conceptually simple feedback control strategy that introduces control in the form of distributed suction and blowing at the wall surface in an attempt to oppose the motion of near-wall turbulent structures [4]. Choi et al. [7] used the physics-based control strategy by employing DNS at a low turbulence Reynolds number. These methods are somewhat limited due to the excessive number of sensors and actuators for flows of engineering importance. Imposing a pre-determined perturbation such as moving or oscillating walls or spanwise periodic perturbations [8–10] are alternative routes for skin friction drag reduction.

For free turbulent shear flows, classical control refers to techniques that are in the frequency domain (as opposed to state-space representations, which are in the time-domain), and often are valid only for linear, single-input, single-output systems (Joslin 2009). Rapoport (2003) applied classical control for vectoring of a turbulent jet. Modern control theory provides a systematic means of constructing closed-loop feedback control laws for multiple-input/multiple-output systems that explicitly account for both model uncertainty and external disturbances [11]. Lee et al. [12] applied modern control, based on reduced-order model (ROM) of the linearized Navier–Stokes equations for Poiseuille flow, to turbulent channel flow and drag reductions in the range of 10–17% were obtained. Other ROM approaches have been developed for turbulent flows, see for example [13]. Most of the strategies are then based on stability–linear or quasi-linear methods. For these methods, the turbulent characteristics of the flow is considered as of secondary importance. However, we will consider later in this text, the possibility to actuate at different scales (time or space) in the objective of exciting extra modes within the turbulent spectrum and then take benefit of the scale interactions such as those occurring within the turbulent cascade.

Recent developments in the fields of micro-machining, piezo or plasmas for designing actuators, high precision flow observation and simulation techniques have motivated some investigation to establish closed-loop flow-control systems. But to date, such kind of work, e.g. on the closed-loop flow-control as applied to active vortex generators (e.g. Benard et al. [14]), is still limited.

For every development of flow control methods, the number of degrees of freedom is large (amplitude and phase of the actuation, location of the actuation, nature of the information collected for closed loop purpose, data processing, ...). Optimization by simple adjustments is limited. Several methods are currently used, such as extremum seeking [15, 16], Genetic Algorithms or related methods. These methods are by nature associated to pre-determined laws of control. An alternative approach that does not require such a-priori choice is the Machine Learning approach [17, 18]. This method finds the control law which minimizes a given cost function. This optimization is performed in a model-free manner directly in the plant. The turbulence characteristics of the flow can then be directly taken into account.

In the next sections we will first present considerations on the impact of the turbulence characteristics of most flows of industrial interest on the flow control interpretations or strategies. In a second part we will give an overview of the CFD approaches. The actuators characteristics will then be presented with the detection methods.

1.1 Control Strategies for Turbulent Regimes

In this section, we recall some turbulent properties, their possible influence on flow control, and illustrate their impact for two flow phenomena, boundary layer and free shear flow control. In this last case we will present two distinct strategies: one is based on Large Scale—Low Frequency Forcing (LFF, addressing more unstable modes), the other is based on Small Scales—High Frequency Forcing (HFF, addressing inertial or dissipative range in the spectral domain). In complement to stability analysis focusing on mean flow properties, the relationships between production domain and dissipative one have to be analyzed from a dual description in frequency and space domains. The possible impacts on flow control strategies will be discussed.

1.1.1 Statistical Description of Turbulent Flows in View of Flow Control

Turbulent flows can be analyzed from their statistical properties. This approach is useful for understanding the energetic equilibria, and for CFD purposes, as initial conditions and for validation. Two main properties of turbulent flows should be recalled for flow control purposes. First, turbulent flows have intrinsically *redistributive properties*. Second, the velocity fluctuations have *3D characteristics* that can be analyzed by considering the Reynolds stress tensor.

1.1.1.1 Redistributive Properties

The equations for the components of the Reynolds Stresses (RS) is usually written as the balance between the different terms. In incompressible form:

$$D\overline{u'_i u'_j}/Dt = P_{ij} + T_{ij} + \Pi_{ij} - \varepsilon_{ij} \tag{1.1}$$

where $P_{ij} = -\left(\overline{u'_i u'_k} U_{j,k} + \overline{u'_j u'_k} U_{i,k}\right)$ is the production rate, $T_{ij} = -\overline{\left(u'_i u'_j u'_k\right)}_{,k}$ the turbulent diffusion rate, $\Pi_{ij} = -\frac{1}{\rho}\overline{p'(\partial u'_i/\partial x_j + \partial u'_j/\partial x_i)}$ the diffusion through velocity pressure-gradient term and $\varepsilon_{ij} = 2\upsilon\overline{(\partial u'_i/\partial x_k \cdot \partial u'_j/\partial x_k}$ the dissipation rate.

Often T_{ij} and Π_{ij} are associated as they represent the diffusion effects linked to turbulence.

In turbulent shear flows, the role of each term can be analyzed by looking at their sign in the plot of the energy balance. Figure 1.1 shows, from the pioneering compilation of Rodi [19], that, depending on the transverse location, the production of shear stress can have several sources such as the production itself plus the turbulent diffusion depending on the transverse location. The sink of energy comes from the dissipation itself but also, still depending on the location in space, from diffusion and convection.

To quote with examples, let us consider 2D (in mean) configurations. The shear stress $\overline{u'v'}$ plays a crucial role in most flows. Indeed, the production of this turbulent shear stress is issued from the combination of the other components of the Reynolds Stresses that appears in the production and, in a nonlinear manner, in the diffusion terms T_{ij} and Π_{ij}. This corresponds to the redistribution of the energy through interactions between turbulent field and mean shear stress, third order correlations and turbulent pressure-strain correlations.

This characteristics should be taken into account to define control strategies. An illustration of the impact of the redistributive effect can be associated to separation control by Vortex Generators (VG). This active or passive method is quite popular for separation on wings (see Siaw and Bonnet, in this volume as an example). The physics of this control method can be interpreted as a global view as the generation of Large

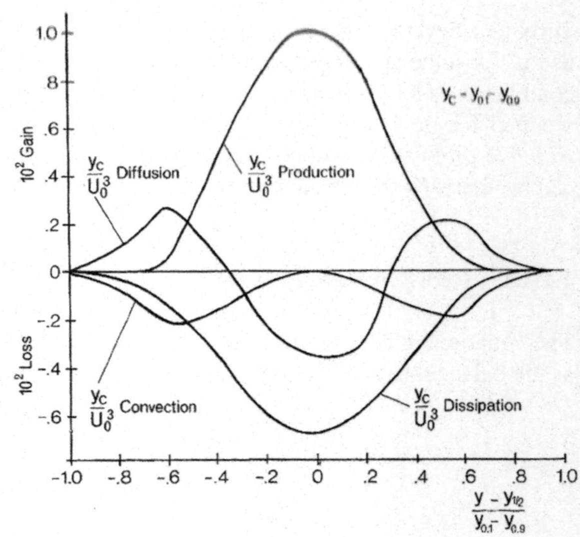

Fig. 1.1 Example of shear stress balance in a self-similar mixing layer after Rodi [19]

Scale *longitudinal* vortices bringing high velocity flow towards the wall and then energizing the boundary layer close to the wall. With this macroscopic interpretation, the turbulent characteristics of the flow can be considered as of secondary importance, even ignored.

However, turbulent process can also be evoked if the *redistributive character* of the turbulent field is considered: the VG effects promotes strong modifications of the Reynolds Stress tensor as evidenced for example by Törblom and Johansson (2007) and Foi (2008). The longitudinal vortices created by the VG generate streamwise vorticity then directly v' and w' components in the Y–Z plane. These fluctuations complement the natural turbulence in creating longitudinal and shear stresses after the redistribution mechanisms described before. The turbulent mechanisms transforms the longitudinal vorticity generated by the VG into transverse and shear stresses. In particular $\overline{u'v'}$ is created, acting on the separation process. Indeed the momentum equation in a 2D approximation can be written as $\frac{\partial U}{\partial x} = -\frac{1}{\rho}\frac{\partial P}{\partial x} + \frac{\partial}{\partial y}\left(-\overline{u'v'}\right)$. From this relation, the streamwise velocity gradient depends not only on the pressure gradient but also of the turbulent shear stress issued from the VG vortex, as outlined by Wallin (2008).

1.1.1.2 3D Characteristics of Turbulence: The Anisotropy Invariant Map and Boundary Layer Examples

The structure of the turbulence, in a statistical description, can also be addressed via the anisotropy of the turbulent field naturally occurring in most of turbulent shear layers. Lumley and Newman [20] introduced the description of the anisotropy between the three velocity components through the invariant parameters IIa and IIIa.

These two invariants are
$IIa = a_{ij} \cdot a_{ji}$ and
$IIIa = a_{ij} \cdot a_{jk} \cdot a_{ki}$,
where $a_{ij} = \langle u'_i u'_j \rangle / q^2 - 1/3\delta_{ij}$

The two invariants can be plotted altogether and stay inside a triangle known as the "Lumley triangle" (see the example in Fig. 1.2). The upper branch corresponds to 2D turbulence: $IIa = 2/9 + 2IIIa$. The two others branches correspond to axisymmetric turbulence: $IIa = 3/2(4/3 + |IIIa|)^{2/3}$. The three summits correspond respectively to a one-component turbulence (upper one), isotropic 2 components turbulence for the left hand one and isotropic turbulence for the lower summit (in which $IIa = IIIa = 0$).

As far as turbulent boundary layers are concerned, this representation is very useful to interpret flow control effects. Fronhpafel et al. [21] presented the evolution of the anisotropy when the turbulent boundary layers is controlled with rigid fibers or sufrantants, showing that close to the wall the flows under control are quite closer to 1D (u′) turbulence than a regular flow. More recently, at ONERA, Bannier (2016) preforms a LES over a ribletted surface. Figure 1.2 shows the impact of the control on the anisotropy, with severe effects particularly, as expected, within the valley

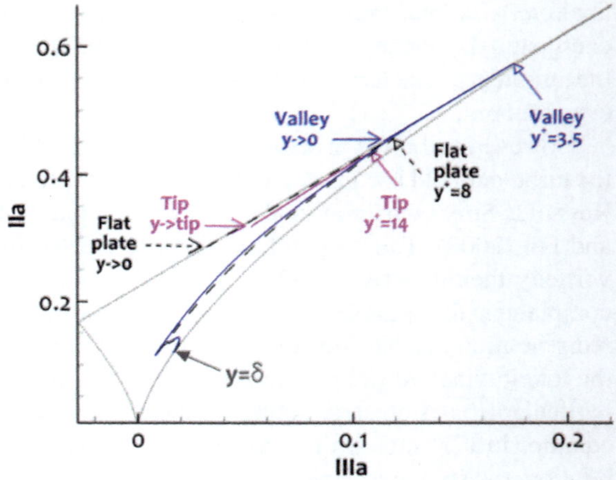

Fig. 1.2 Influence of riblets on the anisotropy of the Reynolds stress after Bannier (2016) with permission

between two riblets. Clearly the description of the anisotropy is an unique tool for the interpretation of boundary layers control mechanisms.

1.1.1.3 Multiplicity of Scales (3D) in the Spectral Domain

From a statistical point of view, the turbulent energy can be described in terms of frequency (time) or wavenumber (space) domains.

Figure 1.3 shows a schematic description of the wave number evolution of the energy of fluctuations in a turbulent flow. The left hand side part corresponds to the production as described in the previous section. The right hand side corresponds to

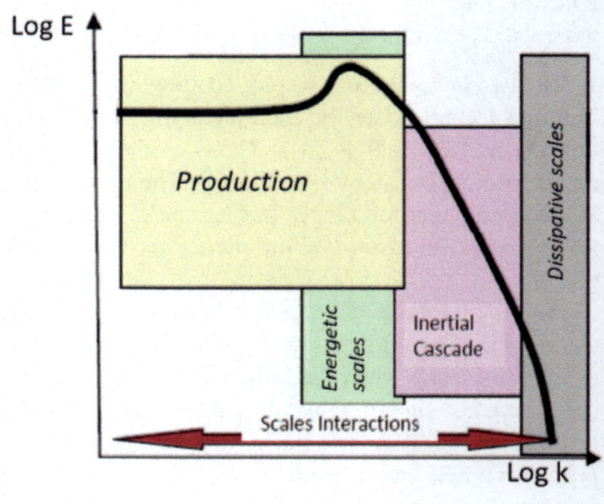

Fig. 1.3 Schematic of the spectral distribution of turbulent energy

dissipative structures. In between, the Kolmogorov-Richardson cascade takes place with energy transfer between decreasing size scales (for considerations on the dissipation scales see Vassilicos [22]). For Homogeneous Isotropic Turbulence (HIT), the dynamics of the scales in the inertial domain are well understood theoretically, numerically and experimentally. As an example, the time T_d required for dissipating an eddy of size L is linked to its own integral (turnover) time T_t via the relation: $T_d \sim T_t(1 - \beta^{2/3})$, where β is the ratio between the two scales [23, 24]. This means that, for spatially evolving flows such as convective ones, an eddy in the inertial domain travels a distance of order of its size before it will be dissipated. This rate can be changed according to the slope of the inertial cascade for out-of-equilibrium flows such as fractal grid arrangement [25, 26].

However for most of free shear flows of engineering importance, a peak of energy at low wavenumbers is observed as schematically plotted on Fig. 1.2. This peak is related to the coherent structures that correspond to the more unstable modes of the flow as we discuss later. In this case, the interpretation of the dynamic of these structures within the inertial domain can be more complex providing they result from a continuous energization from the shear.

1.1.2 Organized Motions in Turbulent Flow Control Strategies

The existence and role of *Organized Motions,* most often *Coherent Structures* (hereafter named CS) or *Large Scales Structures* in turbulent flows is no longer a matter of debate. Since these structures are responsible for noise, vibration, heat transfer, mixing etc., they are natural targets for flow control. The definition of these structures is not unique.

For many purposes, enhancement of CS can be requested via *excitation of unstable modes*. This is the case of separation control, mixing increase …

Conversely, mixing may be required to be reduced, for skin friction reduction, noise abatement, low mixing purposes. *Stabilization of unstable modes* is required in this case and seems to be more difficult to achieve.

In both cases, the dynamics of the CS motions are considered as linear or quasi-linear. The stability theories apply (e.g. the pioneering work of Michalke [27], Ho and Huerre [28]). When linear or quasi-linear methods are invoked, turbulence is not always seen as of primary importance in flow control strategies.

The spectra in Fig. 1.3 shows a relatively narrow peak just before the inertial range that corresponds to the CS signature. From an energetic side, the turbulent field can be separated in three contributions with the triple decomposition proposed by Reynolds and Hussain [29] for an instantaneous velocity, U:

$$U(t) = \bar{U} + \tilde{u}(t) + u'(t) \tag{1.2}$$

The first term on the RHS is the time averaged mean velocity. If we attempt to control this term via flow control then most devices offer little gain in efficiency on a global energy basis i.e. change in energy out equals energy in. Considerations on the energy balance of flow control can be found in Frohnapfel [30] or Seifert [31].

The second term on the RHS is the periodic/dynamic component of the flow and for some specific flow scenarios this can be shown to be dominant in determining the flow state and characteristics. The stresses produced from this term are referred as the periodic stresses or "apparent stresses". It offers some interesting opportunities for demonstrating the way in which to deploy flow control technologies for dynamic environments (responsive environments, smart inputs and sensible control). This also implies that, for statistically steady flows, where the second term disappears, artificial introduction of the periodic term may be necessary for effective control.

The final term on the RHS represents the broadband 'random' turbulent fluctuations, from which the Reynolds stresses are defined. While direct control of the "random" components is the ultimate goal, most of control studies aim to investigate the control of the periodic stresses, the dynamic components of the flow, in order to manipulate the Reynolds stress for the benefit of flow control.

1.1.2.1 Large Scales/Small Scales Actuation

A question then arises, for flow control purposes about the scales that have to be targeted by the actuators. This determines indeed the spatial size and the frequency of the devices.

In most of the demonstrations of flow control, CS or natural instabilities are addressed. The frequencies that are actuated for free shear flows are then of order of the Strouhal numbers based on stability analysis or on the observation of the more unstable modes.

However, some experiments have been devoted to actuation of higher frequencies (higher Strouhal numbers) or higher wavenumbers. This was the case for free shear flows in the experiments by Wistle and Glezer [32]. Luchtenburg et al. [33] address the HFF approach for stabilization of bluff body wakes. Glezer et al. [34] and more recently Barros et al. [35] showed the benefits of HFF on an Ahmed Body wake and Oxalde et al. [36] on an axisymmetric wake. Benton and Visbal [37] addressed the separation on a NACA 0012 airfoil through HFF. In their numerical study, Dandois et al. [38] showed that the major influence of the HFF on a separated ramp was due to a modification of the stability of the inflexion point in the velocity profile. These results corroborate the analysis of Stanek et al. [39]. However, the physical effect of the generation of small, high frequencies scales corresponding to the HFF within the turbulent cascade remains an open question. The stability analysis explaining the effect on the mean flow should be completed with interpretations in terms of energy exchanges within the turbulent cascade.

The spectra obtained by Stanek et al. [39] revisited the HFF of a mixing layer performed Whistle and Glezer (1998). They show drop in low frequency (high energy) domain while the smaller scales gain energy. The dissipation then increases, with typical peaks occurring in the "dissipative scales" shown in Fig. 1.3. This particular efficiency of the HFF led Stanek in 2005 to ask the question on two scenarios: either Production determines Dissipation or Dissipation determines Production.

It should be noticed that the question of the universality of the C_ε coefficient in the relation $\varepsilon = C_\varepsilon U^3/L$ is directly linked with one or the other scenario.

When written in the form of the energy balance as Eq. (1.1), the equilibrium of the turbulent transport equations should be achieved. If the dissipation is increased, the balance should be obtained by a reduction of the production or/and through the other terms such as turbulent diffusion or convection. Luchtenburg et al. [33], states that the small scale fluctuations act as "energy sinks in the evolution equations that an eddy viscosity takes into account in URANS and LES". Stanek states that the drop in turbulent energy is linked to the decrease of production due to the turbulent shear layer stabilization, this stabilization resulting for the flow control on mean shear. Then the decrease of large scales energy seems not directly related to the effects of energizing the small scales, the small scales being decoupled form the stability analysis. This is coherent with the experiments of Vukasinovic et al. [40] in which the effects depends on the streamwise location.

In the recent experiment from Parezanovic et al. [41], a plane mixing layer is controlled by 2D actuation right at the trailing edge through 100 unsteady microjets. The two velocities are 4.7 and 1.3 m/s respectively. Trailing edge momentum thickness is $\theta_0 = 0.7 \times 10^{-3}$ mm. The corresponding Strouhal number after Michalke theory being is 0.013, this corresponds to 87 Hz. Two open loop control have been performed. A first one in which the more unstable frequency is excited (LFF) and another one at higher in frequency (HFF). It should be noticed that, for both cases, the same flow rate was used. The mean velocity profiles are not available, so that it is difficult to apply the stability criteria developed by Dandois et al. [38]. In the case of HFF it can be conjectured that the mean flow is not sensitive to high frequencies and then that the stability can be quite different when LFF and HFF forcing are compared. However, in the case of HFF, there is emission of small scales right at the trailing edge of the splitter plate where the actuators are operating. A discussion on this point is given in Mons et al. [42].

Figure 1.4 shows that open loop control corresponds to a mixing layer increase at a Strouhal number lower than the more unstable one linked to TE boundary layer, corresponding to farther downstream more unstable mode as observed by several authors [41]. However, the more intriguing result is obtained at higher frequency forcing. For HFF at 40 times higher frequencies, the organization of the mixing layer is strongly modified. The large scale organization disappears, small scale turbulence is introduced immediately downstream of the trailing edge although the spreading rate is not so modified. The authors state that "High actuation frequency is not able to synchronize the mixing layer as low actuation frequencies do; instead the momentum injected in this way into the mixing layer diffuses as a part of the turbulence cascade".

It can be conjectured that higher frequency activation can be filtered by the flow resulting in a less receptivity. However, this interpretation is not sufficient to explain the dramatic decrease in large scale growth. A second interpretation is that the coherent structures can synchronize with the actuation in order to take more energy out the

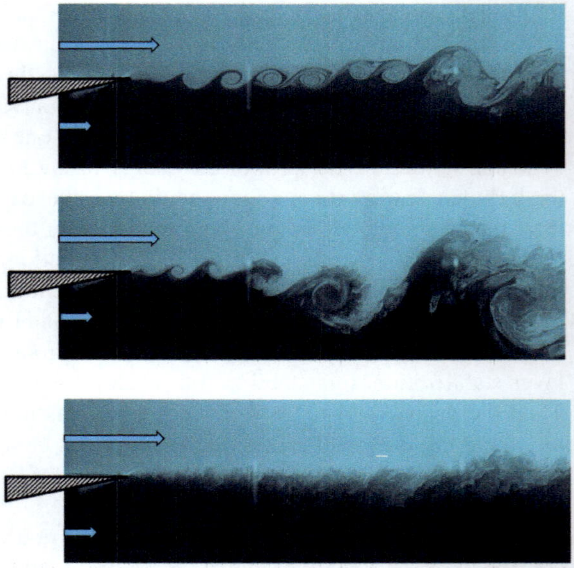

Fig. 1.4 Smoke/laser visualization of the mixing layer by seeding the top (high speed) stream. (Top) un-actuated flow, (Middle) low frequency forcing (10 Hz) and (Bottom) high frequency forcing (400 Hz). Adapted from Parezanovic et al. [41] with permission

mean flow or simply to annihilate the growing process. Firstly, the gradients of the mean flow get reduced via a RANS-type viscosity and the Turbulent Kinetic Energy (TKE) production decreases. Secondly, the excited small scale structures damp the instability of the low frequencies via a Smagorinsky-like viscosity.

HFF appears to be alternative of flow control in free shear flows for specific applications. The physics that are responsible for these effects are not entirely understood.

1.1.2.2 The Cascade "Paradigm" for Flow Control

It is obvious that when the transport equations for turbulence are considered, the production/dissipation values should balance. Indeed the balance is satisfied *locally*. However, the spatial origin of the two terms, namely production and dissipations are different. Typically TKE production will occur locally, for example in the case of a mixing layer, resulting in the Large Scales CS modes issued from the unstable character of the inflexional velocity profiles. This production is governed by the local mean shear. At the other edge of the spectrum, high frequencies or small scales are viewed as the result of a cascade process. At the same given location, the small scales can be considered as the result of a cascade occurring upstream, depending on a mechanism that develops upstream of the actual spatial location.

Schematically, Fig. 1.5 represents a dual schematic description of both a spatial localization of production and associated dissipation scales and local spectral description.

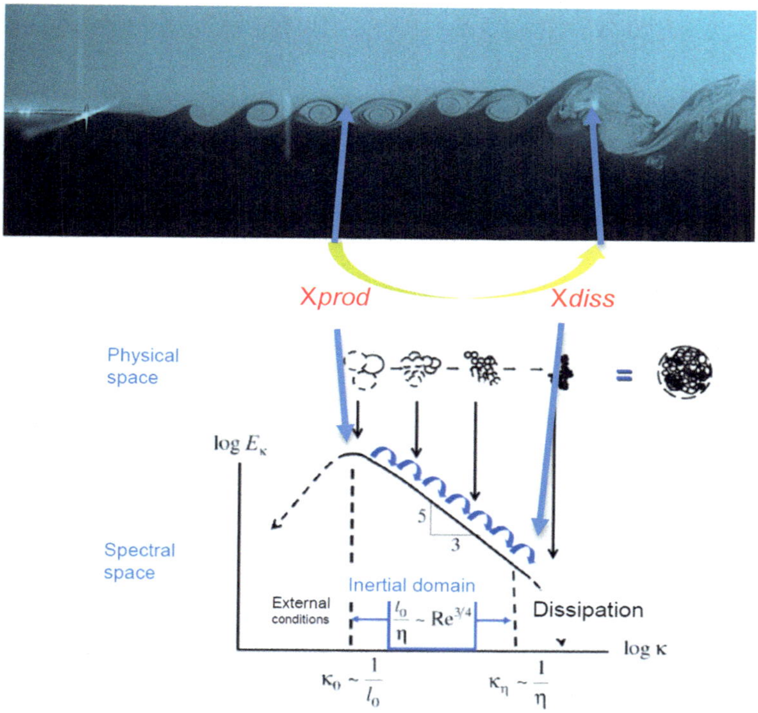

Fig. 1.5 Schematics of the local origin of "energetic scales" and "dissipative scales" and spectral representation

The usual spectral description, corresponds to a statistical local analysis does not take into account the spatial evolution of convective flows. The different locations in space of the two phenomena can be of crucial interest as far as non-equilibrium flows are considered. Particularly for flow control strategies, the actuation can have different local effects in convective flows depending on their frequency/scale domains and on the location within the stream wise direction.

In addition, for HFF, the small scales that are generated will have short life duration, or will stay on short streamwise distances in convective flows (such as the mixing layer in the example) when compared with the characteristics of the large scales (see Sect. 1.1.1.3). However, all the experiments show that the entire spectra is affected, with a lower slope in the inertial domain, even at distances from the actuations where the influence of small scales can be expected to vanish. Then the duration of the influence of the small scales is larger than the one that a pure dynamical analysis can explain. Bos et al. [43] study the spectral imbalance linked to the dissipation rate. Vassilicos [44] recently showed that the modification of the slope of the spectra if artificially forced, out-of equilibrium flows takes longer time—or distance—to relax towards the natural value. Then it can be conjectured that the HFF can act on the spectral imbalance that remains out of equilibrium on longer time scales than the ones linked to the source of the perturbation.

Finally, in the spirit of defining HFF strategies, it will be important to at least evaluate the space/time distances on which the Large Scales Structure will take to promote dissipation scales through the cascade mechanisms. This appears to be still an open question.

1.2 Computational Simulation Approaches for Flow Control

Progress made in computer capabilities (processing speed, memory, storage, and parallel processing) has been accompanied by similar efforts to resolve modelling issues (of turbulence physics and algorithm accuracy). This led to efficient and validated numerical tools for turbulent flow computations at relevant Reynolds numbers. However, these efforts are still far from what the real-world engineering requires, e.g. a trustworthy solution of unsteady turbulent flow control problems on realistic configurations.

1.2.1 RANS, LES, DNS or Hybrid?

It is well-known that, for turbulent flows, RANS solution is unable to produce accurate results when it comes to problems involving unsteady flow separation. As discussed in the previous sections, for active flow control, either the original flow or the flow control actuation or both can involve unsteady separated flows. This leaves us with the choice of large eddy simulation, hybrid RANS-LES, or direct numerical simulation for active flow control problems to resolve the effects of the flow control on various turbulent scales. Direct numerical simulation, if affordable, can give us all the turbulence scales and detailed interaction of these scales with the active flow control activation. Unfortunately, such potential is severely limited by the Reynolds number achievable in the simulation to resolve all the scales and it is beyond our reach for even the laboratory scale flow control problems in the context of this project for the available HPC computing resources.

Regarding LES, it was shown by Spalart et al. [45], that for flow simulations in which Reynolds number was about 10^7, a grid of 10^{11} cells would be necessary in order to obtain a sufficiently accurate solution of the problem. Although computers are in constant development, a grid of such number of cells is still too expensive in the sense of computational requirements. This situation motivated the development of hybrid RANS/LES techniques. Breuer et al. [46] made a comparison of three different approaches such as DES, RANS and LES for a separated flow around a flat plate at high incidence was carried out. It was shown that RANS due to its time-averaged nature lacks the ability to resolve the unsteadiness of massively separated flows and they concluded that RANS is unsuitable to produce reliable results as it cannot reproduce the unsteady characteristics of the separated flow field. LES, on the other hand, can provide reliable results for unsteady separated flows as it resolves large scale turbulent structures of the flow. However, and it is worth mentioning it again, it is an expensive computational technique for practical high Reynolds number flows, as it demands a really fine mesh to predict the resolved turbulent structures.

The hybrid RANS/LES methods, e.g. the detached eddy simulation [47], combines the features of the RANS and LES approaches: near the walls in the attached boundary layer the flow is solved via RANS and a turbulence model is applied and in the rest of the domain where the larger turbulent structures are present LES simulation is performed. Therefore, the application of hybrid RANS/LES is favoured when the problem to solve consists of unsteady turbulent flows with large separated regions. As can been seen in this volume, most authors have adopted the hybrid RANS/LES approach for their study of the active flow control problems. The two baseline cases in the MARS project both involve flow separation: one from a fixed point (in the case of the backward facing step) and the other from a smooth curved surface (in the case of the NACA0015 wing at 11°).

1.2.2 Hybrid RANS/LES Techniques

In this section, different hybrid RANS/LES methods will be briefly reviewed, as they are the dominant simulation tool used by the authors in this volume. More details for specific simulations may be found in the individual contributions embodied in this volume. As mentioned earlier, these techniques were designed to combine the best aspects of RANS and LES approaches. The principle behind the hybrid approach is that, when the boundary layer is attached to the wall RANS can do a good job and as it is separated to form large unsteady separated flow the larger turbulent eddies can be captured via LES.

Detached Eddy Simulation, DES, is the first and one of the most widely used hybrid RANS/LES approaches. It was firstly proposed by Spalart et al. [45]. It was initially developed to be applied in high Reynolds number flows where a massive separation occurs, such as aerospace and ground transportation problems. Essentially, it "*senses*" the grid density and compares the grid spacing in all directions in order to assign RANS to the near wall region to RANS and LES SGS model in the other regions. It is, in effect, "a three-dimensional unsteady numerical solution using a single turbulence model, which functions as a subgrid-scale model in regions where the grid density is fine enough for a LES simulation and as a Reynolds averaged model in regions where it is not" [48]. Therefore, the attached boundary layer would be treated by RANS and regions with massive separation would be treated by LES. The space between these two areas, known as the grey area, may be problematic, as there is no strong physical argument for proper modelling in this area.

DES proposed by Spalart et al. [45] is based in the Spalart-Allmaras one-equation turbulence model [49]. In RANS mode, the length scale is defined as the shortest distance from any point to the closest wall in RANS model and more details may be found in [48]. DES, in RANS mode, uses the Spalart-Allmaras turbulence model and will be operating in LES mode otherwise, using the Spalart-Allmaras model as the subgrid scale model. As mentioned earlier, there is a problem known as the grey area in the hybrid approach. DES only depends on the grid and length scales and when a mesh refinement is performed, it may trigger to LES mode when the boundary layer is fully attached: when the grid spacing parallel to the wall is less than the boundary layer thickness, the LES mode takes over as the length scale is fine enough for the detached eddy simulation to switch into a LES model from RANS mode; however, the grid spacing is partially inside the boundary layer and the resolved Reynolds stresses in LES mode do not completely replace the modelled Reynolds stresses from the RANS mode. This leads to a depletion of the stresses which in turn leads to an over prediction of the separation area producing a reduction of the skin friction computation which is not physically real, [50]. In order to solve the grey area issue, some variants of the DES approach were developed in the recent years.

As mentioned earlier, the transition between RANS and LES for DES is predetermined by wall-distances and grids. Therefore, mesh-refining may trigger the LES mode in attached boundary layers. Considering the drawbacks of DES, Delayed DES was proposed to provide a dynamic RANS/LES transition [51]. However, DDES is sensitive to the initial flow fields. If the initial flow field is far from a physical flow field, the flow-dependent blending function may not work properly. To tackle this problem, an improved version of DDES was proposed by Shur et al. [52]. The improvements mainly contain two aspects. One is the modification of the length scale for LES. The other is that it blends DDES with a wall-modelled LES, depending on whether the inflow conditions have any turbulent content or not.

1.2.3 Numerical Simulation of Flow Control

1.2.3.1 Simulation of Pulse Jet Flow Control

In addition to experimental studies, there are some numerical investigations on the pulsed jet flow control. Deng et al. [53] performed a DNS investigation on pulsed jets applied to NACA0012 at an angle of attack of 4°. With pulsed jets, the separation zone and the drag were both reduced, while the lift maintained approximately the same level as in the baseline flow. The pitch and skew angles were found to be primary control parameters due to the fact that skew angles determine the generation of non-symmetric 3D perturbations containing 3D unstable modes for the early transition. Jewkes and Chung [54] studied the formation of different fluidic vortices generated by different pitched and skewed angles in pulsed jets. A clear vortical "shell" was observed in a low velocity ratio perpendicular pulsed jet, which is not found in other pitch and skew angles. Sau and Mahesh [55] attempted to perform a wide range of optimizations of pulsed jets penetration and spread. Laval et al. [56] also investigated the control parameters in pulsed jets applied to a 2D bump in a converging-diverging channel by LES. The main conclusions are the pitch angle has no significant effect compared with the skew angle. A high jet velocity can increase the control robustness, but with a low energy efficiency. For the optimal pulsing frequency, the low range is found to improve the control efficiency, while the high control frequency may have a negative effect. During the low frequency range, low values of the duty cycle were observed to reduce the mass flux rate consumption with an efficiency as good as for continuous jets. Bobonea [57] numerically studied the impact of different slot/orifice geometries of pulsed jets applied on wind turbine aerofoils.

1.2.3.2 Simulation of Synthetic Jets

An early numerical investigation of synthetic jets was carried out by Kral et al. [58]. A two dimensional and incompressible RANS simulation was conducted, and the synthetic jet was modelled by specifying experiment measured velocity profiles, which means the modelling of the cavity was omitted. Ignoring the whole cavity may sacrifice the accuracy of the curvature of the diaphragm when the oscillating amplitude is not very small. Mittal et al. [59] modelled the diaphragm as a moving boundary in a realistic manner in order to compute the internal cavity flow accurately in a two dimensional RANS configuration. Lee and Goldstein [60] carried out a DNS simulation of an array of synthetic jets driven by pistons in a two-dimensional configuration, which observed the lip and depth of the cavity plays important roles in generating vortices. Three-dimensional DNS simulation of synthetic jets with the cavity flow was carried out by Ravi et al. [61] to investigate the 3D jet flow in quiescent and cross flows. Later, You and Moin [62, 63] carried out a large eddy simulation of the separation control with synthetic jets in NACA0015, involving a small slot across the span connected to a cavity inside of the aerofoil to product the oscillatory synthetic jets. Almost at the same time, Dandois et al. [38] simulated the synthetic jets with the cavity using both DNS and LES in a rounded ramp, which observed that the synthetic frequency similar to the natural shedding frequency can suppressed the flow separation most effectively. Qin and Xia [64] used the dynamic meshing detached eddy simulation to study synthetic jets problems without and with cross flow, in which interesting flow features were found inside the jet cavities. Hong [65] studied synthetic jets flow control on the boundary layer laminar-separation, and it was observed that the synthetic jets were most effective in eliminating the laminar separation bubbles when the forcing frequency was in the lower range of the Tollmien-Schlichting instability. Sawant et al. [66] employed the lumped-element electrotechnical model to study an electrodynamic ZNMF (zero net mass flux) actuators and achieved promising results.

Generally speaking, the main trend in simulation of synthetic jets is to accurately simulate the whole cavity to provide a realizable control effect with LES, DNS or hybrid RANS/LES. Meanwhile, the control frequencies, amplitudes and geometries of the orifice/slot are also hot research areas.

1.3 Experimental Approaches

1.3.1 Overview of Actuators

With the recent introduction of novel actuator concepts, which are locally applied and require only electrical power input, the potential for vehicle control without conventional control surfaces may be within our reach. Seifert et al. [67] showed that separation control using periodic addition of momentum, at a reduced frequency slightly higher than the natural vortex shedding frequency, can save 90–99% of the momentum required to obtain similar gains in performance using steady blowing. Some active flow control applications include stall control using a deformable leading edge, separation control for take-off and landing flight conditions using piezo devices, oscillating vortex generators, and zero net mass oscillators, duct flow separation control and fore-body vortex control using zero net mass suction and blowing, and thrust vectoring with zero net mass oscillatory actuation. The progress in system integration, miniaturization, actuators, sensors and computational techniques enables the utilization of fast-responding, unsteady, flow-control methods into closed-loop system architectures. However promising the technology might look, significant barriers still exist between the capabilities available to the technologist and the successful application. Further comprehensive experimental and computational studies are required to close the gaps between theory, computations, and real-world applications.

A large number of flow control devices have been investigated in the laboratories and, more recently, by numerical simulations. This provides essential understanding of the flow physics at a statistical level and the effectiveness of these devices at the scales under investigation, although few have found their way into practical applications on aircraft. The detailed state-of-the-arts has been provided by Seifert [68], Cattafesta and Sheppard [69]. We recall here of some characteristics of the relevant devices.

Oscillating Surfaces

Early efforts, Wehrmann [70] demonstrated the control of artificially generated Tollmien-Schlichting (T-S) waves through phased single frequency motion of driven flexible walls. This approach has been extended to include artificially generated 3-D disturbances [71] and turbulent burst [72]. The large amplitudes and extremely complicated sensor-actuator feedback control have effectively rendered such approaches difficult for naturally occurring turbulent flows. On the other hand, Flexible Composite Surface (FCS) developed by Sinha (1999), in the form of a 50–100 μm thick micro-flexural tape overcame this hurdle. This can be affixed to selected regions on the surfaces of aircraft wings and stabilizers, where the streamwise flow pressure gradient changes from favourable to adverse, in order to reduce the profile drag under cruise conditions. The membrane of the FCS undergoes extremely small (under 0.1-μm amplitude) flow-induced flexural oscillations, which neutralizes turbulent fluctuations in the near-wall boundary layer airflow at all but a narrow frequency band. The resulting customized turbulent aerodynamic boundary layer, displayed resistance to separation as compared to a laminar boundary layer while exhibiting lower skin-friction induced losses compared to either "naturally occurring" or artificially tripped turbulent boundary layers. Subsequently, the FCS has been used to control separation on airfoils in wind-tunnel and flight tests [73].

Bird et al. [74] introduce an new method by activating a membrane skin for spanwise turbulent boundary layer control through lateral wall oscillation using Kagome lattice structures.

Piezoelectric Actuators

Sakai et al. (2003) developed a method to control two dimensional T-S wave and oblique wave at the early stage of transition of a flat-plate boundary layer. Piezoelectric actuators were used to generate a cancelling wave with an opposite sign, downstream. Kikuchi and Fukunishi [75] carried out an experimental investigation which was aimed at controlling cavity noise using piezo-film actuators. Two pieces of the piezo-film actuators were glued to surface at the upstream edge of the cavity. Choi et al. [7] and Amir et al. [76] conducted experiments to control the flow around an aerofoil using piezoelectric actuators. The study showed that the piezoelectric actuators attached to the aerofoil suction surface effectively increased the lift force at higher angles of attack by delaying separation.

Synthetic Jets

The use of synthetic jets for active control of separation has been studied quite extensively. Ming (1988, 1992) [128, 129] pioneered the work on synthetic jets for flow control. Through experiments using a speaker and a nozzle arrangement, he observed that the velocity at the outlet of the nozzle peaks significantly for a particular frequency set for the speaker. This was named "acoustic streaming" as the actuator is the speaker, through the vibration of the speaker's membrane at a certain frequency. He applied this to the control of separation behind cylinders, either to minimize the separated region or to manipulate the direction of the wake. Glezer and Amitai [77] explored further the idea of synthetic jets for flow control and made the idea popular. Unfortunately, they failed to notice and credit the original work by Ming [128, 129]. In a number of studies by Shah et al. (2008), Rehman et al. (2006), Smith (2002), Hong et al. (2002), Watson et al. [78], Bridges et al. (2003) it has been demonstrated that synthetic jets can indeed reduce the extent of separation over bluff as well as streamlined bodies (see Cattafesta and Sheppard for the references).

Despite these successful demonstrations, it is fair to state that the physical mechanisms through which synthetic jets accomplish this reduction in separation are not completely understood. The unique characteristics of the flow produced by a synthetic jet interacting with a crossflow have also led researchers to suggest other flow features/mechanisms that might play a role in separation reduction in flows where these actuators are employed.

Plasma inside closed cavities can promote synthetic jets, namely Spark Jets [79–81]. These actuators have a high authority and a fast response. However, temperature drifts and the very short duration of the jets can limit their use. In the same spirit, micro-burners are alternative methods when high authority is required (see below). Combustion Powered Actuators (COMPACT) wer developed by Crittenden et al. [82].

Dielectric Barrier Discharge Plasma Actuator

Dielectric barrier discharge (DBD) plasma actuator consists in applying a high voltage between at least two electrodes placed on both sides of a dielectric material. A high electric field results in a weakly ionized region above the dielectric referred as surface non-thermal plasma discharge. The electrochemical interactions between air and plasma. Momentum induced by plasma discharge can be transferred within the boundary layer of bluff body object that being a way to modify aerodynamic and improve the performances as reviewed in Moreau [83] and Corke et al. [84]. The flow induced by the plasma actuator has been also studied both experimentally by Benard and Moreau [85] and by CFD by Maden et al. [86]. Flow control by plasma actuation is appealing due to the absence of mechanical parts, low proper mass, low power consumption and very short time response with easily adjustable command law, particularly useful for closed loop strategies. These features support the current keen interest for DBD plasma actuator as a way to postpone separation over airfoils as demonstrated in Post et al. [87], Goeksel et al. [88], Corke et al. [89] or Little et al. [90] among others contributions. Such flow control actuator was also extensively investigated as skin-friction drag reduction device, Jukes et al. [91], Wong et al. [10], jet mixing enhancement system, Gregory et al. [92], and thrust control, Bohlito and Jacob [93], Benard et al. [94]. All the current studies indicate that plasmas might influence airflow at subsonic speeds but the limited momentum transfer produced by typical DBD actuator restricts the use of plasma discharge to low speed conditions. It is expected that intelligent flow control strategy may extend the effectiveness of DBD to practical flow conditions. Indeed, by using simple, modulated or tonal input waveforms, velocity fluctuations at single, dual or multi-frequencies can be produced by typical DBD, Benard et al. [95].

This process can be linked through a temperature effect to the control scenario observed for localized arc filament plasma actuators (LAFPA) developed by Samimy et al. [96]. This is especially promising as LAFPA has demonstrated a high authority to control vortex dynamic and inherent turbulent stress components by Utkin et al. [97].

A new pulsed jet actuator uses a plasma heating the air passing a sonic throat as demonstrated by Bonnet et al. [98]. The temperature increase allows for 30% modulation of the flow rate of the control jet without any mechanical device.

Nanosecond Pulse DBD Plasma Actuator

A viable alternative to typical non-thermal plasma actuators relates to the use of plasma discharge driven by repetitive nanosecond pulses. The construction of the device is similar to typical dielectric barrier discharge but early reports suggest that nanosecond plasma actuators (NS-DBD) can extend the control authority to practical conditions. Roupassov et al. [99] indicated that leading edge airfoil separation can be suppressed? up to Mach 0.85. The mechanism by which nanosecond plasma actuators produce control is substantially different from typical DBD. As revealed by Unfer and Boeuf [100], numerical modelling of nanosecond discharge does not produce a significant velocity in the neutral species. The numerical results confirmed the experiments performed by Starikovskii and Pancheshnyi [101] that suggested a thermal effect as the source of control. This discharge promotes fast gas heating in the boundary layer. This sudden change in gas heat provokes a local 'micro' shock wave that is highly suspected to be the main parameter of the enlarged control authority. Correale and Kontis [102] showed the efficiency of actuation of NS-DBD in promoting transition from laminar to turbulent flow.

Oscillating Vortex Generators

Flow control by using steady streamwise vortices, such as those generated by mechanical or pneumatic vortex generators, is well established in aeronautical technology. Searching for more effective flow control beyond fixed vortex generators, Mochizuki et al. (2006) investigated experimentally the Reynolds stress field in a turbulent wall jet induced by a streamwise vortex with periodic perturbation. The actuator is an oscillating vortex generator of a delta wing shape. Analysis by triple velocity decomposition [29] and phase averaging shows that the oscillation induces periodic variations in the strength, radius and position of the streamwise vortex centre. The effect of the periodic perturbation manifests itself in the magnitude of the Reynolds stress components $\overline{w^2}$ and $\overline{vw}$, through the generation of an apparent shear stress, or periodic stress, $\overline{\tilde{V}\tilde{W}}$. The study concludes that the periodic stress directly contributes to the Reynolds stress.

Recent developments of sweeping jets have been performed by several authors, see Woszidlo et al. [103]. Schatzman et al. [104, 105] proposes an actuation process that associates continuous suction and oscillatory blowing (SaOB).

High frequency actuators based on the Hartman resonant tubes are developed as individual or groups of actuators by Upadhyay et al. [106]. High frequencies (up to 50 kHz) can be achieved with these devices.

General Considerations for Active Flow Control

Scaling effects: The orders of magnitude between laboratory experiments and in flight demonstrations can be quite different, both for the scales in space and frequencies that are targeted. Indeed, the real full scales applications in road vehicles or aeronautical applications are much larger than the capabilities of most of research laboratories, with the exception when large scale facilities such as NASA (see Andino et al. [107]) or ONERA, DLR etc. are concerned [108]. Laboratory experiments and demonstrations can be much more difficult to perform than the real size ones. Let us consider the example of jet noise control for airplane engines by impinging micro-jets [109–111]. In research laboratories, the typical diameter of the jets to be controlled is $D \approx 5.10^{-2}$ m, the micro-jets diameters being microjet diameter of 0.8–1 mm, the microjet Mach number being 1.5 for a main jet Mach number $M_j = 0.9$. For real engines, the propulsive jet diameter is of order of 2 m, with the same order of magnitude of velocities. A ratio of approx. 40 exists, then full scale control jets will be of order of 3 cm in diameter that are much more easy to machine. In the frequency domain, Strouhal numbers $\left(S_t = fD/M_j\right)$ ranging from 0.2 to 5 are usual targets for active control. This leads to quite different frequency ranges: order of 1–30 kHz for lab experiments and 25–750 Hz for applications. The frequency requirements for lab experiments are extremely demanding and sometimes impossible to satisfy at small scales. Conversely full scales requirements can be easy to achieve but more expansive to test.

To conclude, three characteristics of actuators can be enlightening:

(a) The authority: this is the primary efficiency request. The velocity ratio (when compared to the ambient fluid velocity) or the dynalpy ratio should be large enough for acting on the mean flow field, generating longitudinal vortices in particular. Ratios as large as 3–8 are sometimes requested [112]. For some applications also the momentum ratio ($C\mu$) is often considered as a relevant parameter, in particular in view of estimating the net energy balance required [113]. For flows of engineering relevance the exit jet velocity should be of order of hundred m/s, even transonic or supersonic regimes. Glezer [114], Fronhapfel et al. [30] propose method to evaluate the global energy efficiency of control strategies.

(b) The frequency range: This parameter is requested for open loop active control strategies, in view of exciting unstable modes (even higher frequency modes for high frequency strategies). The requested frequencies can be quite different, for a given flow configuration, depending upon the size of the demonstrator. As outlined before, the laboratory experiments are quite demanding in terms of frequency domain than full scale ones.

(c) The versatility or reactivity: this means that, for closed loop, a device whose characteristics can be adjusted in "real time", at least at frequencies high enough for the time of the loop.

1.3.2 Measurement Technologies

Measurements are required for physical interpretations, CFD validation as well as for input for closed loop control systems.

The in-volume full field measurements are indeed well suited for those purposes. Hot-Wire-Anemometry (HWA) and Laser Doppler Velocimentry (LDV) have been used extensively by various groups world-wide for the dynamic interrogation of the local flow field properties and point measurements of Reynolds stress. Watson et al. [78] investigated the fluid structures exiting from a synthetic jet actuator using smoke, LDV, and hotwire measurement techniques. Kikuchi and Fukunishi [75] used a hotwire probe for measuring velocity fluctuations above a cavity. Fast Fourier Transform (FFT) was used to analyze the data. Wong et al. [115] used HWA and LDV to analyse the three-dimensional flow-field of a supercritical aerofoil with steady and pulsating pneumatic flow control. Amir and Kontis [116] used hot-wire probes and LDV to investigate the effects of oscillation on boundary layer development under the influence of favourable and adverse pressure gradients. PIV (Particle Image Velocitmetry) has been used to compile velocity statistics on a wide variety of flow fields. The data provides unique access to instantaneous velocity vectors that is useful to understand the structural features of different regimes. Two independent PIV systems have also be used in tandem to obtain planar acceleration fields near the regions of interest. Time-resolved wall-pressure measurements can also be obtained simultaneously to compute conditionally averaged velocity fields. Time resolved PIV is widely used for many groups.

The measurement of surface shear stress or skin friction is a non-trivial challenge in aerodynamic applications. A review of the literature yields a variety of techniques including oil films [117], liquid crystals [118], thermal sensors [117], an array of MEMS based sensors [119, 120], balance designs [121], near wall velocity sensors [122] and Shear and Stress Sensitive Films [123]. Of these established techniques for shear stress measurements, liquid crystals or oil films offer non-intrusive and continuous distributions of shear over an area on the surface. Both of these techniques have been utilized in wind tunnels with success; however some limitations exist for each technique. The limitations of the liquid crystal coating technique include the requirement of multiple viewing angles and the fact that the colour change is a function of both shear stress and direction of observation. Some difficulty may be also encountered when working with complex geometries. The oil film technique is well established as a means of measuring shear stress, however it has slow response time and requires periodically recoat the model with oil.

Measurements of pressure on aerodynamic models have traditionally used arrays of pressure taps or surface mounted pressure transducers. As an example, [124] showed that wall pressure sensors can be used for qualifying unsteady transient separation phenomena through stochastic estimation. Non-intrusive and continuous measurements of pressure have been demonstrated using Pressure-Sensitive Paint (PSP) on wind tunnel models, see Liu and Sullivan [125]. PSP is an image-based technique where the luminescent intensity of the coating is a function of the partial pressure of oxygen to which the paint is exposed. Such wall sensors are easy to implement on real configurations.

Acknowledgements JPB thanks fruitful discussions with B. N. Noack and V. Parezanović and C. Vassilicos, with a particular thank to J. Delville, in memoriam. Nicoals Bénard is also acknowledged for the DBD section. NQ would like to thank X Ming, K. Kontis, N Wood for stimulating discussions.

References

1. Schubauer GB, Skamstad HK (1947) Laminar boundary layer oscillation and stability of laminar flow. J Aeronaut Sci 14
2. Poisson-Quinton P, Lepage L (1961) Survey of French research on the control of boundary layers and circulation. In: Lachman GV (ed) Boundary layer and flow control
3. Gad-el-Hak M, Bushnell DM (1991) Separation control: a review. J Fluids Eng 113(1)

4. Gad-el-Hak M, Pollard A, Bonnet JP (1998) Flow control: fundamentals and practices. Springer, Berlin
5. King R (ed) (2007) Active flow control, notes on numerical fluid mechanics and interdisciplinarity design, vol 95. Springer, Berlin
6. Collis SS, Joslinb RD, Seifert A, Theofilis V (2004) Issues in active flow control: theory, control, simulation and experiment. Prog Aerosp Sci 40
7. Choi J, Choi H, Jeon WP (2002) AIAA J 40(5)
8. Choi K-S, De Bisschop J-R, Clayton BR (1998) Turbulent boundary-layer control by means of spanwise-wall oscillation. AIAA J 36
9. Quadrio M, Ricco P, Viotti C (2009). Streamwise-traveling waves of spanwise wall velocity for turbulent drag reduction. J Fluid Mech
10. Wong CW, Zhou Y, Li Y, Li Y (2015) Active drag reduction in a turbulent boundary layer based on plasma-actuator generated streamwise vortices, Paper 9A-6, TSFP9. Melbourne, Australia
11. Zhou K, Doyle JC (1998) Essentials of robust control. Prentice-Hall, Englewood Cliffs
12. Lee WY, Wong, M, Zohar Y (2002) Microchannels in series connected via a contraction/expasion. J Fluid Mech 459
13. Cordier L, Noack BR, Tissot G, Lehnasch G, Delville J, Balajewicz M, Daviller G, Niven RK (2013) Identification strategies for model-based control. Exp Fluids 54(8)
14. Benard N, Pons-Prats J, Periaux J, Bugeda G, Braud P, Bonnet JP, Moreau E (2016a) Turbulent separated shear flow control by surface plasma actuator: experimental optimization by genetic algorithm approach. Exp Fluids 57:22
15. Benard N, Moreau E, Griffin J, Cattafesta L (2009) Plasma flow control—autonomous lift improvement by slope-seeking. AIAA paper 2009-4118
16. Liu SJ, Krstic M (2012) Stochastic averaging and stochastic seeking, communication and control engineering. Springer, Berlin
17. Duriez T, Brunton SL, Noack BR (2016) Machine learning control—taming non linear dynamics and turbulence. In: Series "fluid dynamics and its applications, 116. Springer, Berlin
18. Parezanović V, Cordier L, Spohn A, Duriez T, Noack BR, Bonnet JP, Segond M, Abel M, Brunton SL (2016), Frequency selection by feedback control in a turbulent shear flow. J. Fluid Mech 797
19. Rodi W (1975) A review of experimental data of uniform density free turbulent boundary layers. In: Launder BE (ed) Studies in convection, vol 1. Academic Press, NY, pp 79–165
20. Lumley J, Newman G (1977) The return to isotropy of homogeneous turbulence. J Fluid Mech 82–1:161–178
21. Frohnapfel B, Lammers P, Jovanović J, Durst F (2007) Interpretation of the mechanism associated with turbulent drag reduction in terms of anisotropy invariants. J. Fluid Mech 577:457–466
22. Vassilicos JC (2015) Dissipation in turbulent flows. Annu Rev Fluid Mech 49(95):114
23. Lumley J (1992) Some comments on turbulence. Phys Fluids 4:206
24. Pope S (2000) Turbulent flows. Cambridge University Press, Cambridge
25. Lavoie P, Djenidi L, Antonia RA (2007) Effects of initial conditions in decaying turbulence generated by passive grids. J Fluid Mech 585
26. Valente PC, Vassilicos JC (2015) The energy cascade in grid-generated non-equilibrium decaying turbulence. Phys Fluids 27
27. Michalke A (1965) Vortex formation in a free boundary layer according to *stability* theory. J Fluid Mech 22
28. Ho CM, Huerre P (1984) Pertubed free shear layers. Annu Rev Fluid Mech 16
29. Reynolds WC, Hussain AKMF (1972) The mechanics of an organized wave in turbulent shear flow. J Fluid Mech 54
30. Frohnapfel B, Hasegawa Y, Quadrio M (2012) Money versus time: evaluation of flow control in terms of energy consumption and convenience. J. Fluid Mech 700
31. Seifert A (2015) Evaluation criteria and performance comparison of actuators, instability and control of massively separated flows. In: Theofilis V, Soria J (eds) Fluid mechanics and its applications, vol 107. Springer, Berlin

32. Wiltse J, Glezer A (1993) Manipulation of free shear flows using piezoelectric actuators. J Fluid Mech 249:261–285
33. Luchtenburg DM, Güther B, Noack BR, King R, Tadmor G (2009) A generalized mean-field model of the natural and high-frequency actuated flow around a high-lift configuration. J Fluid Mech 623
34. Glezer A, Amitai M, Onohan AM (2005) Aspects of low and high frequency actuation for aerodynamic flow control. AIAA J 43
35. Barros D, Ruiz T, Borée J, Noack B, (2014) Control of a three-dimensional blunt body wake using low and high frequency pulsed jets. Int J Flow Control 6(1)
36. Oxalde AR, Morrison JF, Qubain A, Rigas G (2015) High-frequency forcing of a turbulent axisymmetric wake. J Fluid Mech 770:305–318
37. Benton S, Visbal MR (2016) Investigation of high-frequency separation control mechanism for delay of unsteady separation. In: 8th AIAA Flow Control Conference paper 2016-4241
38. Dandois J, Garnier E, Sagaut P (2007) Numerical simulation of active separation control by a synthetic jet. J Fluid Mech 574:25–58
39. Stanek MJ, Visbal MR, Rietta DP, Rubin SG, Khosla PK (2007) On a mechanism of stabilizing turbulent free shear layers in cavity flows. Comput Fluids 36(10)
40. Vukasinovic B, Glezer A, Rusak Z (2007) Experimental and numerical investigation of controlled, small-scale motions in a turbulent shear layer. In: 3rd international symposium on integrating CFD and experiments in aerodynamics U.S. Air Force Academy, CO, USA
41. Parezanovic V, Laurentie J-C, Fourment C, Delville J, Bonnet J-P, Spohn A, Duriez T, Cordier L, Noack BR, Abel M, Segond M, Shaqarin T, Brunton SL (2014) Mixing layer manipulation experiment from open-loop forcing to closed-loop machine learning control. Flow Turbul Combust 94(1):155–173
42. Mons V, Chassaing JC, Gomez T, Sagaut P (2014) Is isotropic turbulence decay governed by asymptotic behavior of large scales? An eddy-damped quasi-normal Markovian–based data assimilation study. Phys Fluids 26:115105
43. Bos W, Shao L, Bertoglio JP (2007) Spectral imbalance and the normalized dissipation rate of turbulence? Phys Fluids 19
44. Vassilicos JC (2016) Unsteady turbulence cascades. Phys Rev E. https://doi.org/10.1103/physreve.00.003100
45. Spalart PR, Jou W-H, Michael S, Allmaras SR (1997) Comments on the feasibility of LES for wings and on a Hybrid RANS/LES approach. In: Advances in DNS/LES, 1st AFOSR International Conference on DNS/LES
46. Breuer M, Jovivi N, Mazaev K (2003) Comparison of DES, RANS and LES for the separated flow around a flat plate at high incidence. Int J Numer Methods Fluids 43:357–388
47. Spalart RP (2009) Detached-eddy simulation. Annu Rev Fluid Mech 41:181181evixii, 16, 22, 24, 171
48. Shur M, Spalart PR, Strelets M, Travin A (1999) Detached-eddy simulation of an airfoil at high-angle of attack
49. Spalart PR, Allmaras SR (1992) A one-equation turbulence model for aerodynamic flows. AIAA paper 92-0439
50. Panguluri S, Reasor D, LeBeau RP Jr (2007) Investigation of grey area construction on the performance of detached eddy simulation. AIAA paper 2007–4095
51. Spalart PR et al (2006) A new version of detached-eddy simulation, resistant to ambiguous grid densities. Theoret Comput Fluid Dyn 20:181–195
52. Shur M, Spalart PR, Strelets M, Travin A (2008) A hybrid RANS-LES approach with delayed-DES and wall-modelled LES capabilities. Int J Heat Fluid Flow 29:1638ional
53. Deng S, Jiang L, Liu C (2007) DNS for flow separation control around an airfoil by pulsed jets. Comput Fluids 36(6):1040uidsje
54. Jewkes JW, Chung YM (2010) Low velocity-ratio pitched and skewed jet in a turbulent boundary layer. In: Mallinson GD, Cater JE (eds) 17th Australasian fluid mechanics conference, Auckland, New Zealand
55. Sau R, Mahesh K (2010) Optimization of pulsed jets in crossflow. J Fluid Mechanics 653(365):46

56. Laval JP et al (2010) Large-eddy simulations of control of a separated flow over a 2D bump by means of pulsed jets. J Turbul 11:N52
57. Bobonea A (2012) Impact of pulsed blowing jet on aerodynamic characteristics of wind turbine airfoils. In: AIP conference proceedings, vol 1493, p 170
58. Kral LD et al (1997) Numerical simulation of synthetic jet actuators. AIAA Paper 97-1824
59. Mittal R, Rampunggoon P, Udaykumar HS (2001) Interaction of a synthetic jet with a flat plate boundary layer. AIAA paper 2001-2773
60. Lee CY, Goldstein DB (2002) Two-dimensional synthetic jet simulation. AIAA J 40(3):510al sy
61. Ravi BR, Mittal R, Najjar FM (2004) Study of three-dimensional synthetic jet flowfields using direct numerical simulation. AIAA paper 51
62. You D, Moin P (2006) Large-eddy simulation of flow separation over an airfoil with synthetic jet control. In: Center for turbulence research annual research briefs, 337surbu
63. You D, Moin P (2007) Study of flow separation over an airfoil with synthetic jet control using large-eddy simulation. In: Annual research briefs, center for turbulence research, Stanford University, 311lence
64. Qin N, Xia H (2008) Detached eddy simulation of a synthetic jet for flow control. Proc Inst Mech Eng Part I: J Syst Control Eng 222(5):373–380
65. Hong G (2012) Numerical investigation to forcing frequency and amplitude of synthetic jet actuators. AIAA J 50(4):788estig
66. Sawant SG et al (2012) Modeling of electrodynamic zero-net mass-flux actuators. AIAA J 50(6):1347amic Z
67. Seifert A, Darabi A, Wygnanski I (1996) Delay of airfoil stall by periodic excitation. J Aircraft 33(4)
68. Seifert A (2009) Closed-loop active flow control systems: actuators. In: King R (ed) Notes on numerical fluid dynamics and multidisciplinary design, active flow control, vol 95. Springer, Berlin
69. Cattafesta LN III, Sheplak N (2011) Actuators for active flow control. Annu Rev Fluid Mech 43:247–272
70. Wehrmann O (1965) Tollmien—Schlichting waves under the influence of a flexible wall. Phys Fluids 1389–1390
71. Breuer KS, Haritonidis JH, Landahl MT (1989) The control of transient disturbances in a flat plate boundary layer through active wall motion. Phys Fluids A 1:574
72. Wilkinson SP, Malik MR (1985) Stability experiments in a flow over a rotating disk. AIAA J 23
73. Sinha NK, Ananthkrishnan N (2000) Level flight trim and stability analysis using continuation methods. In AIAA atmospheric flight mechanics conference, Paper 2000-4112, Denver, CO, USA
74. Bird J, Santer M, Morrison J (2015) Turbulent boundary layer control through spanwise wall oscillation using Kagome lattice structures. In: 68th Annual Meeting of the APS Division of Fluid Dynamics, vol 60, no 21
75. Kikuchi S, Fukunishi Y (1999) Active flow control technique using piezo-film actuators applied to the sound generation by a cavity. ASME FEDSM99-7232
76. Amir M, Kontis K (2008b) Application of piezoelectric actuators at subsonic speeds. J Aircr 45, 1419–1430. https://doi.org/10.2514/1.35630
77. Glezer A, Amitai MA (2002) Synthetic jets. Annu Rev Fluid Mech 34:503–529. https://doi.org/10.1146/annurev.fluid.34.090501.094913
78. Watson M, Jaworski AJ, Wood NJ (2003) Contribution to the understanding of flow interactions between multiple synthetic jets. AIAA J 41(4):747–749
79. Caruana D, Rogier F, Dufour G, Gleyzes C (2013) The plasma synthetic jet actuator, physics, modelling and flow control. Application to separation. ONERA J Aerosp Lab 6
80. Emerick T, Ali MY, Foster C, Alvi FS, Popkin S (2014) Spark jet characterizations in quiescent and supersonic flow fields. Exp Fluids 55

81. Grossman KR, Ossman KR, Cybyk BZ, Wie BZ (2003) Spark jet actuators for flow control. AIAA paper 2003-0057
82. Crittenden TM, Woo GTK, Glezer A (2012) Combustion powered actuators for separation control. AIAA paper 2012-3135
83. Moreau E (2007) Airflow control by non-thermal plasma actuators. J Phys D: Appl Phys 40(3)
84. Corke TC, Enloe CL, Wilkinson SP (2010) Dielectric barrier discharge plasma actuators for flow control. Annu Rev Fluid Mech 42(2010):505–529
85. Benard N, Moreau E (2014) Electrical and mechanical characteristics of surface AC dielectric barrier discharge plasma actuators applied to airflow control. Exp Fluids 55:1846
86. Maden I, Maduta R, Kriegseis J, Jakirlić S, Schwarz C, Grundmann S, Tropea C (2013) Experimental and computational study of the flow induced by a plasma actuator. Int J Heat Fluid Flows 41:80–89
87. Post ML, Corke TC (2003) Separation control on high angle of attack airfoil using *plasma* actuators. AIAA paper 2003-1024
88. Goeksel B, Greenblatt I, Nayeri C, Paschereit C (2006) Steady and unsteady plasma wall jets for separation and circulation control. AIAA paper 2006-3686
89. Corke T, Post M, Orlov D (2009) Single dielectric barrier discharge plasma enhanced aerodynamics: physics, modeling and applications. Exp Fluids 46:1–26
90. Little L, Nishihara M, Adamovich I, Samimy M (1999) High-lift airfoil trailing edge separation control using a single dielectric barrier discharge plasma actuator. Exp Fluids. https://doi.org/10.1007/s00348-009-0755-x
91. Jukes TN, Choi K-S, Johnson GA, Scott SJ (2006) Characterisation of surface plasma induced wall flows through velocity and temperature measurement. AIAA J 44(4):764–771
92. Gregory JW, Ruotolo JC, Byerley AR, McLaughlin TE (2007) Switching behavior of a plasma-fluidic actuator. In: 45th AIAA aerospace sciences meeting (AIAA 2007-0785)
93. Bohlito M, Jacob J (2009) Active vortex generators using jet vectoring plasma actuators. SAE Int J Aerosp 1(1):610–618
94. Benard N, Bonnet JP, Touchard G, Moreau E (2008) Flow control by dielectric barrier discharge actuators: jet mixing enhancement. AIAA J 46(9)
95. Benard N, Sujar-Garrido P, Bonnet JP, Moreau E (2016b) Control of the coherent structure dynamics downstream of a backward facing step by DBD plasma actuator. Int J Heat Fluid Flow 61(Part A):158–173
96. Samimy M, Adamovich I, Webb B, Kastner J, Hileman J, Keshav S, Palm P (2004) Development and characterization of plasma actuators for high-speed jet control. Exp Fluids 37:577–588
97. Utkin YG, Keshav S, Kim JH, Kastner J, Adamovich IV, Samimy M (2007) Development and use of localized arc filament plasma actuators for high-speed flow control. J Phys D: Appl Phys 40(3)
98. Bonnet JP, Acher G, Benard N, Lebedev A, Moreau E (2016) Sonic flow control by plasma: a new pulsed jet actuator. ICTAM Montreal, Canada
99. Roupassov DV, Nikipelov AA, Nudnova MM, Starikovskii AY (2009) Flow separation control by plasma actuator with nanosecond pulsed-periodic discharge. AIAA J 47(1):168–185
100. Unfer T, Boeuf JP (2009) Modelling of a nanosecond surface discharge actuator. J Phys D Appl Phys 42(19)
101. Starikovskiy A, Pancheshnyi S (2013) Dielectric barrier discharge development at low and moderate pressure conditions. AIAA Paper 2013-0902
102. Correale G, Kontis M (2015) Control of backward facing step flow using NS-DBD plasma actuators. 9C-4 paper, TSFP9, Melbourne, Australia
103. Woszidlo R, Nawroth H, Raghu S, Wygnanski IJ (2010) Parametric study of sweeping jet actuators for separation control. In: AIAA 5th Flow Control Conference paper 2010-4247
104. Schatzman D, Wilson J, Arad E, Seifert A, Shtendel T (2014) Drag reduction mechanisms of suction-an-oscillatory-blowing flow control. AIAA J 52(11)
105. Schatzman D, Wilson J, Marom L, Palei V, Seifert A, Arad E (2015) Suction and oscillatory blowing interaction with boundary layers. AIAA Paper

106. Upadhyay P, Gustavsson JPR, Alvi FS (2016) Development and characterization of high-frequency resonance-enhanced microjet actuators for control of high-speed jets. Exp Fluids 2016:57
107. Andino MY, Lin JC, Washburn E, Whalen EA, Graff EC, Wygnanski IJ (2015) Flow separation control on a full scale vertical tail model using sweeping jet actuators. AIAA Sci Tech paper 2015-0785
108. Jacquin L (2009) Scales in turbulent motions. ONERA J Aerosp Lab 1
109. Arakeri V, Krothapalli A, Siddavaram V, Alkislar MB, Lourenco LM (2003) On the use of microjets to suppress turbulence in a Mach 0.9 axisymmetric jet. J. Fluid Mech 490:75
110. Castelain T, Sunyach M, Juvé D, Béra J-C (2008) Jet-noise reduction by impinging microjets: an acoustic investigation testing microjet parameters. AIAA J 46(5)
111. Laurendeau E, Jordan P, Bonnet JP, Delville J, Parnaudeau P, Lamballais E (2008) Subsonic jet noise reduction by fluidic control: the interaction region and the global effect. Phys Fluids 20(1)
112. Johari H, Rixon GS (2003) Effects of pulsing on a vortex generator jet. AIAA J 41
113. Stanlov O, Seifert A (2008) On amplitude scaling for active separation control. In: International conference on jets, wakes and separated flows TU Berlin
114. Seifert A (2015) Evaluation criteria and performance comparison of actuators. Fluid Mech Its Appl 107:59–64. https://doi.org/10.1007/978-3-319-06260-0_8
115. Wong WS, Qin N, Sellars N, Holden H, Babinsky H (2008) A combined experimental and numerical study of flow structures over 3D shock control bumps. Aerosp Sci Technol 12:436–447
116. Amir M, Kontis K (2008a) Oscillation effects on boundary-layer development under the influence of favourable pressure gradients. J Aircr 45(6):1955–1968
117. Goldstein RJ (1996) Fluid Mechanics measuremensts. Taylor and Francis, Washington, DC
118. Reda DC, Muratore JJ Jr (1994) A new technique for the measurement of surface shear stress vectors using liquid crystal coatings. AIAA paper 94-0729
119. Hall JW, Tinney C, Ausseur JM, Pinier JT, Hall AM, Glauser MN (2008) IUTAM symposium on flow control and MEMS. In: Morrison JF, Birch DM, Lavoie P (eds) Springer, Berlin
120. Sheplak M, Cattafesta L, Nishida T, Mcginlevet C (2004) MEMS Shear Stress Sensors: Promise and Progress. In: 24th AIAA Aerodynamic measurement technology and ground testing conference (AIAA 2004-2606, Portland)
121. Baars WJ, Squire DT, Talluru KM, Abbassi MR, Hutchins N, Marusic I (2016) Wall-drag measurements of smooth- and rough-wall turbulent boundary layers using a floating element. Exp Fluids 57(90):1–16
122. Hochareon MB, Fontaine A (2004) Wall shear-rate estimation within the 50cc Penn State artificial heart using PIV, J. of Biomech. Eng. 126:430–437
123. Tarasov VN, Orlov AA (1990) Tarasov V N and Orlov A A 1990 Method for determining shear stress on aerodynamic model surface. Pat Russ 4841553/23/1990
124. Siaw WL, Bonnet JP (2017) Transient phenomena in separation control over a NACA 0015 airfoil. Int J Heat Fluid Flow http://www.sciencedirect.com/science/journal/0142727X
125. Lui T, Sullivan JP (2005) Pressure and temperature sensitive paints, experimental fluid mechanics. Springer, Berlin
126. Bolitho M, Jacob JD (2008) Thrust vectoring flow control using plasma synthetic jet actuators. AIAA paper 2008-1429
127. Debien A, Aubrun S, Mazelier N, Kourta A (2015) Active separation control process over a sharp edge ramp, 3D-1 paper, TSFP9. Melbourne, Australia
128. Ming X, Dai CH (1991) A new phenomenon of acoustic streaming. Acta Mech Sinica. Proceedings of the international conference on fluid dynamics measurement and its applications Oct 1989. vol. 7(3). Beijing China
129. Ming X (1992) New phenomenon of rectifying effect, Chinese. J Theor Appl Mech 24(1):52–60

Chapter 2
Description of the Test Cases

Gabriel Bugeda, Jacques Périaux, Ning Qin and Jordi Pons-Prats

Abstract The high-level objective of MARS project was to understand the formation and behaviour of turbulent structures which affects the Reynolds stress and skin friction. The aim was, once understood, to apply flow control techniques in order to control these structures and reduce the overall drag derived from the Reynolds stress and mainly from the skin friction. Active flow control devices were the main interest; DBD plasma, Synthetic jets, Micro Blowing and Suction, Moving Surfaces were included on the list. To test all these devices, two test cases were defined, and a database and file repository were established in the project webserver. The present chapter is aimed to describe the test cases, including the set-up of the flow control devices, as well as to describe the file repository were all the data was stored.

Keywords MARS · Reynolds stress · Skin friction · Drag reduction · Separation control · Flow control · DBD plasma · Synthetic jets · Blowing and suction · Slots · Synthetic Vortex Generator · Fluidric Vortex Generator · Moving Surface · Backward facing step · NACA0015

2.1 Introduction

Drag reduction and separation control are directly related to more efficient air transportation and less emission of harmful gases into the environment. While the aerospace industry is striving to have more and more optimised designs, there is still some way away from the targets set out in the ACARE 2020 vision for 50% reduction in aircraft emissions. Separation control and drag reduction contribute directly towards this target and active flow control could play an important role in achieving

G. Bugeda · J. Périaux · J. Pons-Prats (✉)
CIMNE, Barcelona, Spain
e-mail: jpons@cimne.upc.edu

G. Bugeda
Universitat Politècnica de Catalunya, Barcelona, Spain

N. Qin
University of Sheffield, Sheffield, UK

N. Qin et al. (eds.), *Advances in Effective Flow Separation Control for Aircraft Drag Reduction*, Computational Methods in Applied Sciences 52,
https://doi.org/10.1007/978-3-030-29688-9_2

it. Active flow control provides an additional dimension for further improving aircraft performance and in particular, for performance at different operational points, such as at cruise and take-off and landing.

The importance of reducing skin friction on meeting the global fuel burn targets is obvious. At cruise approximately one half of the total drag of a modern commercial transport aircraft is attributed to skin friction. The importance of improved separation control is less obvious unless we acknowledge the influence of aircraft mass on fuel burn. In fact the sensitivity of fuel burn to mass is greater than that to skin friction. Therefore if the structural mass of the aircraft can be reduced by more efficient low speed configurations or improved load alleviation off-design then this carries a significant direct benefit on cruise fuel burn.

The turbulence Reynolds stress is the most important dynamic quantity affecting the mean flow as it is responsible for a major part of the momentum transfer in the wall bounded turbulent flow. There is a lack of current understanding of the interrelationship between the various flow control devices and the Reynolds stresses in the flow field they produced. An improved understanding can potentially significantly improve the effectiveness of flow control as the Reynolds stresses are closely related to the flow behaviour at the surface for effective separation control or drag reduction. A variety of control devices are available and new ones are invented but which one for what purpose is an open question yet to be fully answered.

The focus of MARS is on the effects of a number of active flow control devices on the discrete dynamic components of the turbulent shear layers and the Reynolds stress. From the application point of view, the current proposal provides a positive and necessary step in the right direction wherein it will demonstrate the capability to control individual structures that are larger in scale and lower in frequency compared to the richness of the time and spatial scales in a turbulent boundary layer. The project will investigate active flow control means rather than passive controls.

An important characteristic of MARS project is the exploration of the possibility of influencing the mean flow via the Reynolds stress by direct manipulation of discrete structures that contribute to the stress. This is in contrast to most previous work in which the actuators change the mean flow directly. Therefore, the MARS approach offers the potential for higher system gain.

2.2 MARS Project Test Cases

Two fundamental and distinct flow cases have been chosen for the study of the effects of various flow control devices. The methodology will be based on both wind tunnel experimental investigation and numerical simulation, complementing each other for extracting flow details regarding the dynamic components and the Reynolds stresses in the shear layers, in addition to the surface properties.

The backward facing step presents a separated turbulent shear layer from a fixed sharp edge. Depending on the step height in relation to the incoming flow boundary layer thickness, various separated flow characteristics can be generated. At certain conditions, periodic unsteady flows can be achieved and the effects of the periodic component on the turbulence Reynolds stresses can be investigated in detail. For the step flow, the reattachment zone exhibits a characteristic frequency as a function of the external fluid velocity, the flow state and the step geometry. Flow control actuators will be investigated actuators regarding their influence on the characteristic frequency, amplitude and coherence of the periodicity. The capability of the flow control on increasing and decreasing the Reynolds stresses downstream of the step will be studied. The project will also investigate whether a more responsive environment can be created for flow control.

The second basic test case is closer to a realistic wing regarding the pressure gradient but the separation from a smooth surface is more difficult to handle for unsteady flows. For the NACA0015 wing, extensive studies have been carried out with different flow control devices, including the control of training edge separation and stall characteristics. Previously, fluidic vortex generators and synthetic jets were studied for suppressing trailing edge separation and wing stall. Unsteady flow separation was observed and the responsiveness of the flow to the control input was investigated. At some flow conditions, periodicity of the dynamic components was observed. However, the detailed behaviour of the Reynolds stresses near and downstream of the control devices is not very clear so far. The project will focus on the control of the periodic component to manipulate the Reynolds stresses. The interaction of the underlying flow periodicity and the control device periodicity will be identified. For the wing case, the effects of the flow control on the Reynolds stresses can be more directly related to separation control and drag reduction.

Summarizing, the paper presents the setup of the following test cases:

- Backward Facing Step; coupled with the control devices Synthetic Vortex Generator, Synthetic Jets, Plasma and Slot
- NACA0015 coupled with Plasma, Synthetic Jets, Blowing and suction, Moving Surface and Fluidic Vortex Generators.

Regarding the optimization test cases, the ones described are:

- Backward Facing Step with Plasma and with Synthetic Jets
- NACA0015 with Synthetic Jets and with Pulsed Jets.

2.3 Backward Facing Step (BFS)

The main parameters of BFS case are shown as follow:

- Length of test section L: 4 m
- Height of test section H: 0.5 m
- Width of test section W: 0.3 m
- Height of step h: 0.01 m (adaptable)
- Location of step: 3 m from inlet of test section
- Velocity: 10–30 m/s (adaptable)

Several flow control devices can be installed on a BFS, such as SJ, movable VG, and fluid VG, etc. Some considerations for this BFS case:

A. To be separated from conventional BFS flow;
B. To get full developed turbulence;
C. To be convenience for measurements and simulations;
D. To be easy for installation of FC devices (The height of BFS can be changed).

2.3.1 Synthetic Vortex Generator (SVG)

Experimentally, the SVG consisted of a metal tape stretched over a strip of magnets placed along the span of the wind tunnel. The magnets had a width of 0.005 m and were located at 0.002 m away from the wall. AC current is applied to the metal tape generating a Lorentz force that moves the tape up and down. The frequency and magnitude of the movement is controlled by the applied current. The experimental frequency of 280 Hz produced oscillation of the strip and the amplitude was 0.002 m.

All the experimental work for this study was carried out at NUAA facilities. Figure 2.1 shows a picture of the wind-tunnel where the experiments took place.

The wind tunnel had a total height of 0.47 m and a span width of 0.30 m. The step height was 0.03 m. Experiments were carried out at a free stream velocity of 19.8 m/s and the Reynolds number based on step height was 39,000. In this case, the boundary layer thickness at step edge location was around 2H according to the experimental readings.

Fig. 2.1 Wind-tunnel facilities at Nanjing University of Aeronautics and Astronautics

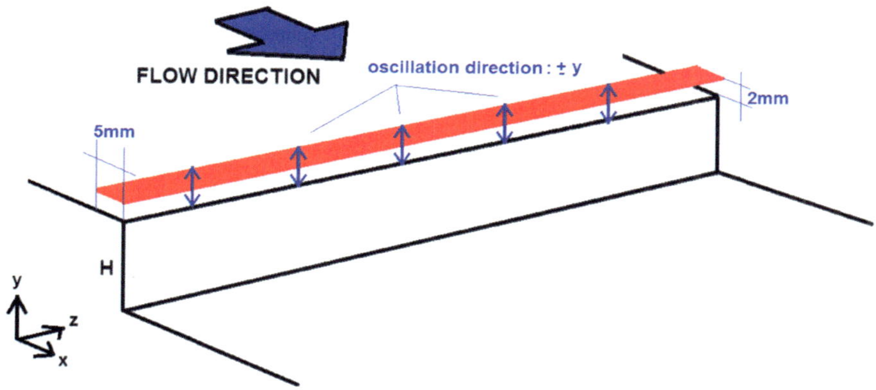

Fig. 2.2 Schematics of the simulated SVG

In order to perform the simulations, it was assumed that the SVG performed as a flat surface 0.005 m long located just before the step edge at 0.002 mm over the wall with a frequency of oscillation as the experimental, 280 Hz and maximum amplitude of 0.002 m. See Fig. 2.2.

Regarding the computational mesh, an 8 million cells provided with very reliable results. Figure 2.3 shows the computational mesh and an instantaneous capture at 0.29 s of a detail of the step region when SVG were activated in simulations.

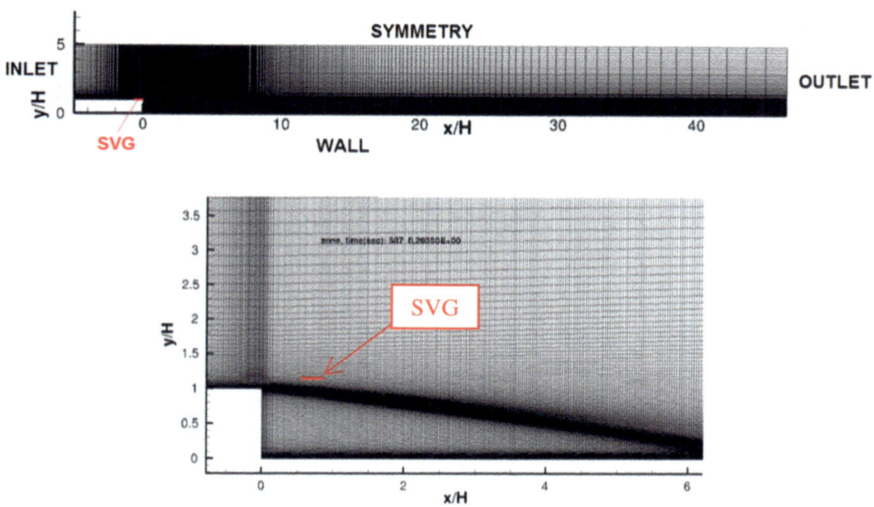

Fig. 2.3 Computational mesh and detail of step region when SVG are in operation

2.3.2 *Synthetic Jets (SJ)*

For the controlled NPU-BFS case with synthetic jets, as shown in Fig. 2.4, five jet gaps are set up in sequence along the span from the centre line and located at 65 mm upstream the step. The hole of the jets in experiment is in square shape, and distances between jets are about 40 mm. The detail parameters for this case are given in Table 2.1.

The wind tunnel in NUAA has 4 m total length, 0.5 m height and 0.3 m width. The Reynolds number is $Re_h = 8.0 \times 10^4$, based on the free-stream velocity of $U_1 = 40$ m/s and the step height $h = 30$ mm. On the other hand, the wind tunnel in DLR has an open test section of 0.7 m height and 1.05 m width. The Reynolds number is $Re_h = 2.0 \times 10^4$, based on the free-stream velocity of $U_2 = 10$ m/s and the step height $h = 30$ mm. The incoming boundary layer was artificially tripped upstream of the step to generate a turbulent boundary layer. A two-dimensional coordinate system has its origin point at the corner on the wall, a horizontal X-axis and a vertical Y-axis.

The actuator consists of a square chamber box, a thin horizontal slot, a pipe and a piston. The slot is 2 mm wide close to the step edge and facing to the downstream direction (Fig. 2.5). The square box has a cross-section of 30×30 mm^2. The fre-

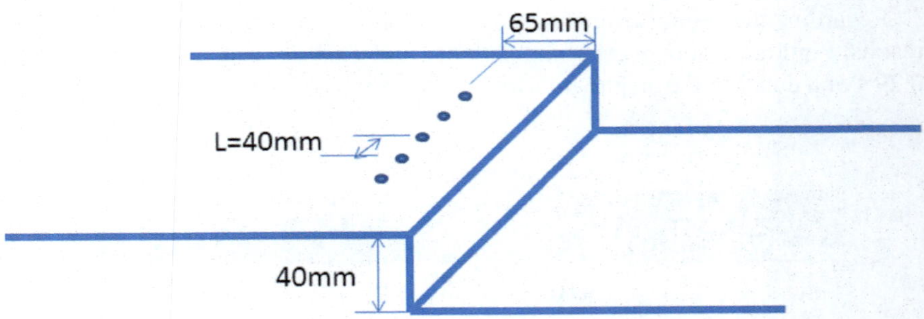

Fig. 2.4 Schematics of NPU-BFS synthetic jets test case

Table 2.1 Boundary conditions

Parameter	Value
Number of jets	5
Total span length	123 mm
Distance between jets	40 mm
Jet length	3 mm
Jet width	0.18 mm
Jet height	1 mm
Maximum jet velocity	5 m/s
Jet frequency	3000–5000 Hz
Jet density	1.17663 kg/m^3

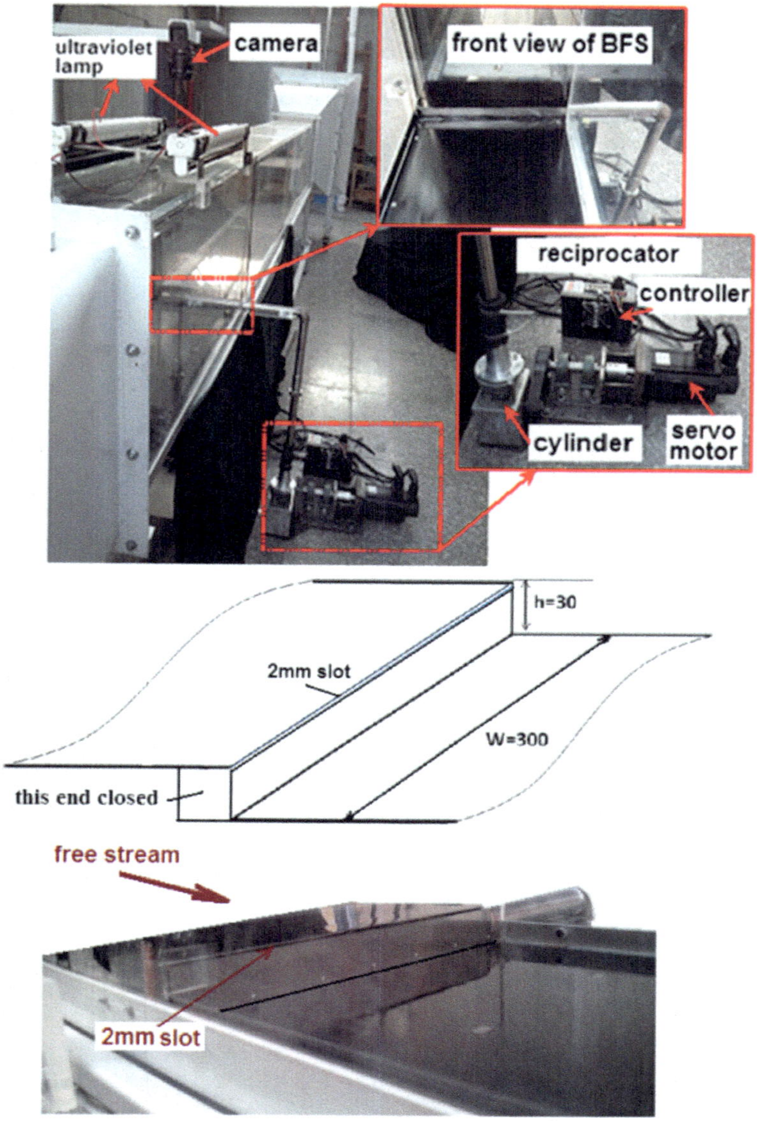

Fig. 2.5 Photo and schematic of the synthetic jet

quency of the reciprocating piston can be controlled by the servo motor. Synthetic jet is generated through the thin slot at the upper corner of the step. Three actuation frequencies of f = 0, 35 and 50 Hz were tested in NUAA by using fluorescence oil-film, while one frequency f = 100 Hz was tested in DLR by using PIV. The corresponding Strouhal number, based on the free-stream velocity and step height, are $St_h = 0, 0.026, 0.0375$ and 0.30, respectively.

2.3.3 Plasma

Experiments performed at the wind-tunnel in Poitiers facilities, Fig. 2.6, were characterised by the following configuration: the step height, H, was 0.03 m with a spanwise length of 0.3 m. The wind tunnel test section is $0.3 \times 0.3 \times 1$. Measurements were taken at the middle section covering $x/H = \{-2, 7\}$ and $y/H = \{-1.5, 2\}$.

Experiments were carried out for a free stream velocity of 15 m/s, therefore the Reynolds number based on step height is 30,000. The boundary layer thickness was calculated by means of velocity profiles measurement at various locations upstream of the step and it was determined to be approximately 13 mm at 1H before the step edge.

The actuator was mounted as shown in Fig. 2.7.

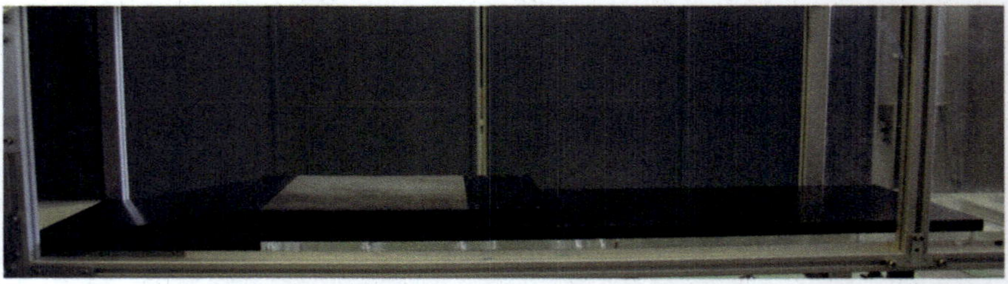

Fig. 2.6 BFS model in wind-tunnel (CNRS PPRIME Poitiers)

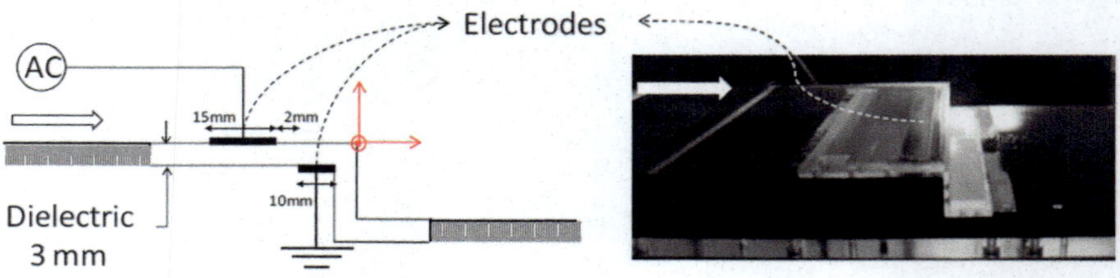

Fig. 2.7 DBD plasma actuator configuration

As it can be seen in Fig. 2.7, the exposed electrode is 0.015 m long. The embedded electrode, on the other hand, is located 0.002 m downwards the exposed and it is 0.01 m long. The dielectric thickness in this case is 0.003 mm.

2.3.4 Slot

The current simulation refers to the experiment of Driver et al. [1] in a low-speed wind tunnel, the size of whose inlet section is 1.000, 151 and 101 mm in the streamwise, spanwise and wall-normal directions, respectively. The step height, h, is 12.7 mm. The aspect ratio (the ratio of the spanwise size of wind tunnel to step height) is 12, which is large enough to ensure a quasi-2D flow in the mid-span of the wind tunnel [2]. The Reynolds number based on the free-stream velocity, $U_0 = 44.2$ m/s, and the step height, h, is $Re_h = 3.7 \times 10^4$. The boundary layer thickness is $\delta = 1.5$ h at 4 h upstream of step.

The slot harmonic actuator is located at the step edge, as shown in Fig. 2.8 with a width, d, of 0.1 h. The jet is prescribed by a velocity profile, which is uniformly distributed in spanwise direction and harmonically oscillates in time.

$$u_{ex} = U_{ex} f(\xi, \eta) \sin(2\pi f_{ex} t + \varphi)$$

where $\varphi = 0$ is the initial phase angle, f_{ex} is the excitation frequency, $U_{ex} = VRU_0$ is the amplitude and VR is velocity ratio.

The velocity profile is a laminar-like profile. An additional coefficient 1.5 is applied to ensure that the mass flow rate equal the one of the uniformly distributed profile. The profile read,

$$f(\xi, \eta) = 1.5\left[1.0 - 4.0(\xi + 0.5)^2\right]$$

where $\xi = x/d \in [-1, 0]\eta = z/d =$ x/d. The jet angle is 45° deviating from the streamwise direction.

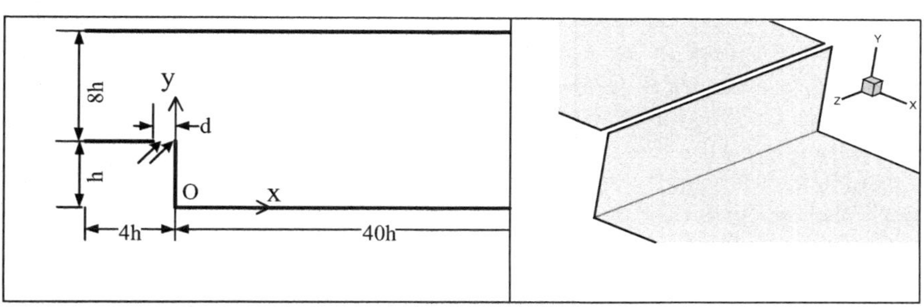

Fig. 2.8 Schematic of the slot harmonic actuator (dimensions and perspective view)

We consider the full height of the wind tunnel, which is 8 h upstream of step, so a non-slip boundary condition is applied to the upper and lower wall. The dimensions of 2D geometry are shown in Fig. 2.8. The inlet is located at x/h = −4 h, where the time-averaged velocity is available and the turbulence information is not. Hence, we conducted a 2D RANS of the wind tunnel and extracted the time-averaged velocity and turbulence information at the corresponding location to the inlet, feeding the 3D simulation. The outlet is located at x/h = 40 h, which is downstream enough to apply a convective boundary condition. Considering that the BFS flow is quasi-2D, the spanwise boundary condition is set as periodic condition. And the length of spanwise computational domain is 4 h, which is large enough to weaken the spanwise correlation [3].

2.4 NACA0015

NACA0015 geometry is well known, based on the NACA definition. It can be found on NASA archives. Following the link, a description of the geometry and aerodynamic characteristics can be found; https://ntrs.nasa.gov/archive/nasa/casi.ntrs.nasa.gov/19990047898.pdf.

2.4.1 Plasma

The experiments were conducted in one of the subsonic wind tunnels at the Aero-Physics Laboratory in the school of MACE at The University of Manchester. The facility is an open-return low subsonic 'blower' with a 0.455 m square cross section by 1.4 m long test section. The ceiling and side walls are made of optical grade perspex to allow optical access for photography and visualization. Turbulence in the test section is reduced by the presence of honeycombs located upstream of the test section. The turbulent intensity in the test section over the range of velocities used in the experiment is approximately 0.24%. The wind tunnel speed was monitored using a pitot-static tube placed upstream of the models mounted within the test section.

The aerofoil used in the experiments was a NACA0015 with a chord length, c, of 0.24 m and span of 0.40 m. For minimizing the end effects, end plates are mounted on the sides of the aerofoil made from optical grade perspex to allow visual access for ow visualization and measurements. The end plates were 0.685 × 0.457 × 0.01 m (length × height × thickness). The leading and trailing ends of the end plates were machined to have a 5 mm radius. The maximum blockage of the aerofoil which occurs at the highest incidence was estimated to be 3.9%. Therefore, tunnel blockage effect on the aerodynamic coefficients considered minimal and no correction in the measurements taken were necessary.

The configuration of plasma actuator examined in the current investigation is shown in the schematic of Fig. 2.9 [4]. The actuator is placed at the leading edge

	Width [mm]	Depth [mm]	Offset [mm]
Exposed electrode	5	0	0
A	5	0.24	0
B	40	0.48	5
C	5	0.18	45

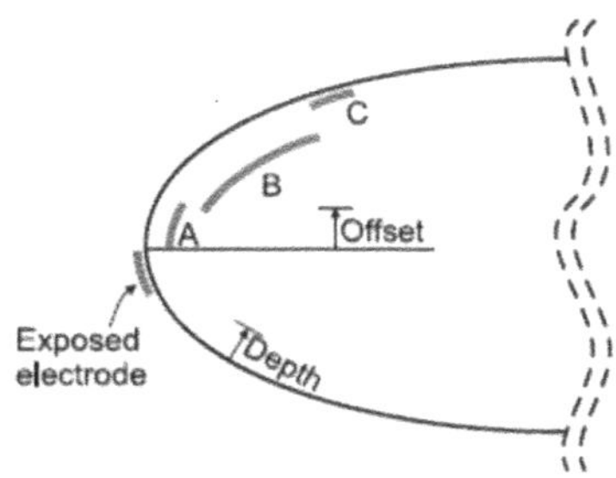

Fig. 2.9 Aerofoil actuator configuration at the leading edge (figure not scale)

while the interface of the exposed electrode and the first covered electrode is located at x/c = 0. The encapsulated electrodes are aligned so that there is no offset between the edges of successive electrodes. All the electrodes are tinned copper foils, 74 μm thick and 200 mm in length, in the spanwise direction. Layered Kapton tape was used as a dielectric material with each layer having a 60 μm thickness. Dimensions and placement of the electrodes are also provided in Fig. 2.9. The offset listed in figure is the distance measured from the downstream edge of the exposed electrode. To have a uniform plasma along the span, a small amount of overlap is applied between the downstream edge of the exposed electrode and upstream edge of the first encapsulated one.

2.4.2 Synthetic Jets (SJ)

The flow under consideration is that on a NACA0015 at $v_\infty = 10$ m/s and $\alpha = 18°$. The chord length of the airfoil is 0.5 m, and the far field boundary is 20c away from the airfoil. The width of slot is 1.5 mm, and it is located at 10%c on the upper surface of the airfoil. A structured/unstructured hybrid mesh is generated, the total number of cells is 665,000 for a 2D mesh. The airfoil is surrounded by C-type structured mesh, other part is filled by triangle mesh. There are 12 points at the slot of jet, 280 points on the upper surface totally and 120 points on the lower surface of airfoil. The corresponding 3D mesh is generated by translating the 2D mesh in span wise direction with 40 sections, and span wise length is 0.2c, as shown in Figs. 2.10 and 2.11.

On the outer boundary, the left (inlet) boundary is fixed with a uniform dimensionless inlet velocity of unity, the upper and lower boundary condition are free-stream boundaries which satisfy the Neumann condition, and the right (outer) boundary condition is set to a zero velocity gradient condition. For the airfoil, the inner boundary condition is a no-slip wall boundary condition. The near wall y^+ values of the airfoil were generally kept within 1, well within the viscous sublayer region. Three parameters (Fig. 2.12) for each jet are selected for the search investigation, namely

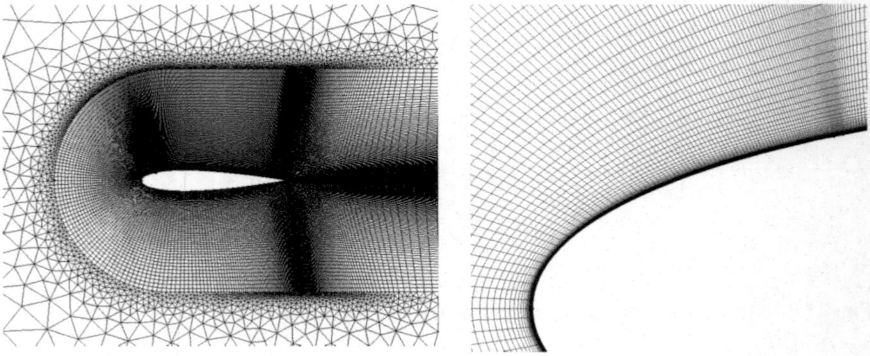

Fig. 2.10 Partial view of the mesh and mesh close to the jet

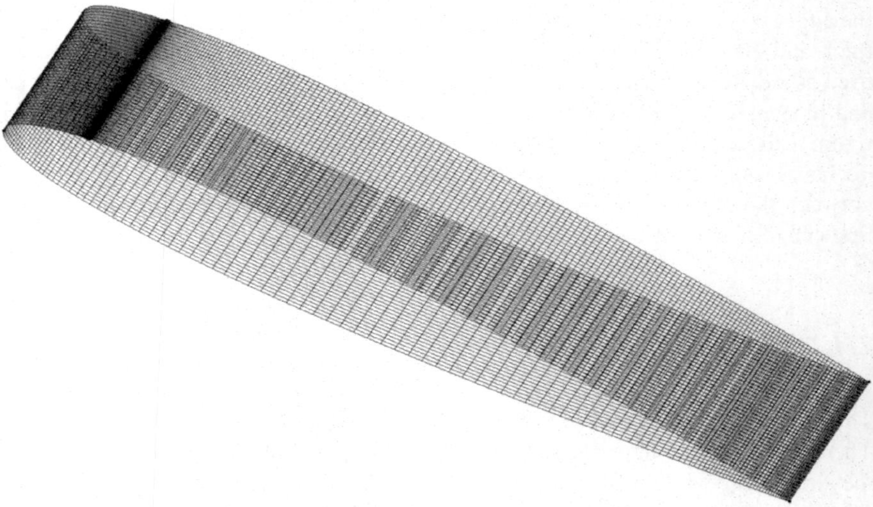

Fig. 2.11 Three dimensional mesh on the surface of wing section

Fig. 2.12 Schematic of the jet parameters for a single jet and the baseline flow conditions

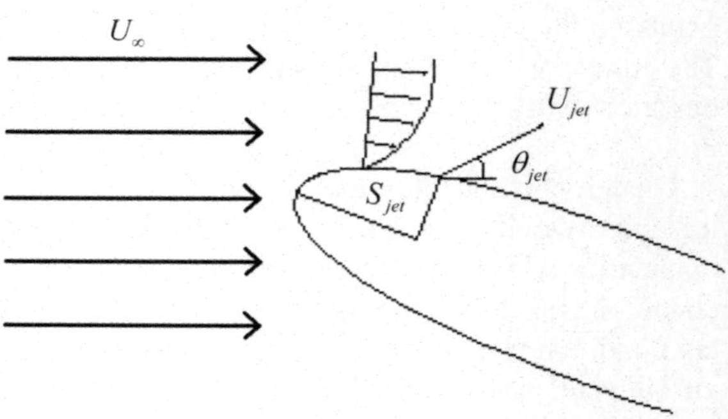

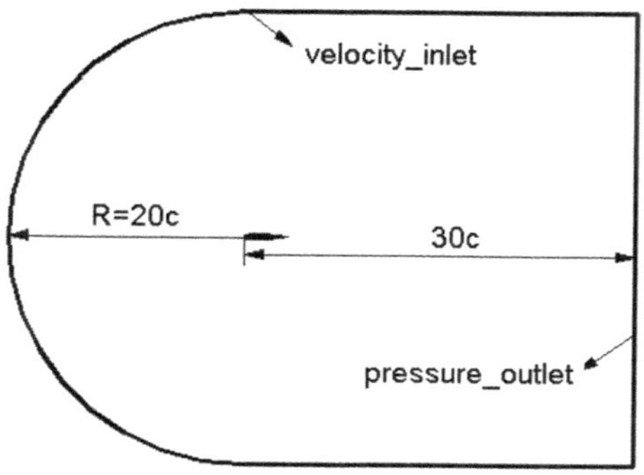

Fig. 2.13 NAC0015 airfoil computational domain

suction/blowing amplitude U_{jet}, frequency N_{jet} of jet flow and suction/blowing angle θ_{jet}. In the numerical investigation, the jet entrance velocity is set as where θ_{jet} is the angle between the chord direction and the local jet surface.

2.4.3 *Blowing/Suction (MBS)*

Considering a NACA0015 airfoil, the flow mechanism of separation flow control with blowing/suction approaches have been investigated. According to the experiment configuration by Giarranz et al. [5], the inflow velocity of 35 m/s and the control slots width of 0.53%c are set in present simulations, and the Reynolds number based on chord is approximate to 8.96×10^5. The flow control positions are located at 12, 30 and 70%c, respectively. The time step is $\Delta t = 5.357 \times 10^{-5}$ s, then the dimensionless time is $\Delta t^+ = 5 \times 10^{-3}$ so that every flow period is guaranteed to contain enough time steps. The flow is initialized by uniform inflow, lift and drag coefficients are averaged after a quantity of time steps with 30 iterations per time step when the periodicity of the flow is achieved.

The computational domain is shown in Fig. 2.13, the velocity inlet and pressure outlet conditions are separately utilized in inlet and outlet boundaries, while no slip wall condition is enforced on the airfoil surface.

The structured mesh used in present study contains 79,000 nodes and the airfoil surface is resolved with 472 points (Fig. 2.14). The local mesh refinements are applied near the jet location. The minimum grid size near the wall is 1×10^{-5} c, and the value of y^+ approximate to 0.3.

2.4.4 *Moving Surface (MS)*

A NACA0015 airfoil used in the test has a chord length 0.35 m. The MS are implemented with installation of rotating cylinders at leading edge or leeward side of the

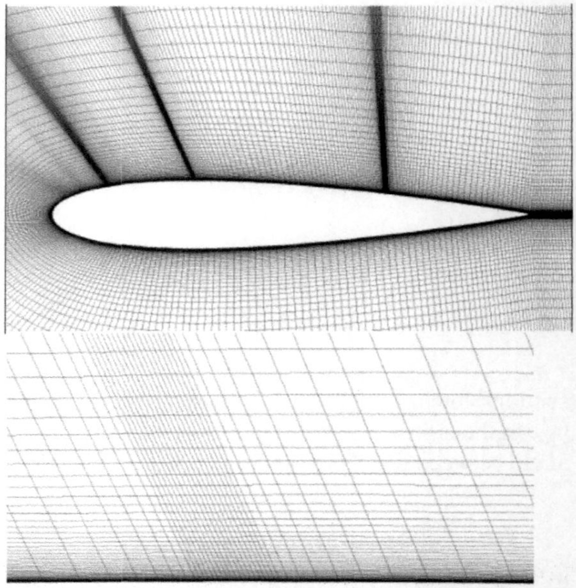

Fig. 2.14 Computational grid for NACA0015 airfoil and mesh near control device

airfoil. The front cylinder at leading edge is located at $C_1(0.05\text{c}, 0)$ with a radius $R_1 = 0.05\text{c}$, while the back cylinder at leeward side is located at $C_2(0.52\text{c},\ 0.036\text{c})$ with a radius $R_2 = 0.036\text{c}$. As noticed, since there are two different rotating cylinders, one can use them alone or together for boundary layer flow control. The cases include: single cylinder control at leading edge, single cylinder control at leeward side, and a combined control with cylinders at the leading edge and the leeward side both.

For the baseline case with a clean airfoil, the mesh is generated with a C-type topology, where the far field is with 50 times chord length, and a total number of volume elements about 90 thousands. The first level height from the wall is kept an order of magnitude around 10^{-6}c. For the control cases, all meshes are generated with the same strategy and based on the baseline case to ensure a similar size and a similar distribution for all meshes (see example in Fig. 2.15).

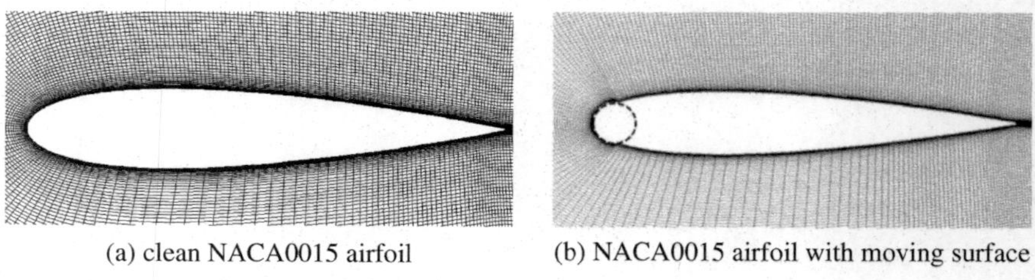

(a) clean NACA0015 airfoil (b) NACA0015 airfoil with moving surface

Fig. 2.15 Illustration of computational meshes

The flow control effect of cylinder is investigated under the flow condition at free-stream inflow velocity $U_{\infty} = 40$ m/s, angle of attack $\alpha = 11°$, and the Reynolds number Re/c $= 1 \times 10^6$.

2.4.5 *Fluidic Vortex Generators (FVG)*

NACA0015 airfoil model with a chord length of 0.35 m. We analyse the impact of the transient attachment and separation process at a Reynolds number of 1 million. To control the flow, angled fluidic vortex generators (FVG) through 44 orifices were positioned in a single array at a position of 30% from the leading edge of the airfoil.

2.5 Optimization Test Cases

2.5.1 *BFS with Plasma*

Backward Facing Step

The backward facing step geometry with DBD plasma actuator used in this study is modelled after the configuration in Sujar-Garrido et al. [6]. The BFS geometry and fully structured grid can be seen in Fig. 2.16. In this case, the top and bottom walls are included in the grid with a no-slip boundary condition. The inlet velocity profile applied at −10 h has been based according to experimental LDV measurements, and the velocity profile can be seen in Fig. 2.17. The 2D model has been extended in the spanwise direction by 1.4 h, with a grid resolution of 50 cells, resulting in a total cell count of approximately 4.5 million cells.

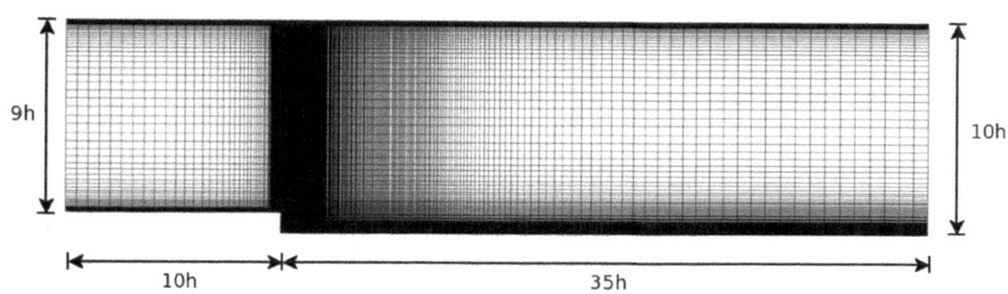

Fig. 2.16 The BFS geometry based on Sújar-Garrido et al. [6], and fully structured grid

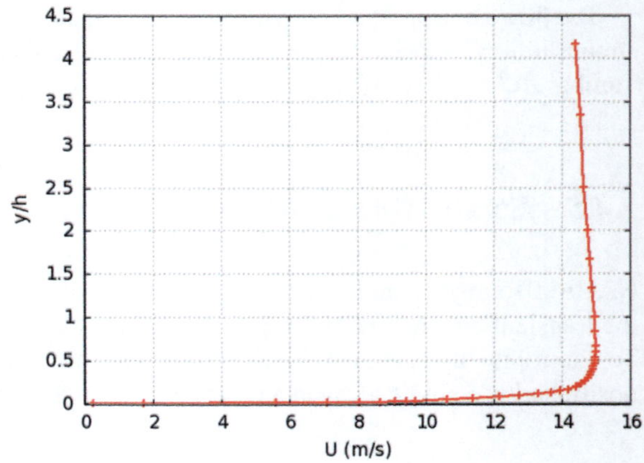

Fig. 2.17 Velocity profile at −10 h according to experimental LDV measurements by Sújar-Garrido et al. [6]

2.5.2 *BFS with Synthetic Jet*

Base Line Setup
The NUAA BFS wind tunnel test case

- Step height h = 30 mm
- Spanwise length = 10 h
- Wind tunnel height = 16 h before the step, the origin of the x–y axis is at the top of the step.
- Free stream velocity is 13.8 m/s.
- Reynolds number (based on h) = 30,000.
- According to experimental observations, the reattachment length is 5 h without control.

The NPU BFS wind tunnel test case

- Step height h = 40 mm
- Spanwise length = 18 h
- Wind tunnel height = 5 h before the step, the origin of the x–y axis is at the top of the step.
- Free stream velocity is 14.8 m/s.
- Reynolds number (based on h) = 40,000.
- According to experimental observations, the reattachment length is 6.5 h without control (Figs. 2.18, 2.19, 2.20, 2.21 and 2.22).

Optimisation Problem
The aim for the optimisation of the actuation is to change the Reynolds Stress distribution in the shear layer downstream the step.

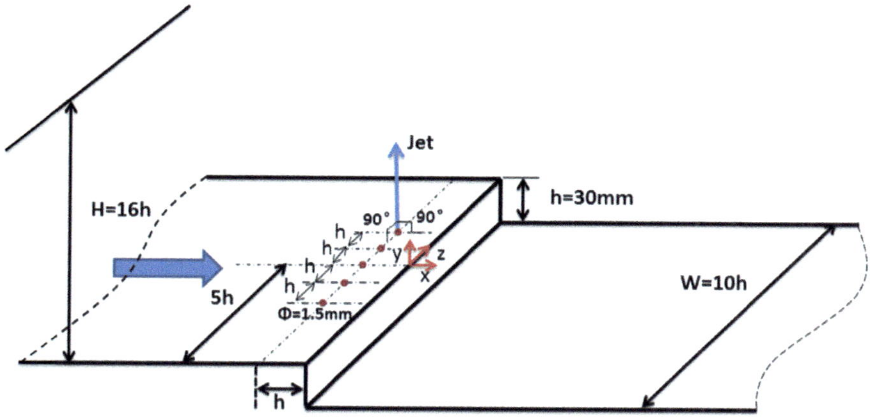

Fig. 2.18 NUAA experimental case 1

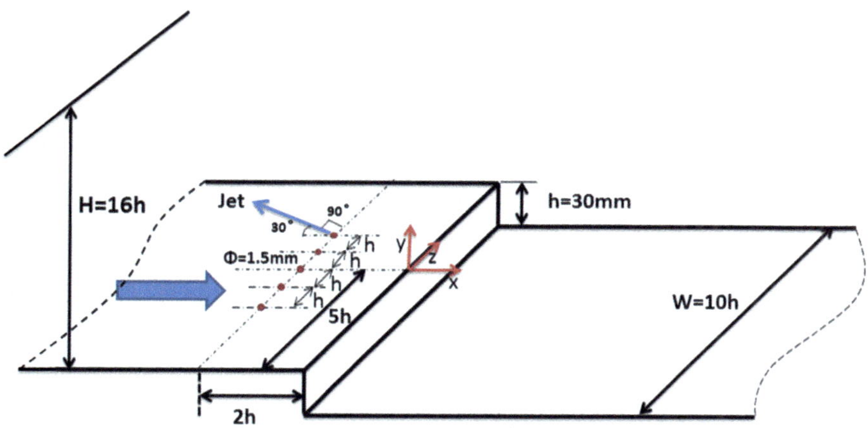

Fig. 2.19 NUAA experimental case 2

Optimization objective: The displacement of the reattached line (backward or forward).

Design Parameters

Parameter	Range
Frequency (f)	50 < f < 200 Hz
Angle of jets (based on free stream)	0°, 30°, 90°, 150°
Peak of jet velocity	3–15 m/s
Position in the stream-wise of jets	

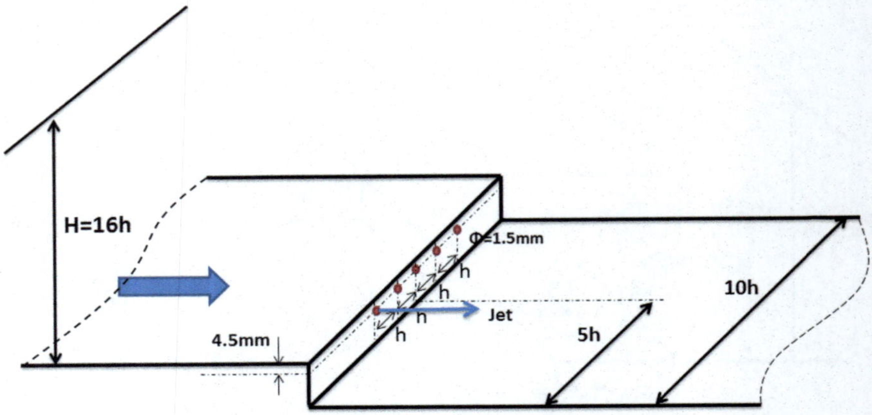

Fig. 2.20 NUAA experimental case 3

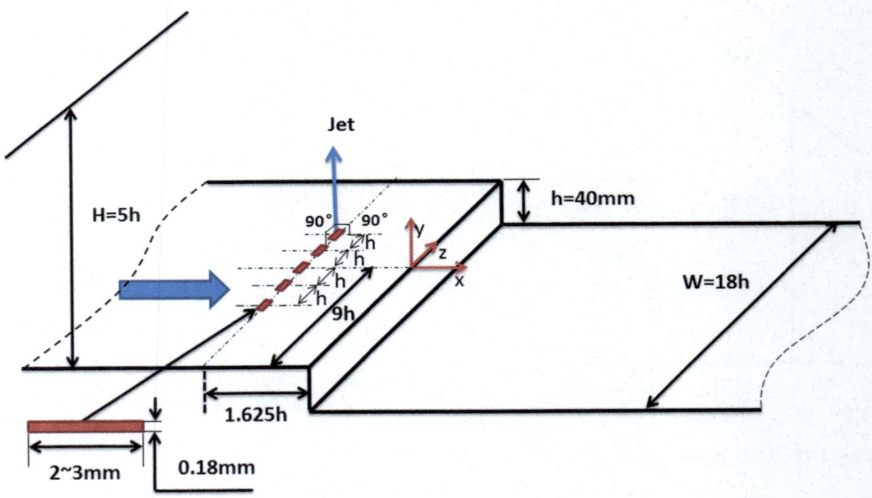

Fig. 2.21 NPU experimental case 4

2.5.3 NACA0015 with Synthetic Jet

The computational domain is defined as: upstream: 20 * C far away from leading edge, downstream: 20 * C far away from trailing edge, top boundary: 20 * C far away from chord line, bottom boundary: 20 * C far away from chord line. C is the chord length, C = 500 mm. Spanwise length = 200 mm (Fig. 2.23).

While the boundary condition; left and right boundary: periodic boundary condition, upstream, down stream, top and bottom boundary: far field boundary condition, free stream velocity = 10, 20, 30 m/s, angle of attack = 18°. No-slip boundary condition applied to the wall. Width of slot is 1.5 mm.

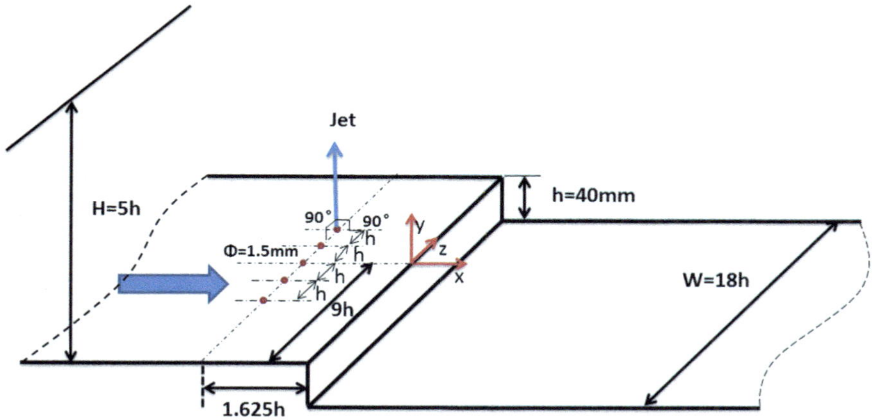

Fig. 2.22 NPU experimental case 5

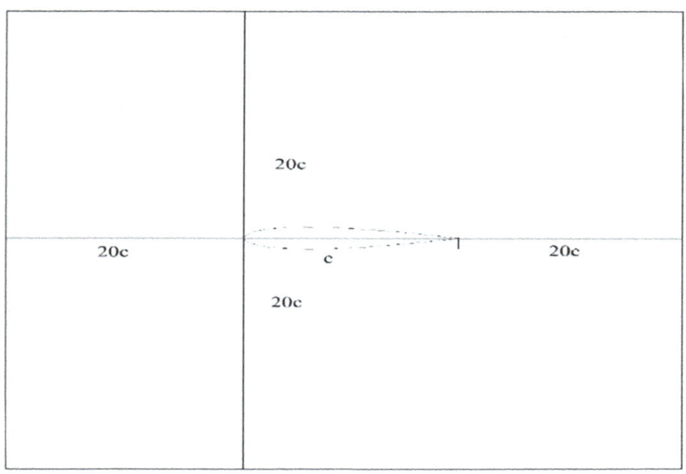

Fig. 2.23 Diagram of computational domain

The design variables are:

- frequency of jet: f, [30, 60 Hz]
- amplitude of jet flow: V_Jet, [30, 80 m/s]
- direction of jet flow: Theta, [30°, 160°]
- position of jet flow exit: p. [5–7%c].

Objective function: J is defined as the averaged lift, which should be maximized. No constraints are applied. The Optimizer is a Genetic algorithms with adaptive surrogate model.

2.5.4 NACA0015 with Pulsed Jets

Base Line Setup

The configuration of the wind tunnel and airfoil is specified in Table 2.2 and schematics with dimensional information can be seen in Figs. 2.24, 2.25 and Table 2.3.

Pulsed Jets Setup

The pulsed jets experimental configuration is described in Fig. 2.26 and Table 2.4. From Fig. 2.26 it can be seen that the orifice shape is elliptic, rather than circular. The elliptic shape of the pulsed jets can be derived from the intersection of a plane (i.e. wing surface) and a cylinder of 1 mm diameter (i.e. orifice) orientated according to Fig. 2.26. An illustration of this can be seen in Fig. 2.27.

Due to the span of the wing and number of pulsed jets, the numerical simulation has to be carried out on a smaller spanwise length to reduce computational costs. The resulting simulation model specifications that differ from the experimental configuration are listed in Table 2.5. Here, it has to be noted that the spanwise length is calculated so that periodic conditions can be correctly applied.

Table 2.2 Wing tunnel and airfoil geometric definition

Parameter	Value
Wind tunnel height	2.6 m
Wind tunnel width	2.4 m
Wind tunnel length	6 m
Aerofoil chord length (c)	0.35 m
Angle of incidence	11°

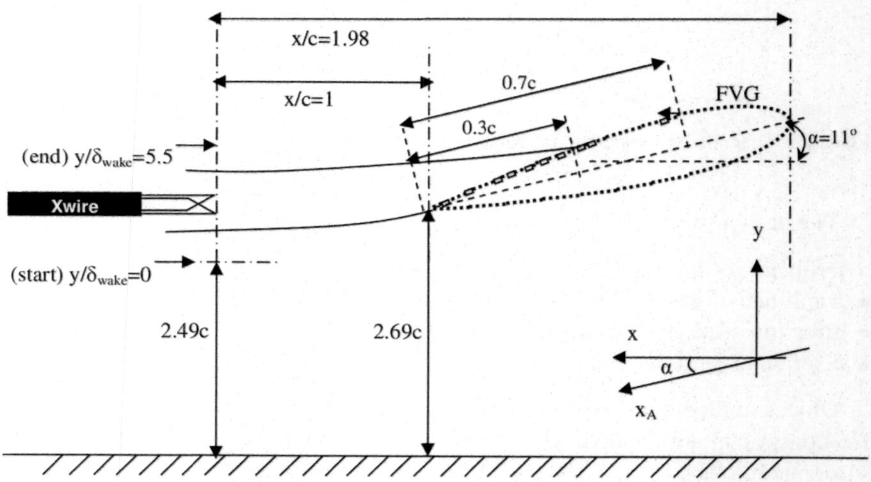

Fig. 2.24 Vertical positioning of the wing on the wind tunnel

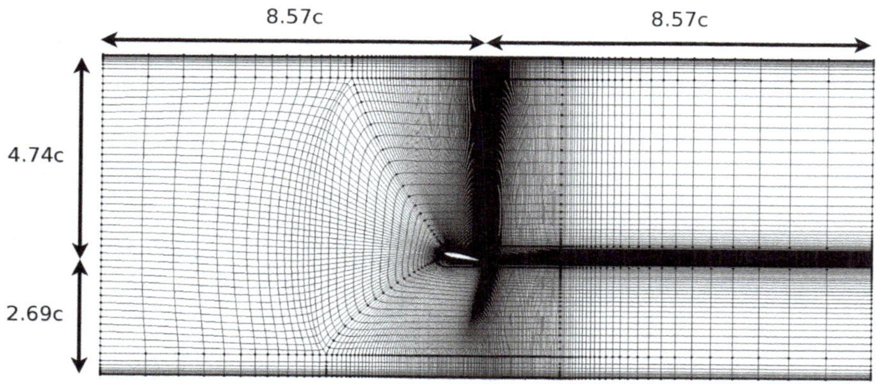

Fig. 2.25 Baseline mesh and positional information with respect to the wing's TE

Table 2.3 Boundary conditions

Parameter	Value
Velocity inlet	u = 40 m/s
Top and bottom	No slip wall
Airfoil	No slip wall
Outlet	Pressure outlet
Sides (0.2c width)	Periodic

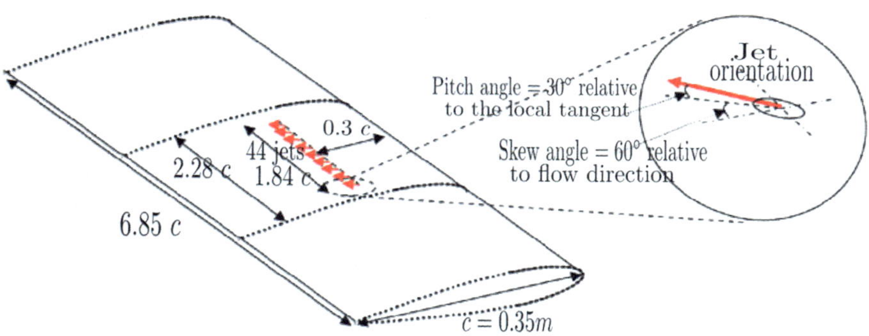

Fig. 2.26 Schematic of the pulsed jets configuration

2.5.5 *Optimisation Problem*

The aim for the optimisation of the pulsed jets operating parameters is to minimise the coefficient of drag C_D of the aerofoil.

The parameters for optimisation have been chosen due to their feasibility of being changed during an experimental campaign. The work duty here is defined as the

Table 2.4 Experimental fluidic vortex generators configuration

Parameter	Value
Total number of jets	44
Total span length	1.84c
Distance between jets	15 mm
Orifice diameter	1 mm
Jet velocity	200 m/s
Jet density	1.16376 kg/m^3
Pulsation frequency (on/off)	1 Hz
Work duty	50%

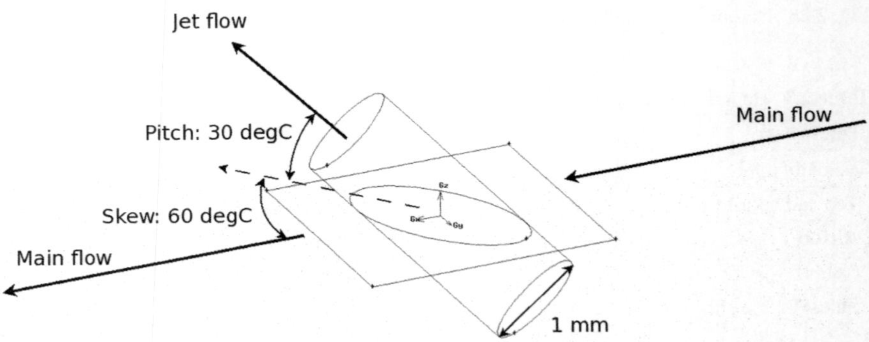

Fig. 2.27 Illustration of intersection between plane and cylinder to create the orifice elliptical shape

Table 2.5 Model parameters specific to numerical simulations

Parameter	Value
Number of FVG's	4
Span length	0.1714c
Distance between orifice centre and wing edge	0.0214c

ratio of the time where the jets are on, divided by the time for one cycle (1 Hz in the original operating parameters).

Design Parameters

Parameter	**Range**
Jet velocity, V_J	$50 < V_J < 300$ m/s
Pulsation frequency, f	$0.1 < f < 10$ Hz
Work duty, w	$20\% < w < 80\%$

2.6 Test Cases Contributors

	e——EXPERIMENTAL SIDE							WP3——NUMERICAL SIDE								
PARTNER	EU PARTNERS			CHINESE PARTNERS				EU PARTNERS					CHINESE PARTNERS			
CASE	UNIMAN	POITIER	DLR	AVIC-ARI	NUAA	BUAA	NPU	USFD	FOI	NUMECA	INRIA	CIMNE	AVIC-ARI	THU	ACTRI	BUAA
BFS			Vol PIV													
Oscillating Surfaces	X				X			X					X		X	
Plasma	X	X	X				X	X								X
Vortex Generators					X				X		X	X	X	X		
Synthetic Jets					X		X		X		X			X	X	
micro-blowing/suction					X							X				X
NACA 0015			Stereo PIV													
Oscillating Surfaces	X		X					X					X		X	
Plasma				X												X
Vortex Generators									X		X		X	X		
Synthetic Jets		X		X					X		X			X	X	
Moving Surfaces				X										X		
micro-blowing/suction						X										X

2.7 MARS website and File repository

To help communication among partners, but also to communicate the progress of the project to the general public, as part of the dissemination task within the project, a website was created. The web site was separated into two parts, a public and a private part. The private server was password protected. It was being used to assemble and distribute all project relevant information and results including agendas and minutes of meetings, reports, results and deliverables. It also includes a file repository and database to share the results of the test cases and related information.

Public Part of the MARS Web Page
The public part of the MARS web page can be found at http://www.cimne.com/mars. It contains the following sections:

- About MARS
  - Introduction
  - Objectives & Expected Results
  - Workplan
  - Consortium
- Diffusion
  - Public Docs: it contains all those documents produced by the project and considered of public access
  - Public Dissemination Activities: this is for the dissemination of the different dissemination activities organized from the project
- Members Area: this give access to the restricted part of the project file server

The first version of the web page was available in January 2011.

Restricted Part of the MARS File Server
The restricted part of the project file server contains the following sections/folders:

- Files
  - Documents
    3rd Party Publications
    Partner Publications
    Proposal and Documents
  - Project Repository
    Deliverables: here is where all files corresponding to project deliverables will be uploaded
    Meetings: here is where all documents corresponding to the different project meetings (agenda, minutes, presentations, etc.) can be uploaded.
    Technical Documents: here is where all the different working documents can be uploaded. There is a folder for each of the project tasks.

- Forum: this section allows the project participants to follow discussions about any topic related with the MARS project.
- Members: this section contains information of all participants in the project (telephone, address, photograph, e-mail, etc.). Each participant can modify and actualize his/her information when needed.
- News: this intends to be a repository of all news related with the MARS project.

All the different folders where files can be uploaded contain the corresponding self-explanatory facilities for performing the necessary operations.

Short Guide for the Website and File Repository

The website, and in particular the file repository, is organized in the most simple way. It is similar to a standard desktop environment, where folder are created to store and organize the files. The file repository was located within the MARS website (Fig. 2.28), in a private area protected with password (Fig. 2.29).

The private area of the website contains the same information as the public section, but the user can found a link to the documentation database (Fig. 2.30). The documentation is understood as the whole set of relevant data for the project and the members of the consortium. Organized under a set of folders, one can easily identify the documents related to each work package, as well as the data from each test case (Fig. 2.31). Regarding the test case folders, they were organized by test case and by contributors. In this way, all the contributors and consortium members easily located a specific data related to the test case and the contributors. The interaction between contributors were smooth and direct.

Fig. 2.28 MARS website front page and URL

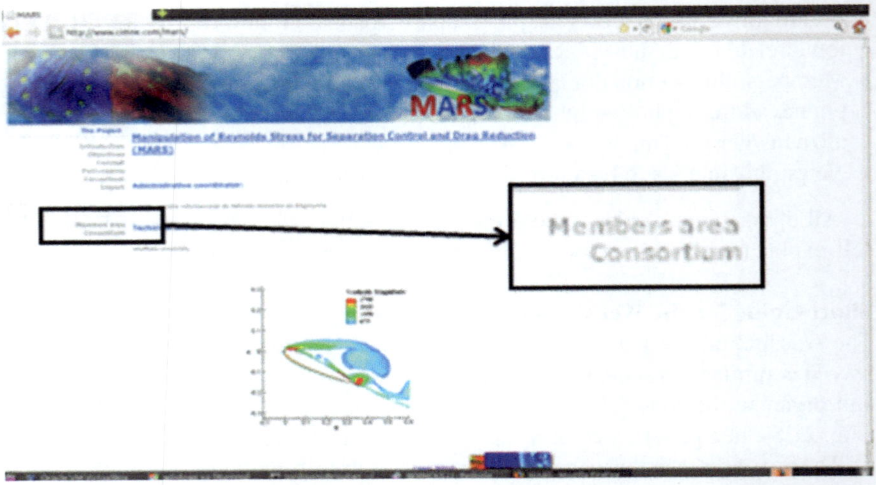

Fig. 2.29 Access to the restricted area

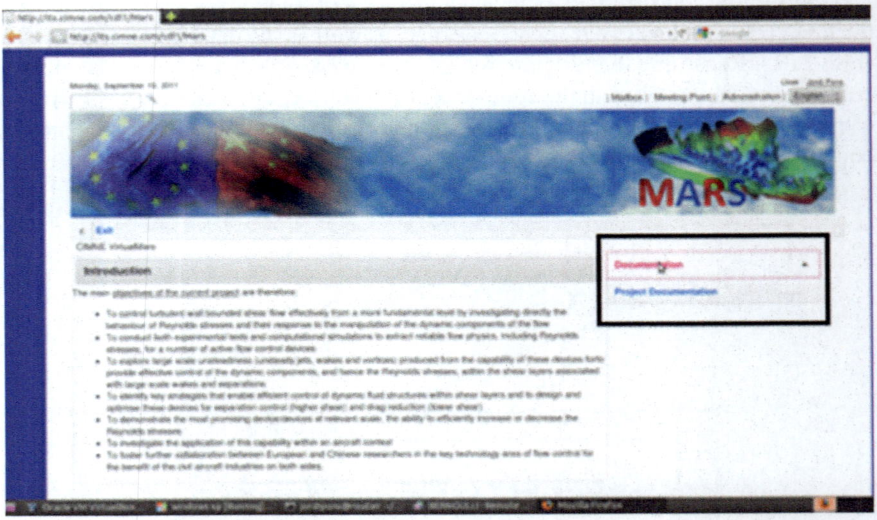

Fig. 2.30 Access to the documentation section

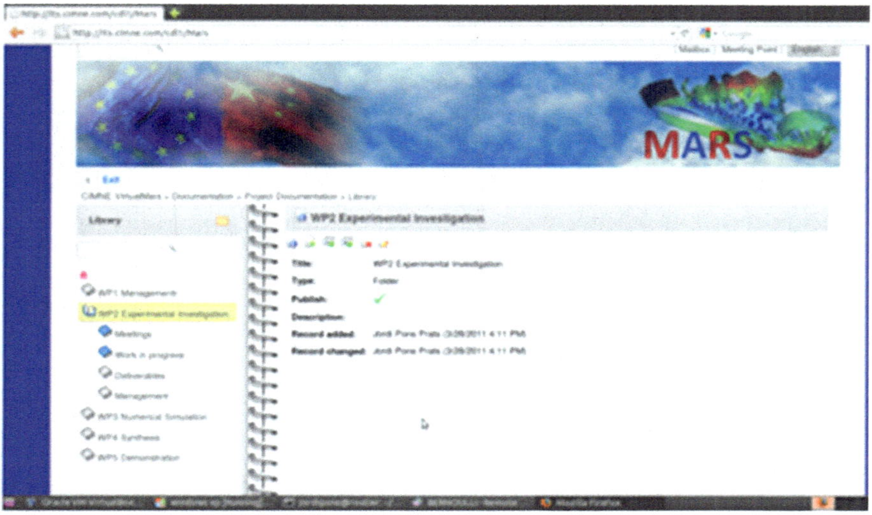

Fig. 2.31 Folders on the file repository

References

1. Driver DM, Seegmiller HL, Marvin JG (1987) Time-dependent behavior of a reattaching shear layer. AIAA J 25(7):914–919
2. De Brederode V, Bradshaw P (1972) Three-dimensional flow in nominally two-dimensional separation bubbles: flow behind a rearward-facing step. I. Department of Aeronautics, Imperial College of Science and Technology
3. Le H, Moin P, Kim J (1997) Direct numerical simulation of turbulent flow over a backward facing step. J Fluid Mech 330:349–374
4. Erfani R, Erfani T, Utyuzhnikov SV, Kontis K (2013) Optimisation of multiple encapsulated electrode plasma actuator. Aerosp Sci Technol 26(1):120–127
5. Zdravkovich MM (1997) Flow around circular cylinders: fundamentals. Oxford Science Publications. Oxford University Press
6. Sujar-Garrido P, Benard N, Moreau E, Bonnet JP (2013) Modifications of the shear layer downstream a backward facing step by dielectric barrier discharge plasma actuator. In: TSFP Digital Library Online. Begel House Inc

Chapter 3
Shear Layer and Shedding Modes Excitations of a Backward-Facing Step Flow by Surface Plasma Discharge

Nicolas Benard, P. Sujar-Garrido, Jean-Paul Bonnet and E. Moreau

Abstract The present experimental study interests in determining the influence of a linear plasma actuator (dielectric barrier discharge) on the development of a separated turbulent shear layer. More specifically, the plasma actuator is used to impose periodic perturbations at the step corner of a backward-facing step. Two different modes of excitation are explored. One concerns the shear layer mode of instability, a mode whose amplification leads to a minimization of the recirculation bubble. The present investigation shows how a dielectric barrier discharge plasma actuator can impose periodic perturbations that excite the shear layer mode and result in a strong regularization of the vortex street. The case of excitation at the shedding mode is also experimentally investigated using a DBD actuator. The measurements show the increase in Reynolds stress caused by this excitation as well as the specific growing mechanism of the shear layer. Indeed, phase-averaged flow measurements highlights the difference in the mechanism of development of the shear layer regarding the type of excitation used, the shear layer mode promoting a growing mechanism by fluid entrainment while the shedding mode enhancing the pairing of successive vortical flow structures.

Keywords Flow control · Plasma actuator · Turbulent shear layer · Triple decomposition · PIV

3.1 Introduction

Separated flows and the associated vortical flow dynamics are involved in many industrial problematics and they are sources of performance penalties, structural vibrations, and pressure losses among many detrimental effects. They are caused

N. Benard (✉) · J.-P. Bonnet · E. Moreau
Institut PPRIME (CNRS UPR3346, Université de Poitiers, ISAE-ENSMA), Bd Marie and Pierre Curie, Futuroscope, BP 30279, 86962 Poitiers, France
e-mail: nicolas.benard@univ-poitiers.fr

P. Sujar-Garrido
KTH, Linné FLOW Centre, Stockholm, Sweden

N. Qin et al. (eds.), *Advances in Effective Flow Separation Control for Aircraft Drag Reduction*, Computational Methods in Applied Sciences 52,
https://doi.org/10.1007/978-3-030-29688-9_3

by an adverse pressure gradient developing on a wall or by an abrupt change in the geometry. This latter flow separation scenario can be simplified to the case of a flow developing downstream of a backward-facing step. This simple geometry leads to the formation of a complex flow dynamics including a region of free sheared flow, a contra-rotating recirculation bubble, a shear layer in interaction with a reverse flow caused by the reattachment and an unsteady reattachment point. Several instability wavelengths coexist in the flow but some of them can be excited in order to modify the characteristics of the mean flow. Regardless of the manner to introduce periodic perturbations in the initial region of the shear layer formation (pulsed jet, synthetic jet, moving wall, loudspeaker or plasma actuators), the organization of the vortical flow structures that dominate the separated shear flow can be imposed by exciting specific instability modes.

The backward-facing step flow presents a primary periodic organization that highly resembles to a typical free plane mixing layer and that is consequently highly sensitive to an excitation at the shear layer mode of instability [1, 2]. Indeed, a separated shear layer is composed of vortical flow structures formed due to the roll-up of the separated boundary layer caused by an initial Kelvin Helmholtz instability. These structures are organized in a periodic manner but they present higher wavelengths close to the separation point than close to the mean reattachment position. In fact, these structures often pair while they are convected by the mean flow and the pairing mechanism reduces their frequency signature [3]. The pairing process implies the amplification of the subharmonic signature while the fundamental amplification rate is reduced. By imposing periodic perturbation at the step corner, the pairing mechanism can be partially or totally inhibited [4]. When scaled on the momentum thickness in the initial region, the shear layer mode corresponds to a Strouhal number $St_\theta = 0.012$ [1, 2]. As it is explicitly demonstrated in Hudy et al. [5], amplification of the shear layer instability mode results in a larger spreading rate of the shear layer that promotes a shortened reattachment. The minimization of the recirculation bubble has been observed by many authors over the years, this for a periodic excitation at a Strouhal number $St_h = 0.2$–0.4, where h refers to the height of the step [6]. Pulsed jet [1], Gautier and Aider [7], oblique periodic jet [8], and pulsed jet resulting from plasma discharges [9–12] have also demonstrated a great control authority to reduce the mean reattachment point location. In all of these studies, a minimization of the recirculation bubble when the control is conducted in order to excite the shear layer mode of instability ($St_\theta = 0.012$), sometimes also referred as step mode of instability ($St_h = 0.2$–0.4). Minimization of the recirculating region corresponds to a basic optimization problem and some recent studies have introduced optimization algorithms in feedback control loop including actuator and sensors. Genetic algorithms, flow sensor and actuators have been coupled in order to determine the optimal forcing parameters [13] or to identify the best performing control law [14]. The former study confirms the minimization of the recirculation length providing that the flow is forced at the shear layer mode excitation while the latter investigation finds an original control law (related to the low-frequency flapping motion of the shear layer) performing better than pulsed jet injection at the shear layer most instable frequency.

Another periodic excitation can affect the development of the free shear layer. Indeed, periodic perturbations whose primary frequency corresponds to the first subharmonic of the shear layer mode can enhance the pairing mechanism [15]. Subharmonic forcing can promote the merging of vortices by vortex pairing mechanism and it can produce large periodic fluctuations at the forcing frequency in the reattachment position [11]. This control mechanism is based on the importance of the first subharmonic of an instability wave in the development and growth rate of a separated shear layer [16]. It corresponds to a forcing at $St_h = 0.12$ [11] that can be rescaled based on the reattachment position X_r. In this case, the Strouhal number St_{Xr} corresponding to the subharmonic excitation is of about 0.6, a value correlated with the frequency of the vortical flow structures when they impact on the bottom wall and are periodically shed in the wake of the backward-facing step [17–19]. Thus, periodic forcing at the first subharmonic of the shear layer mode can be interpreted as an excitation of the shedding mode of the backward-facing step [20]. Excitation at the shedding mode is similar to the 'subharmonic forcing mode' used by Ho and Huang [15], Mansour et al. [21] or Paschereit and Wygnanski [22] and it can be used for promoting the merging of vortices in order to produce large periodic fluctuations in the reattachment region, these fluctuations being driven by the forcing frequency [11].

The present investigation concerns the manipulation of a separated shear layer by plasma actuator. A single dielectric barrier discharge is installed at the corner of a backward-facing step. The ionic wind produced by this type of discharge [23, 24] is used to impose periodic perturbations at both the shear layer and shedding modes of instability. Experimental measurements by particle image velocimetry are conducted in order to differentiate the influence of both control modes on the mean and dynamic flow fields, this for Reynolds numbers from 10,000 up to 60,000.

3.2 Experimental Setup

3.2.1 Step Model and Wind-Tunnel

The backward facing step configuration is composed of an initial ramp guiding the flow from the inlet of the test section to a plate where the turbulent boundary layer develops (Fig. 3.1). This plate is 400-mm long to reach a fully developed boundary layer while a zig-zag tripper is installed 360 mm upstream the step corner (its thickness is 90 μm). This device is used for a fully homogeneous turbulent boundary flow regardless of the velocity of the incoming flow. Indeed, by using a forced laminar-to-turbulent flow regime transition, the mean and dynamic properties of the incoming flow become both independent of the Reynolds number. In the present experiments, the step has a height, h, of 30 mm, this leading to an expansion ratio (ratio between test section and test section minored by the step height) equal to 1.1. This value is high enough for reducing the influence of the streamwise pressure

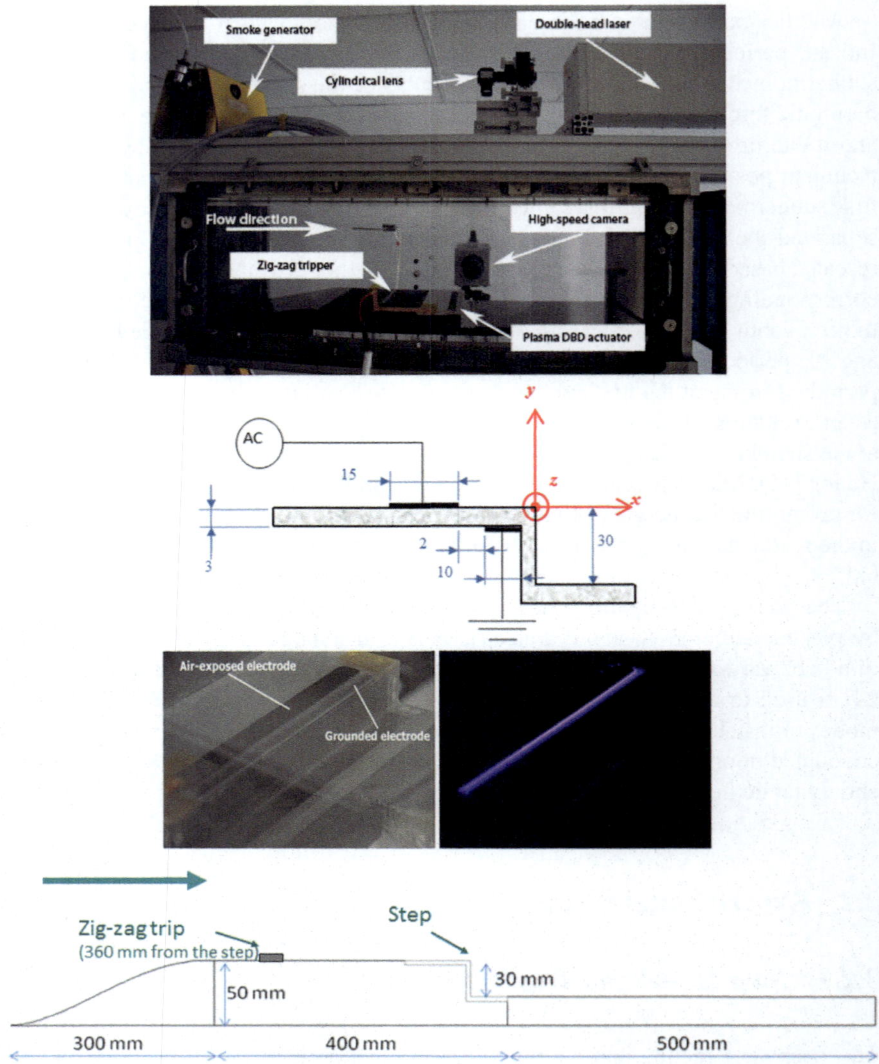

Fig. 3.1 Images of the test section equipped for PIV measurements, sketch of the linear DBD plasma actuator placed just upstream of the step corner and sketch of the backward facing step model

Table 3.1 Characteristics of the boundary layer and forcing conditions

U_0 (m s^{-1})	Re_h	δ (mm)	δ^* (mm)	θ (mm)	H	Shear layer mode	Vortex shedding mode
5	1×10^4	19	2.4	1.5	1.6	f = 42 Hz $St_h = 0.25$ $St_\theta = 0.012$	–
15	3×10^4	13	2.04	1.4	1.45	f = 125 Hz $St_h = 0.25$ $St_\theta = 0.012$	f = 65 Hz $St_h = 0.13$ $St_\theta = 0.006$
30	6×10^4	13	1.8	1.2	1.5	f = 250 Hz $St_h = 0.25$ $St_\theta = 0.01$	–

gradient developing along the wall of the wind-tunnel on the mean flow reattachment position. The aspect ratio (channel width on step height) is larger than 10 in order to have a two-dimensional flow in the centre of the wind tunnel. Here, the model spans the entire width of the test section. This section is made of transparent side walls (glass) in order to capture flow properties by optical diagnostics with lasers and cameras. This test section, whose size is 300 × 300 × 1000 mm^3, is connected to a closed-loop wind tunnel. In the present investigation, the freestream flow ranges from 5 to 30 m/s. This corresponds to Reynold numbers based on the step height, Re_h, from 10,000 to 30,000. Preliminary measurements by LDV with the backward facing step in place have shown that the meshes and grids used for flow homogenization lead to a moderate turbulent intensity less than 1.5% at $Re_h = 30{,}000$ [25]. The characteristics of the boundary layer for the three investigated freestream velocities are given in Table 3.1.

3.2.2 *Description of the Dielectric Barrier Discharge Actuator*

A non-thermal plasma actuator is accommodated on the step model for manipulation of the separated flow. The actuator design is linear in order to impose 2D periodic perturbations that may be effective for affecting the instability developing in the streamwise direction. It is made of two electrodes spanning the entire width of the test-section and arranged asymmetrically on both sides of a dielectric material. Here, the model has been designed in order to have a removable part (see Fig. 3.1) that corresponds to the step geometry. This removable part is made of a 3-mm thick machined PMMA piece and it is used as dielectric barrier directly. The air-exposed (top) and the grounded (bottom) electrodes have a width of 15 and 10 mm respectively. An inter-electrode gap of 2 mm is used. The electrode thickness is 70 μm, this low electrode protuberance into the flow reduces the influence of the actuator on the boundary layer characteristics when the actuator is not activated.

In all experiments, the air-exposed electrode is powered by a high voltage power amplifier used as voltage source (Trek 30/40). This amplifier generates a maximum output voltage of $\pm$ 30 kV (tension gain 3000 V/V, 600 W maximum power, slew rate 500 V/μs). The AC signal sends to the air-exposed electrode corresponds to the modulation of a sine wave by a gate function.

The base sinusoidal signal (24 kV amplitude and 1 kHz frequency) is used to ionize air molecules and to produce the electric field that accelerates the ionized species. For a voltage peak amplitude of 24 kV and an AC frequency (f_{AC}) of 1 kHz (consumed power of 1.3 W/cm, see Fig. 3.2b), the base signal can produce a tangential wall jet whose maximum mean velocity reaches 4 m/s (Fig. 3.2a, measurements by laser Doppler velocimetry from Sujar-Garrido [25]. However, the present investigation interests in exciting two specific modes of the separated shear layer, these two modes corresponding to low frequencies. In order to take advantage of the mean flow produced by an AC signal at 1 kHz, the AC electrical signal is alternatively turned *off* and *on* at a low frequency ($f_{BM} = 125$ Hz in Fig. 3.2a) by using a modulation with a gate function (Burst Modulation, BM). The duty-cycle is fixed at 50%. This type of modulation does not really affect the spatial distribution of the produced ionic wind; it still resembles a tangential wall jet with a thickening of the region with maximal velocity as it is shown in Fig. 3.2a. However, the amplitude of the produced mean flow velocity is reduced. For instance at 24 kV and 1 kHz, the maximal velocity is reduced from 4 to 2.7 m/s, but the produced flow presents a strong periodic oscillation at the forcing low frequency [26, 27]. This type of unsteady forcing allows us to impart predefined perturbation at low frequencies, in a range closes to the one of the preferred instability mode of the separated shear layer. As it is shown in Fig. 3.2b, the control system presents a maximal consumption of 1.8 W/cm when operated in

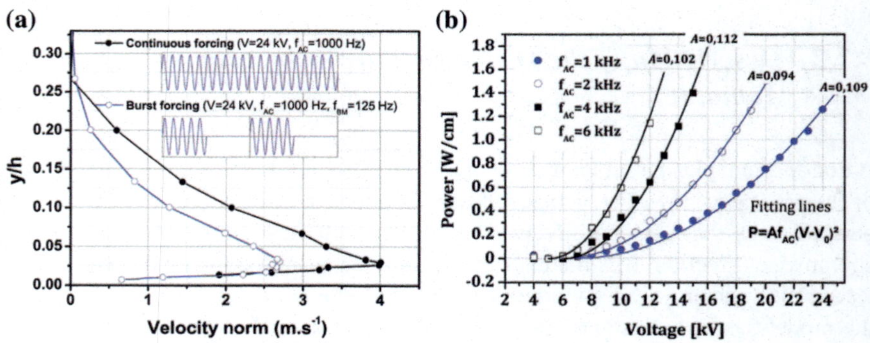

Fig. 3.2 Electromechanical conversion of the plasma DBD actuator with **a** velocity profiles (at x/h = −0.2) of the mean flow produced in case of continuous actuation (24 kV, $f_{AC} = 1000$ Hz) and burst modulation (24 kV, $f_{AC} = 1000$ Hz, $f_{BM} = 125$ Hz) and **b** the electrical power consumption of the actuator (in meter of linear discharge along the spanwise direction, i.e. normalized here by 300 mm)

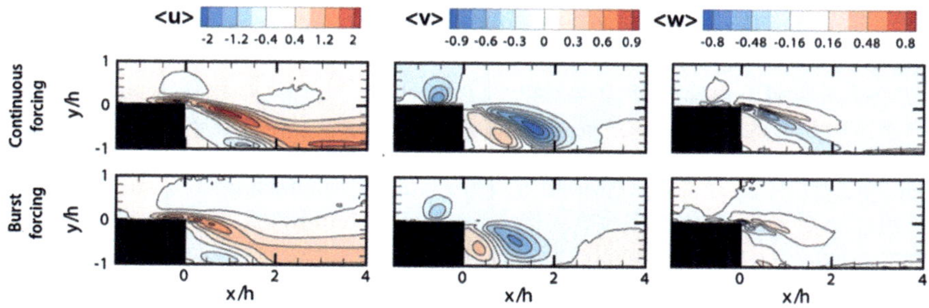

Fig. 3.3 Large field of view of the ionic wind flow produced by the actuator. The three velocity components (in m s^{-1}) are obtained by stereoscopic PIV [25]

continuous mode. However, this consumption is halved by using a burst modulation and changing the burst frequency f_{BM} has no influence on the consumed power [23].

The measurements proposed in Fig. 3.2a are focused on the small region where the plasma layer propagates. The flow produced by the actuator for initial quiescent flow conditions can also be observed from a large field of view, at the scale of the step model (Fig. 3.3). One can note that in this case the time delay between two PIV images has to be adjusted and the region of ionic wind production is no more accurately resolved. The mean flow velocity components proposed in Fig. 3.3 show that the jet penetration is high. The primary velocity component can be observed up to $x/h = 4$ but the pressure gradient at the wall forces the produced wall jet to reattach at $x/h \approx 2$. One can notice that the secondary and third velocity components are of a same order, especially when continuous forcing is achieved. In this case, the perturbation superimposed by the plasma actuator cannot be considered as fully 2D despite of a linear geometry. The burst modulation produces a more 2D flow, but at the price of a lowered maximal velocity for all the three velocity components.

3.2.3 Optical Flow Diagnostic

The turbulent flow field from the separation point to the earlier stage of the boundary layer developing downstream of the reattachment point is measured by a high-speed particle image velocimetry system. The acquisition system includes a CMOS sensor (Photron® FASTCAM SA1, 12 bits and 20 μm pixel) with maximum resolution 1024 × 1024 pixel^2 at 5400 frames per second. A 2D view of the flow is obtained thanks to a thin laser plan (1 mm) located at the mid-span of the step model. This visualization plan is performed by using cylindrical lens in the optical path of the laser beams produced by a dual laser (Quantronix® Darwin-Duo) having a wavelength 527 nm and maximum repetition frequency 10 kHz. The flow is seeded with atomized cosmetic oil (ATM210 TOPAS atomizer and Shell Ondina 917 oil). Measurements indicate that 99% of the particles have a size smaller than 0.9 μm and their mean diameter is 0.4 μm. These particles deflect the light emitted by the laser sheet on the CMOS sensor (placed perpendicularly to the laser sheet). After a noise reduction procedure by background subtraction, an adaptive correlation from interrogation areas of 64 × 64 pixels^2–16 × 16 pixels^2 with overlap 50% is applied to two consecutive images. Finally, a detailed view of the flow can be obtained with a spatial resolution of one vector each 2.4 mm. Due to the strong correlated signal during a time-resolved flow sequence, a minimum of 1500 image pairs were required for adequate statistical convergence of time-averaged velocity. It was also observed that more than 4000 image pairs are required for a converged measurement of the flow fluctuations. It was decided to conducted acquisitions by series of 5000 images recorded at a frequency of 1 kHz, this for each flow condition.

3.3 Shear Layer Forcing Mode Results

This section summarizes the results observed by forcing the separated flow at the shear layer mode. For this purpose, the applied electrical signal corresponds to burst modulation where the f_{BM} frequency is adjusted to produce a Strouhal number equal to 0.25 (Table 3.1). The influence of the plasma discharge is investigated for Reynolds numbers from 9927 to 60,000.

3.3.1 Mean Flow Field Modifications

The velocity norm and the positive/negative mean velocity component u are both shown in Figs. 3.4, 3.5 and 3.6 ($Re_h \approx 10{,}000$, 30,000 and 60,000, respectively). This latter representation is used because the frontier between positive and negative u velocity is a simple manner to identify the reattachment point at the bottom wall. In addition to the flow under the periodic forcing by burst modulation, results for

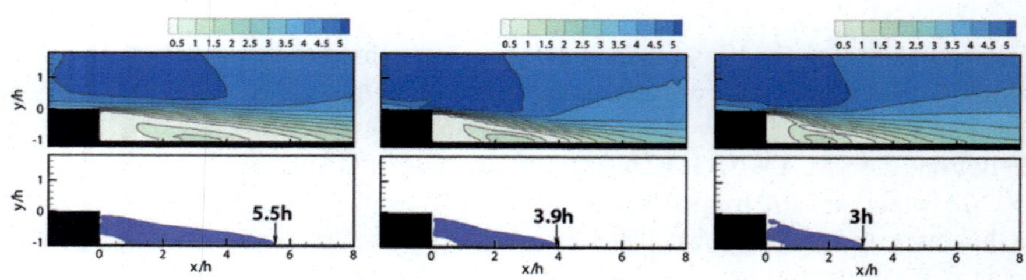

Fig. 3.4 Isolines of the velocity norm on top and reattachment position by positive and negative first velocity component u on bottom plots. Measurements for freestream flow velocity $U_0 = 5$ m s^{-1} ($Re_h \approx 10{,}000$)

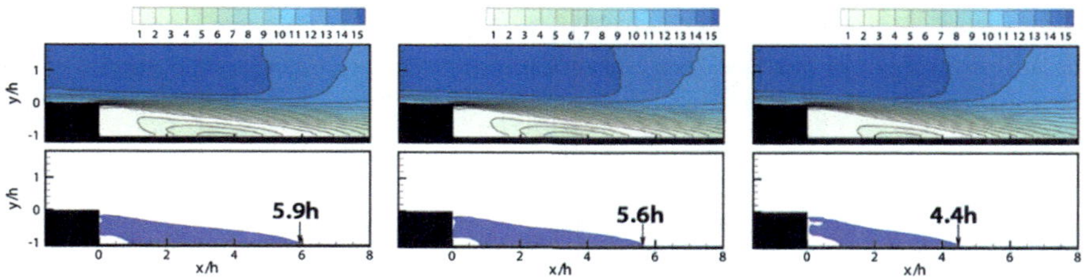

Fig. 3.5 Isolines of the velocity norm on top ($m\ s^{-1}$) and reattachment position by positive and negative first velocity component u on bottom plots. Measurements for freestream flow velocity $U_0 = 15\ m\ s^{-1}$ ($Re_h \approx 30{,}000$)

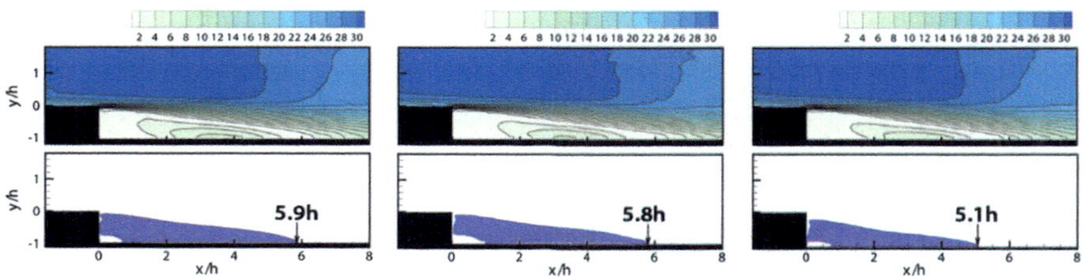

Fig. 3.6 Isolines of the velocity norm ($m\ s^{-1}$) on top and reattachment position by positive and negative first velocity component u on bottom plots. Measurements for freestream flow velocity $U_0 = 30\ m\ s^{-1}$ ($Re_h \approx 60{,}000$)

the unforced flow (baseline) and the flow forced by a continuous perturbation are proposed.

The flow fields at $U_0 = 5\ m\ s^{-1}$ are presented in Fig. 3.4. The baseline flow is characterized by a large recirculating region whose mean reattachment position is at 5.5 h from the step corner. Both forcing conditions result in a significant shortening of the recirculating region. The larger reduction (−45%) is obtained for the burst forcing at the shear layer mode (mean reattachment at 3 h), but the continuous forcing (whose action can be considered as steady regarding the flow time scales) shortens the recirculation by 29% (mean reattachment at 3.9 h). At such low freestream velocity, dielectric barrier discharges have demonstrated their high control authority in many situations [24, 28]. The ionic wind peak velocity is of the same order than the freestream flow, this for continuous and burst forcing. In such a favorable scenario, the baseline flow can be easily manipulated. The control mechanism is dominated by the mean momentum transfer produced by the discharge and a specific excitation at the shear layer mode performs only a little better.

The case of a flow at velocity of $15\ m\ s^{-1}$ ($Re_h \approx 30{,}000$) is shown in Fig. 3.4. The baseline flow reattaches at 5.9 h. This is close to the mean reattachment position observed at $Re_h \approx 10{,}000$ because in this Reynolds number range the reattachment is nearly Reynolds independent. Here, the velocity ratio U_0/U_{plasma} (with U_{plasma} the peak mean velocity produced by the linear DBD in quiescent flow) is of about 3. The momentum transfer caused by the EHD force does not dominated anymore compared to a flow control conducted at a freestream velocity of 5 m s^{-1}. It results that the continuous forcing has only a minored influence on the reattachment point (mean reattachment at 5.6 h). However, forcing at the shear layer mode leads to a reduction of the recirculating region by 25%. One can notice that this is slightly better than the results reported for the same model, but with a voltage amplitude lowered at 20 kV and AC frequency increased to 2000 Hz [11].

A final series of measurements have been conducted for a freestream velocity of 30 m s^{-1} ($Re_h \approx 60{,}000$, see Fig. 3.6). The baseline flow has a mean reattachment position 5.9 h, a same position than for freestream velocity of 15 m s^{-1} confirming the Reynolds independent character of the mean flow in the proposed Re range. Here, the continuous forcing mode has almost no effect. The mean reattachment position is located at 5.8 h. Better results are obtained with burst forcing; the recirculation length can be reduced by 13%. This reduction is still appreciable, but it also shows the reduction in control authority for an increasing Reynolds number.

3.3.2 *Effects on Turbulent Kinetic Energy*

Forcing the flow at the shear layer mode supposes that the actuation will primarily act on the organization and strength of the coherent vortical flow structures of the turbulent separated flow. In this section, the spatial distribution of the mean turbulent kinetic energy, k, measured by the rms of the 2D velocity fluctuations $\left(k = \frac{1}{2}\left(\overline{u'^2} + \overline{v'^2}\right)\right)$ is evaluated for natural and forced flows (Fig. 3.7). It highlights how the actuation can globally change the production and dissipation of turbulent energy. The spatial distribution of TKE is typically dominated by the region related to the shear layer development as for a plane mixing layer, but there is also a large region of high TKE amplitude near the mean reattachment location. The continuous forcing does not increase the amplitude of the TKE but it produces a small

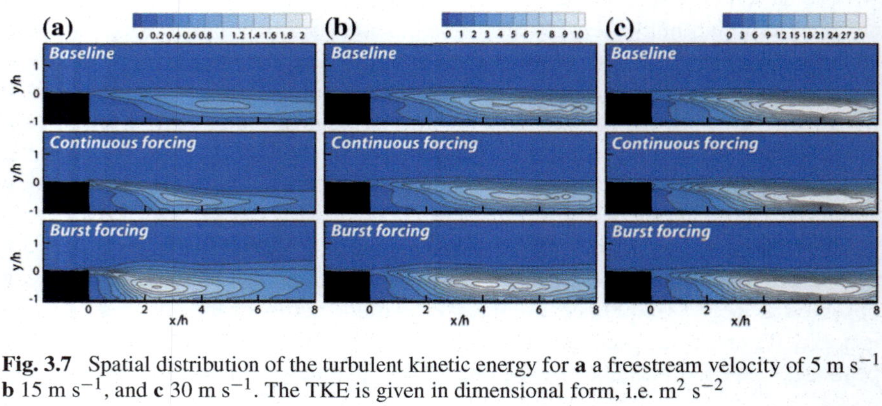

Fig. 3.7 Spatial distribution of the turbulent kinetic energy for **a** a freestream velocity of 5 m s^{-1}, **b** 15 m s^{-1}, and **c** 30 m s^{-1}. The TKE is given in dimensional form, i.e. m^2 s^{-2}

redistribution of the energy over the spatial domain. This means that production and dissipation of turbulent energy are not amplified by the continuous forcing, however the momentum transfer caused by the plasma discharge redistributes the turbulent energy. In fact, the TKE can be amplified by applying a forcing at the shear layer mode. In this case, the TKE is spatially redistributed accordingly to the change in the mean flow reattachment but a significant increase in the TKE amplitude is also observed. The TKE increase can reach almost 100% for a freestream velocity of 15 m s^{-1} ($Re_h \approx 30{,}000$), this in agreement with the turbulent production increase by 100% observed by Yoshioka et al. [4] when they imposed perturbations at $St_h = 0.19$ by loud speakers. The present results confirm the larger influence of a periodic forcing at the correct frequency than a control mechanism based on pure momentum addition using a continuous forcing signal.

3.3.3 *Effects on Reynolds Stress Tensor*

This section concerns the influence of the plasma discharge operated by burst modulation at the shear layer mode on some components of the Reynolds stress tensor. This is of particular interest for turbulent separated flows as Reynolds stress is responsible for a major part of the momentum transfer. The three components R_{xx}, R_{yy} and R_{xy} are shown in Figs. 3.8, 3.9 and 3.10 for freestream flow velocities of 5, 15 and 30 m s^{-1}, respectively. The baseline flow has a similar spatial distribution of its RS components for an increasing Reynolds number. The only change is an increase in the amplitude of R_{xx}, R_{yy} and R_{xy} with Re. As for the turbulent kinetic energy, the continuous forcing can affect the topology of RS components, but it has almost no influence on their amplitudes. At low freestream velocity, the redistribution of the momentum transfer is clearly evidenced in Fig. 3.8. However, when the Reynolds number is increased, changes on RS components are not significant. The burst modulation is more effective in modifying both spatial distribution and amplitude of the

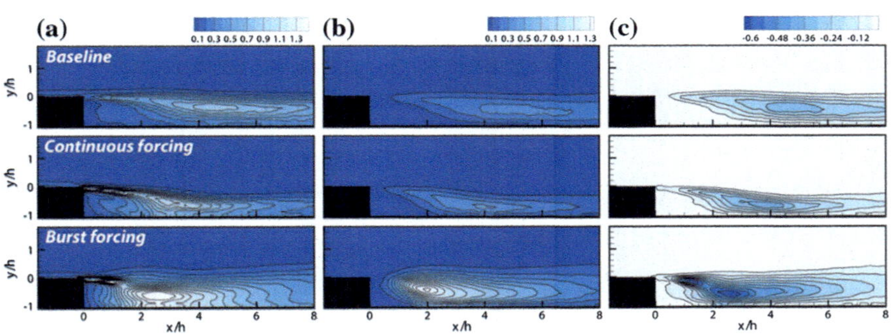

Fig. 3.8 Spatial distribution of the Reynolds stress component **a** R_{xx}, **b** R_{yy} and **c** R_{xy} for a freestream velocity of 5 m s^{-1} ($Re_h \approx 10{,}000$). Data given in dimensional form, i.e. $m^2\ s^{-2}$

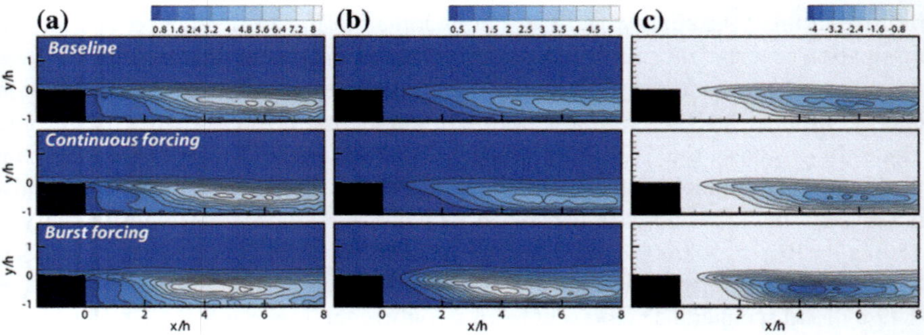

Fig. 3.9 Spatial distribution of the Reynolds stress component **a** R_{xx}, **b** R_{yy} and **c** R_{xy} for a freestream velocity of 15 m s^{-1} ($Re_h \approx 30{,}000$). Data given in dimensional form, i.e. $m^2\ s^{-2}$

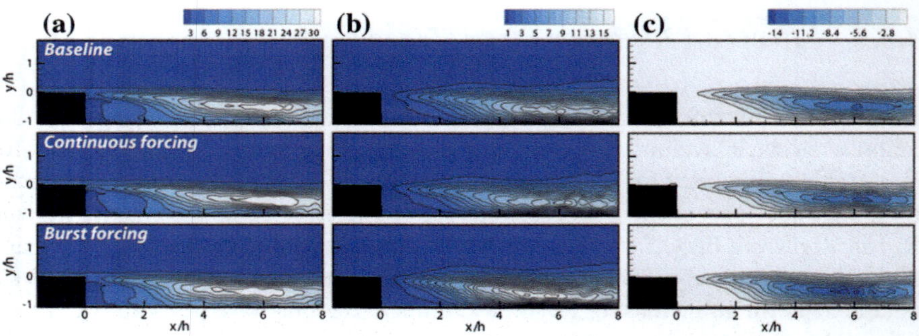

Fig. 3.10 Spatial distribution of the Reynolds stress component **a** R_{xx}, **b** R_{yy} and **c** R_{xy} for a freestream velocity of 30 m s^{-1} ($Re_h \approx 60{,}000$). Data given in dimensional form, i.e. $m^2\ s^{-2}$

three RS components. Up to a freestream flow velocity of 15 m s^{-1}, the plasma actuator can drastically increase the momentum transfer in the normal, transverse and oblique directions due to its action on the turbulent flow structures. This leads to the large increases in the amplitude of the three RS components shown in Figs. 3.8 and 3.9. At the higher speed flow (30 m s^{-1}, $Re_h \approx 60{,}000$), the ionic wind produced by the discharge can still slightly increase the amplitude of the RS components, but it is no more able to totally redistribute their topologies.

3.3.4 Dynamics of the Flow

A phase-averaging procedure applied to the velocity vector fields is now used in order to provide useful information about the mechanisms and processes underlying the flow control when using an excitation at the shear layer mode. The time-resolved character of the PIV measurements helps in reconstructing the flow field classified by phase angle, this in a posteriori processing procedure. A phase angle value is assigned to each of the PIV snapshots and a F90 code (Intel Pro Fortran) ranks the data accordingly to their phase value and it performs an ensemble average of all the flow fields at a same phase angle. This procedure corresponds to a triple flow decomposition as originally proposed by Reynolds and Hussain [29]:

$$u(x,t) = \bar{u}(x) + \tilde{u}(x, \phi(t)) + u'(x,t)$$

where u is one of the two measured velocity components, $\bar{u}$ is its mean value, $\tilde{u}$ is the phase dependent part (with ϕ, the phase angle) while u' corresponds to the stochastic turbulent part of the instantaneous flow field u. Here, phase-resolved averaged vector fields at discrete phase angles can be easily identified only for the controlled flow cases where a time reference exists.

The influence of the forcing conditions on the vortical flow structures $(\bar{u}(x) + \tilde{u}(x, \phi(t)))$ at the selected phase angles Φ is depicted in Fig. 3.11 for freestream flow velocities of 5, 15 and 30 m s^{-1}. The λ_2-criterion, as it was originally defined by Jeong and Hussain [30], has been chosen among other techniques (see in Bonnet et al. [31]) for an effective extraction of the region of low pressure, a region that can be supposed to be a vortex center. The separated shear layer exhibits a strong organization of the vortex motion when it is forced at the shear layer mode. The large scale flow structures form at x/h $\approx$ 0.05 and they are convected at a mean velocity roughly equal to $0.5U_0$ as it was observed by Bhattacharjee et al. [6]. One can notice that the convection speed is not strictly constant (see the different slopes

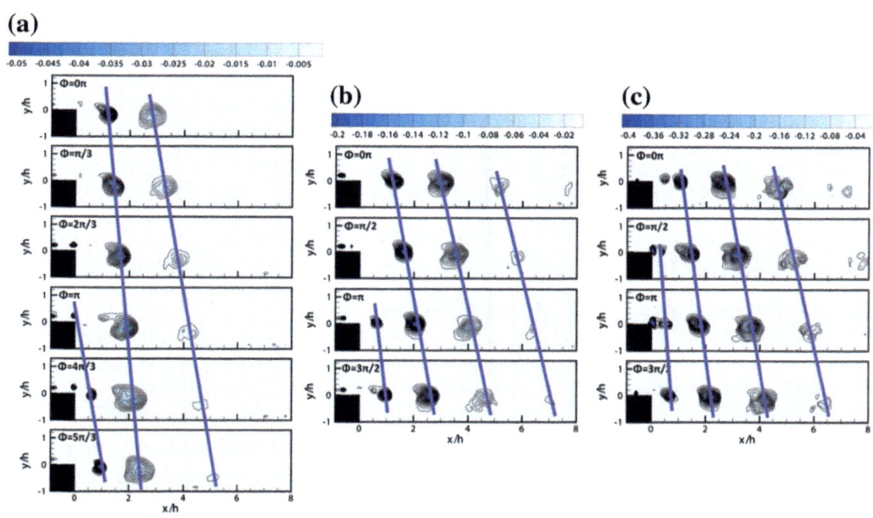

Fig. 3.11 Phase-averaged description of the vortical flow structures identified by a negative λ_2 criterion for plasma actuator operated at the shear layer mode and for freestream velocity of **a** 5 m s^{-1}, **b** 15 m s^{-1}, and **c** 30 m s^{-1}

of the line connecting a same flow structure in Fig. 3.11) because it depends on the measurement location as it is discussed in Hudy et al. [32] or Benard et al. [11]. The vortical flow structures grow in size and strength up to position x/h $\approx$ 3. Beyond this position, the vortical flow structures reach their maximal size that is roughly equal to the height of the step. One can also notice that despite inherent dissipation in turbulent flow, the flow structures survive past the mean reattachment point as in Troutt et al. [33]. The main result from these phase-averaged flow fields is the type of growing mechanism involved when shear layer forcing mode is applied. Indeed, there is almost no vortex pairing occurring during the development of the separated shear layer. A few can be observed before position x/h = 2 (see for instance the phase angle $\Phi = 3\pi/2$, at x/h $\approx$ 1, for freestream velocity of 15 m s^{-1}), in all situations the pairing involves two flow structures of different size. It can be concluded that the development of the separated shear layer primarily occurs by a fluid entrainment mechanism rather than vortex pairing when shear layer mode excitation is used. The last remark on Fig. 3.11 relates to the presence of vortical flow structures in all the tested flow velocities. Despite a lack of control authority regarding the capability of the plasma discharge to reduce the reattachment position at 30 m s^{-1}, it is shown here that the actuator has still a strong control authority on the shear layer development and periodicity of the embedded flow structures.

3.4 Shedding Forcing Mode Results

Now, the actuator is operated in a manner to excite the shedding mode of the backward facing step flow. The measurements have been conducted at only one freestream velocity fixed at 15 m s^{-1}. The sinusoidal wave at 1000 Hz is modulated by a gate function at a frequency f_{BM} of 65 Hz. In such a flow condition, this physical frequency corresponds to a Strouhal number $St_h = 0.125$. When scaled by the initial momentum thickness at the step corner, this physical frequency of 65 Hz corresponds to a Strouhal number $St_\theta = 0.006$ (see Table 3.1).

3.4.1 Mean Flow Field Modifications

The mean velocity norm and flow with positive and negative u velocity component (discriminated by color) are illustrated in Fig. 3.12a for both baseline and controlled flow whose actuation characteristics are made to excite the shedding mode. This mode of forcing seems not really effective in reducing the recirculation bubble; the reduction is limited to 12%. This actuation mode is not the best manner to minimize the recirculation length (shear layer mode performs better with a reduction by 25%). However, when the flow response to the forcing frequency is scanned manually [13] or autonomously with a genetic algorithm [11], it shows a clear reduction at the first subharmonic of the shear layer mode and it is the second better performing frequency.

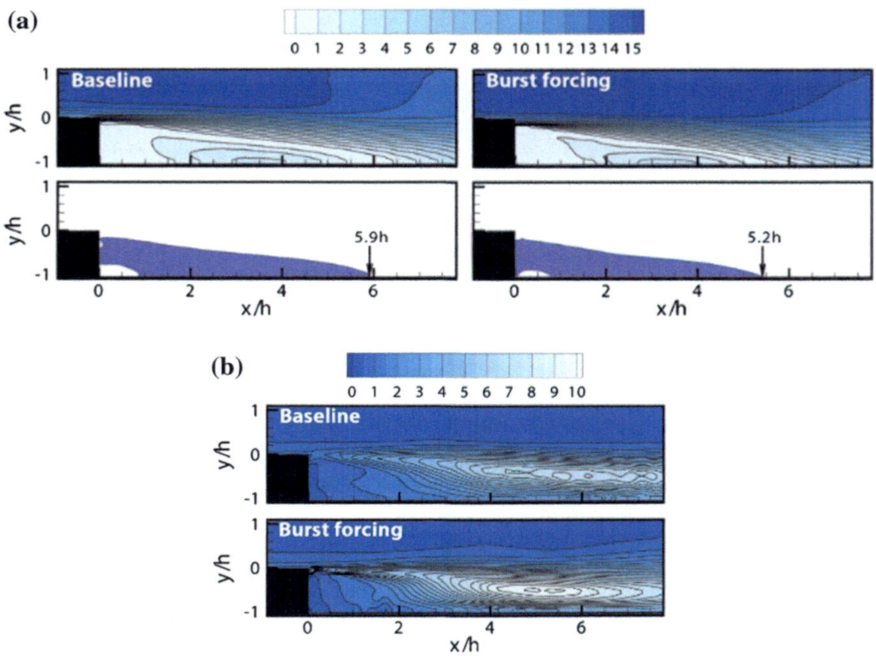

Fig. 3.12 Spatial distribution of **a** the mean velocity norm and **b** turbulent kinetic energy for baseline flow and actuation by burst modulation at the shedding mode

3.4.2 Effects on Turbulent Kinetic Energy

The spatial distribution of the turbulent kinetic energy is plotted in Fig. 3.12b in order to determine the region of high turbulent energy. Because an actuation at the shedding mode can impose strong periodic fluctuations to the reattachment point, it is expected that this type of forcing strongly affects the turbulent quantities. In fact, the spatial distribution of TKE can be divided into three regions. One is related to the first instant of the shear layer development (x/h < 2) where the actuation seems to render thinner and more intense the beginning of the shear layer. The second one expands from x/h > 2 to x/h < 3, where the flow structures are developed and can contribute to the reattachment of the separated shear layer. This region seems not to be really modified by the plasma actuation. The third region corresponds to the flow reattachment (x/h > 3.5). Here, the forcing at the shedding mode alters both spatial distribution and amplitude of TKE. The region of high TKE is not limited to the mean reattachment location but it also expands laterally. This lateral expansion is caused by the reattachment position oscillating between different locations along the bottom wall when shedding mode is applied [11].

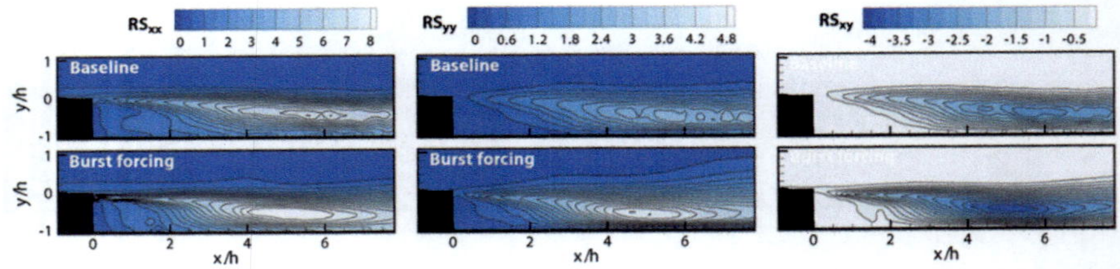

Fig. 3.13 Spatial distribution of the Reynolds stress components R_{xx}, R_{yy} and R_{xy} for a freestream velocity of 15 m s^{-1} ($Re_h \approx 30{,}000$) and burst forcing at the shedding mode. Data given in dimensional form, i.e. m^2 s^{-2}

3.4.3 Effects on Reynolds Stress Tensor

The Reynolds stress components RS_{xx}, RS_{yy}, and RS_{xy} contribute to the turbulent kinetic energy and they should be reinforced by the actuation in agreement with the TKE increase reported in the previous section. The spatial distributions of the mean value of these three components are shown in Fig. 3.13. The Reynolds stress components are dominated by RS_{xx} as typically observed in convective flows. The actuation has a severe effect on the three components of the Reynolds stress tensor. The RS_{xx} amplitude is reinforced at the beginning of the shear layer, this without changes in the shear layer width compared to the baseline flow. Beyond position x/h = 3, the amplitude and spatial distribution of this component are both increased, with a maximum amplitude located above the mean reattachment point. When shedding forcing is used, the reattachment oscillates periodically with a large amplitude, this along the bottom wall [32]. These oscillations are caused by the periodic shedding of large scale flow structures in rotation and convected by the mean flow [11]. The periodic passage and shedding of flow structures are responsible for the increase in RS_{xx}, while both increase in RS_{yy} and RS_{xy} are caused by the entrainment of irrotational fluid between two consecutive flow structures.

3.4.4 Dynamics of the Flow

Forcing the flow by periodic fluctuations at the shedding mode weakly shortens the recirculation bubble but it has a strong effect on the Reynolds stress components. Here, a phase-average procedure is used to determine the dynamic of the convected flow structures when shedding forcing mode is used (see Fig. 3.14). Flow structures, identified here by the negative λ_2 criterion, are observed in the earlier stage of the shear layer development ($x/h = 1$). They gain energy from entrainment mechanism (see for instance the vortical flow structure at x/h = 1 for $\Phi = 2\pi/5$ and convected up to position x/h = 2 at phase angle $\Phi = \pi$), however multiple places of vortex pairing are identified from the phase-averaging procedure. For instance, the merging of two successive vortices

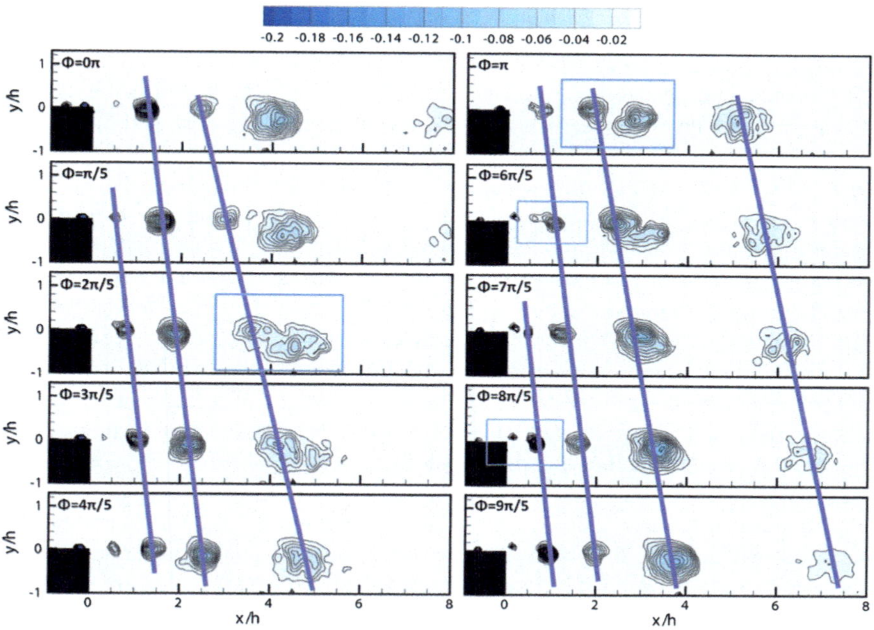

Fig. 3.14 Phase-averaged description of the vortical flow structures identified by a negative λ_2 criterion for plasma actuator operated at the shedding mode (freestream velocity of 15 m s^{-1})

is observed in the first stage of the shear layer development (x/h = 1 at $\Phi = 6\pi/5$), in the region of maximal amplification of instability (x/h = 3, $\Phi = 6\pi/5$) and just upstream of the mean reattachment point (x/h = 4, $\Phi = 2\pi/5$). The forcing at the shedding mode conduces to a shear layer which vortical activity is highly periodic and with a growing mechanisms primarily based on the pairing of successive vortical flow structures. The successive pairing of flow structures conduces to a progressive reduction in the frequency signature of the shear layer all along its expansion. In fact, the flow structures merge by two, without preference in the size of the paired flow structures, this up to having a frequency signature equal to the one imposed by the plasma actuator at the step corner. Finally, the vortical flow structures are shed at the imposed frequency. In fact, as it was suggested by Dejoan and Leschziner [34] when the large-scale flow structure detaches from the bottom wall, this leads to a sudden shrinkage of the recirculation as it was described by Benard et al. [11]. This periodic shrinking of the recirculation bubble generates strong fluctuations in the pressure coefficient measured at the wall [13]. These pressure fluctuations are also caused by the flow structures formed by pairing whose the size, and particularly their height, can exceed the height of the step. They present a strong interaction with the bottom wall as recently confirmed by numerical simulation in Hu et al. [35]. They are responsible for strong fluctuations in the size of the recirculating bubble and they interact with the bottom wall beyond of the mean reattachment position.

3.5 Conclusions

The present experiments illustrate how a turbulent separated flow reacts to an excitation at the shear layer or shedding mode. Here the excitation is imposed by a plasma actuator whose effects on the mean flow, Reynolds stress and flow structure dynamics are determined by PIV measurements. It is shown that the plasma discharge can impose perturbations that trigger shear layer or shedding mode. The benefit of an excitation at the shear layer mode has been demonstrated from Reynolds number of 10,000 up to 60,000. In this type of flow forcing, the recirculation length is minimized and the actuator imposes a highly periodic street of vortical flow structures. The formation of vortical flow structures is locked-in by the forcing frequency and the mechanism of growing by pairing is mostly inhibited. It is shown here that the control authority of the linear DBD actuator is reduced while increasing Reynolds number as already observed in many applications [24]. However, it is shown for all the tested Reynolds number that the actuator keeps its capability to impose a regular vortex street.

The plasma actuator can also be used for exciting the shedding mode of the backward-facing step. In this case, the mean position of the reattachment point is not significantly affected but a strong increase in the amplitude of the Reynolds stress components have been observed, this for positions all along the shear layer development. This type of forcing conduces to a vortex street growing by pairing of vortices at different locations along the shear layer trajectory. At the final stage of pairing, the size of the flow structures is larger than the height of the step and they are shed periodically in the wake of the backward-facing step.

The capability to impose mono-frequency flow fluctuations at the origin of large changes in the turbulent flow has been demonstrated. Plasma actuators are recognized as very fast system susceptible to impart perturbations at single and multi-frequencies. A full research space is open regarding the flow modifications that can be obtained by using simultaneously excitation of the shear layer by fundamental and subharmonic perturbations. Indeed, the concept of subharmonic resonance with dual-frequency perturbations identified by Mansour et al. [21] or Raman and Rice [36] can be easily verified using a linear plasma actuator.

Acknowledgements This work was supported by FP7/2010-2013, MARS (grant agreement no. 266326). A part of the equipment has been funded by the French Government program "Investissements d'Avenir" (LABEX INTERACTIFS, reference ANR-11-LABX-0017-01).

References

1. Chun KB, Sung HJ (1996) Control of turbulent separated flow over a backward-facing step by local forcing. Exp Fluids 21:417–426
2. Hasan MAZ (1992) The flow over a backward-facing step under controlled perturbation: laminar separation. J Fluid Mech 238:73–96

3. Winant CD, Browand FK (1974) Vortex pairing: the mechanism of turbulent layer growth at moderate Reynolds number. J Fluid Mech 63:237–255
4. Yoshioka S, Obi S, Masuda S (2001) Organized vortex motion in periodically perturbed turbulent separated flow over a backward-facing step. Int J Heat Fluid Flow 22:301–307
5. Hudy LM, Naguib AM, Humphreys WM (2003) Wall-pressure-array measurements beneath a separating/reattaching flow region. Phys Fluids 15
6. Bhattacharjee S, Scheelke B, Troutt TR (1986) Modification of vortex interactions in a reattaching separated flow. AIAA J 24
7. Gautier N, Aider JL (2015) Frequency-lock reactive control of a separated flow enabled by visual sensors. Exp Fluids 56:1–10
8. Yoshioka S, Obi S, Masuda S (2001) Turbulence statistics of periodically perturbed separated shear over a backward-facing step. Int J Heat Fluid Flow 22:393–401
9. Pouryoussefi GS, Mirzaei M, Hajipour M (2014) Experimental study of separation bubble control behind a backward-facing step using plasma actuators. Acta Mech 226:1153–1165
10. d'Adamo J, Sosa R, Artana G (2014) Active control of a backward facing step flow with plasma actuators. J Fluid Eng 136
11. Benard N, Sujar-Garrido P, Braud P, Bonnet JP, Moreau E (2016) Control of the coherent structure dynamics downstream of a backward facing step by DBD plasma actuator. Intern J Heat Fluid Flow 57:1–16
12. Sujar-Garrido P, Benard N, Moreau E, Bonnet JP (2015) Dielectric barrier discharge plasma actuator to control turbulent flow downstream of a backward-facing step. Exp Fluids 56:70
13. Benard N, Pons-Prats J, Periaux J, Bugeda G, Braud P, Bonnet JP, Moreau E (2016) Turbulent separated shear flow control by surface plasma actuator: experimental optimization by genetic algorithm approach. Exp Fluids 57:1–17
14. Gautier N, Duriez T, Aider JL, Noack B, Segond M, Abel M (2015) Closed-loop separation control using machine learning. J. Fluid Mech 770:442–457
15. Ho CM, Huang LS (1982) Subharmonics and vortex merging in mixing layers. J Fluid Mech 119:443–473
16. Martin RA, Kaul UK (2014) Optimization of perturbation parameters for simulated free shear layer flow. AIAA 2014-2223
17. Cherry NJ, Hillier R, Latour P (1984) Unsteady measurements in a separated and reattaching flow. J Fluid Mech 44
18. Driver DM, Seegmiller HL, Marvin JG (1987) Time-dependent behavior of reattaching shear layer. AIAA J 25
19. Oster D, Wygnanski I (1982) The forced mixing layer between parallel streams. J. Fluid Mechanics 123:91–130
20. Dandois J, Garnier E, Sagaut P (2007) Numerical simulation of active separation control by a synthetic jet. J Fluid Mech 574:25–58
21. Mansour NN, Hussain F, Buell C (1988) Subharmonic resonance in a mixing layer. In: Proceedings of the summer program, Center for Turbulent Research
22. Paschereit CO, Wygnanski I (1991) Instabilities in the axisymmetric jet: subharmonic resonance. In: Unger Y, Branover H (eds) Advances in turbulence studies. Progress in Astronautics and aeronautics. https://doi.org/10.2514/4.866227
23. Benard N, Moreau E (2014) Electrical and mechanical characteristics of surface AC dielectric barrier discharge plasma actuators applied to airflow control. Exp Fluids 55:1846
24. Moreau E., 2007, 'Airflow control by non-thermal plasma actuators,' J Phys D: Appl Phys 40
25. Sujar-Garrido P (2014) Active control of the turbulent flow downstream of a backward facing step with dielectric barrier discharge plasma actuators. Ph.D. thesis, University of Poitiers
26. Aono H, Sekimoto S, Sato M, Yakeno A, Nonomura T, Fujii K (2015) Computational and experimental analysis of flow structures induced by a plasma actuator with burst modulations in quiescent air. Mech Eng J 2
27. Benard N, Moreau E (2010) Capabilities of the dielectric barrier discharge plasma actuator for multi-frequency excitations. J Phys D Appl Phys 43:145201

28. Wang J-J, Choi K-S, Feng L-H, Jukes TN (2013) Recent developments in DBD plasma flow control. Prog Aerosp Sci 62:52–78
29. Reynolds WC, Hussain AKMF (1972) The mechanics of an organized wave in turbulent shear flow. Part 3. Theoretical models and comparisons with experiments. J Fluid Mech 54:263–288
30. Jeong J, Hussain F (1995) On the identification of a vortex. J Fluid Mech 285:69–94
31. Bonnet JP, Delville J, Glauser MN, Antonia RA, Bisset DK, Cole DR, Fiedler HE, Garem JH, Hilberg D, Jeong J, Kevlahan NKR, Ukeiley LS, Vincendeau E (1998) Collaborative testing of eddy structure identification methods in free turbulent shear flows. Exp Fluids 25:197–225
32. Hudy LM, Naguib AM, Humphreys WM (2007) Stochastic estimation of a separated-flow field using wall-pressure-array measurements. Phys Fluids 19:024103
33. Troutt TR, Scheelke B, Norman TR (1984) Organized structures in a reattaching separated flow field. J Fluid Mech 143:413–427
34. Dejoan A, Leschziner MA (2004) Large eddy simulation of periodically perturbed separated flow over a backward-facing step. Int J Heat Fluid Flow 25:581–592
35. Hu R, Wang L, Fu S (2016) Investigation of the coherent structures in flow behind a backward-facing step. Int J Numer Meth Heat Fluid Flow 26:1050–1068
36. Raman G, Rice EJ (1989) Subharmonic and fundamental high amplitude excitation of an axisymmetric jet. AIAA paper 1989-0993

Chapter 4
MEE-DBD Plasma Actuator Effect on Aerodynamics of a NACA0015 Aerofoil: Separation and 3D Wake

R. Erfani and Kostas Kontis

Abstract Dielectric barrier discharge (DBD) plasma actuators have received considerable attention by many researchers for various flow control applications. Having no moving parts, being light-weight, easily manufacturable, and their ability to respond almost instantly are amongst the advantages which has made them a popular flow control device especially for application on aircraft wings. The new configuration of DBDs which uses multiple encapsulated electrodes (MEE) has been shown to produce a superior and more desirable performance over the standard actuator design. The objective of the current study is to examine the effect of this new actuator configuration on the aerodynamic performance of an aerofoil under leading edge separation and wake interaction conditions. The plasma actuator is placed at the leading edge of a symmetric NACA 0015 aerofoil which corresponds to the location of the leading edge slat. The aerofoil is operated in a chord Reynolds number of 0.2×10^6. Surface pressure measurements along with the mean velocity profile of the wake using pitot measurements are used to determine the lift and drag coefficients, respectively. Particle image velocimetry (PIV) is also utilised to visualise and quantify the induced flow field. The results show improvement in aerodynamic performances of aerofoil under leading edge separation and also facing the wake region.

Keywords Plasma actuator · Dielectric barrier discharge (dbd) · Flow control · Leading edge separation · Wake interaction

---

R. Erfani
Department of Engineering, Manchester Metropolitan University, Manchester M1 5GD, UK
e-mail: r.erfani@mmu.ac.uk

K. Kontis (✉)
School of Engineering, University of Glasgow, Glasgow G12 8QQ, UK
e-mail: kostas.kontis@glasgow.ac.uk

N. Qin et al. (eds.), *Advances in Effective Flow Separation Control for Aircraft Drag Reduction*, Computational Methods in Applied Sciences 52,
https://doi.org/10.1007/978-3-030-29688-9_4

4.1 Introduction

Based on the current trend of aircraft development, environmental factors such as noise and emission limits will play a more vital role in future transport aircraft design, driving the need for greener more efficient aircraft. While the aerospace industry is striving to have more and more optimised designs, it is still some way away from the targets set out in the ACARE 2020 vision for 80% cut in NO_x emissions and 50% reduction in CO_2 emissions. Of the primary objectives of the Clean Sky initiative is achieving a total reduction in aircraft drag of 10% by reducing the wing drag by 25%. Another goal is the reduction of fuel burnt by 20%. Both the reduction in drag and fuel savings are intertwined since through innovative flow control mechanisms the need for heavy and bulky devices along the wing would be eliminated, leading to lighter aircraft and the lower consumption of fuel and hence, lower emissions. In comparison, although passive flow control methods are relatively simpler to design and manufacture, they are only effective over a small range. Therefore, when dealing with unsteady motion, such as wake interaction with aerofoils, active flow control is the dominant choice. One of the disadvantageous of active flow control is the requirement of additional power.

The first recorded use of plasma for active flow control was by Velkoff and Ketcham [1] who used a corona discharge to manipulate the transition point on a flat plate. The use of the 'ionic wind' generated by a corona discharge was the main focus of research attention until the late 1990s. In 1998 a new configuration of electrodes was presented by Roth et al. [2] that produced a One Atmosphere Uniform Glow Discharge Plasma (OAUGDP). This configuration was able to produce a jet in still air and manipulate the boundary layer of a flat plate.

Plasma actuators require no moving parts in converting electrical energy into kinetic energy, use a simple system structure, can operate over a broad frequency range and can be used instantaneously, making them ideal for flow control purposes. The simple actuator design and ease of construction allows the actuator to be retrofitted to existing aircraft and are also easy to manufacture and service. Previous studies have shown that dielectric barrier discharge (DBD) plasma actuators are efficient in different flow control applications: boundary layer manipulation [3], lift augmentation on a wing section [4], transition point manipulation [5], and in particular in the field of separation control [6–10].

Their benefits have not been fully utilised on a large scale since the applicability of plasma actuators is limited by the maximum induced velocity it can achieve. Dielectric temperature has significant effect on the strength of the induced velocity [11]. It was initially shown that breaking down the encapsulated electrode and burying each of the electrodes in a different depth respect to the surface can affect the maximum induced velocity and jet characteristics of the plasma actuator [12, 13]. This type of actuator is known as the multiple encapsulated electrode (MEE) configuration. MEE-DBD plasma actuator has shown its improved performance over the conventional DBD plasma actuator by increasing the induced jet velocity at a lower power consumption [14].

Two topics which have attracted a great deal of attention from aerodynamicists are: (i) aerofoil leading edge flow separation leading to stall, and (ii) aerofoil-wake interaction. These two test cases will be used to demonstrate the performance of a MEE-DBD plasma actuator in manipulating the aerodynamic coefficients of a NACA 0015 aerofoil with a chord Reynolds number of 0.2×10^6. The description of the optimum design of MEE-DBD which has been used in the current study has been presented by Erfani et al. [15, 16].

4.1.1 Leading Edge Stall

The stall condition is an undesirable phenomenon where controlling it will improve the aircraft performance by enabling it to take off and land at higher incidences and lower speeds. Controlling the stall angle can be achieved through delaying the leading edge separation on the aerofoil which is usually associated with achieving higher lift magnitudes and lower drag. There are several mechanisms in the field of passive and active flow control to prevent leading edge separation, such as: roughness and trips, vortex generators, 2D and 3D bumps, and dimples are amongst the passive methods (but not limited to) while boundary layer suction, leading edge flap and tangential steady/pulsating blowing piezoelectric actuators, synthetic jets, Lorentz force separation control, oscillating surfaces and membranes are amongst the active ones. All of these methods are designed to enhance the momentum of the flow or remove the low momentum flow close to the surface in order to overcome adverse pressure gradients.

Post and Corke [7] illustrated the advantage of plasma actuators compared to vortex generators. The aerofoil used in their experiment was a NACA 66_3-018 with a maximum Reynolds number of 0.15×10^6. The DBD plasma actuator found to delay the separation by 6° past the stall angle. He et al. [6] showed that the leading edge separation control by the use of conventional DBD plasma actuator resulted in an increase in both the maximum lift coefficient and the stall angle of attack by 4°, for NACA 0015 aerofoil with chord Reynolds number of 0.2×10^6. Rethmel et al. [10] carried out a study with an identical aerofoil and chord Reynolds number flow and observed that the nanosecond pulse driven DBD can extend the stall angle.

4.1.2 Aerofoil-Wake Interaction

In aeronautics, most open and closed turbo-machineries, and other engineering disciplines the flow over an aerofoil is influenced by a wake which for instance is originated from the preceding aerofoil. Such a wake can reduce the lift and increase loads on the downstream aerofoil which can create dangerous conditions for flying vehicles. This makes the study of the wake influence on aerodynamic characteristic of the aerofoil necessary. The disturbances in external flow initiate and force transition of boundary layers from laminar to turbulence. Extensive analytical, numerical and experimental studies have been conducted on boundary layer-wake interaction [17–19].

The aerodynamic characteristics under the influence of the preceding wake have also been studied in relation to the laminar-turbulent transition of the boundary layer over an aerofoil. It is expected that the boundary layer over the aerofoil becomes turbulent through a transition process caused by the interaction between the wake and the boundary layer [20–24]. In addition, there are some investigations regarding the studies of aerodynamic forces of flying vehicle facing the wake of an aircraft [25, 26].

Kornilov et al. [27] discussed the interaction between an incompressible two-dimensional turbulent wake produced by a symmetrical aerofoil at incidence and a boundary layer formed on a similar aerofoil immediately downstream. A significant reduction in the level of turbulence was found in the boundary layer of the downstream aerofoil when it is located in the wake periphery. They identified that the outer region of this interaction can be described by simple correlations traditionally used for the wake behind a circular cylinder. Fujisawa et al. [20] discussed the influence of the circular cylinder wake on the aerodynamic performance of an aerofoil. Their results indicated that due to the interactions between the fluid forces and the cylinder wake the drag force decreases and the lift force increases at relatively large angles of attack typically beyond $\alpha = 12$ degrees. Their results show an improvement in aerodynamic performance of the aerofoil due to the influence of cylinder wake except for the lift coefficient at $\alpha = 9$ to 12 degrees.

4.2 Plasma Physics

The typical plasma actuator is a linear asymmetric arrangement of two electrodes separated by a dielectric material. One electrode is exposed to the air, while the other is encapsulated in the dielectric. A schematic illustration of such an actuator is depicted in Fig. 4.1 for a standard and MEE-DBD actuator. A high voltage alternating current (a.c.) input, with typical voltages of 2 kV_{p-p} to 60 kV_{p-p} (peak to peak voltage) and frequencies of 300 Hz to 1 MHz is supplied to the exposed electrode while the encapsulated electrode is earthed. Typically, the plasma actuator is long and thin and is placed spanwise on the aerodynamic surface.

Air ionisation occurs at the edge of the exposed electrode on top of the insulated surface where a region of high electric field potential exists, resulting in the formation of a visible plasma. The plasma is limited to the extent of the covered electrode [28]. Energetic, relatively high mass ions impart momentum into the neutral air, forming an induced jet. This jet is directed away from the exposed electrode, due to the asymmetric placement of electrodes, and in turn introduces momentum into the flow.

Although the plasma appears as a relatively uniform diffuse discharge to the unaided eye, similar to that shown in Fig. 4.2, optical measurements of the plasma

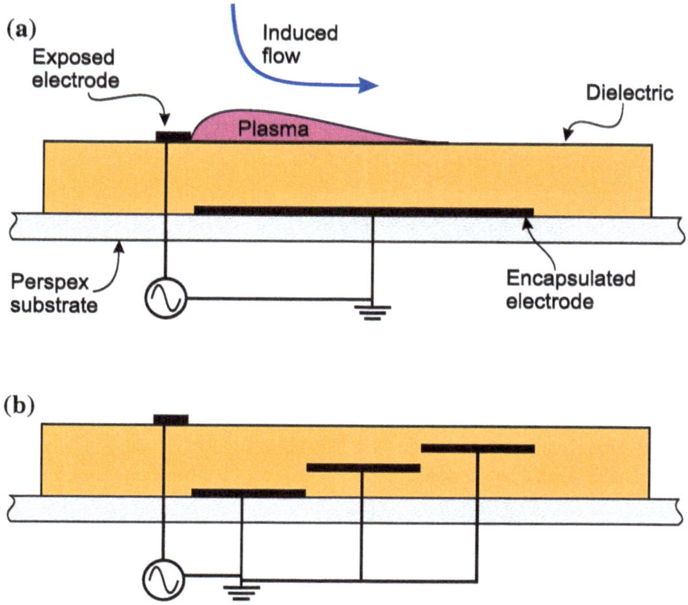

Fig. 4.1 Typical configurations of standard DBD (**a**), and MEE-DBD plasma actuators (**b**)

Fig. 4.2 Photograph of plasma created on a DBD plasma actuator

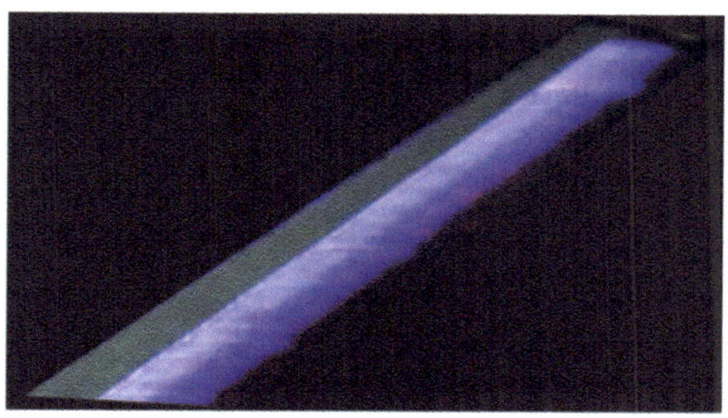

indicate that each half of the voltage cycle consists of the ignition, expansion and quenching of many microdischarges [29, 30]. The charge deposited on the surface reduces the applied electric field at that location. It also shuts down the discharge at macro scales, therefore it is necessary to supply a alternating voltage to maintain a high enough potential gradient to sustain the plasma. The presence of dielectric prevents the formation of an electric arc and spreads the discharges along the dielectric surface almost uniformly.

Figure 4.3a shows the current and alternating voltage trace for a DBD plasma actuator. As the voltage of the exposed electrode becomes more negative electrons are deposited upon the dielectric surface, as shown schematically in Fig. 4.3b. This

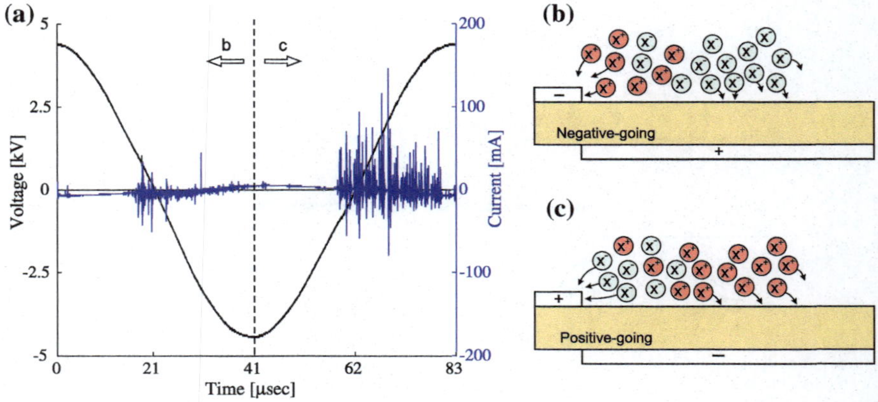

Fig. 4.3 Typical Voltage and current trace of a DBD plasma actuator (**a**), when electrons leave the exposed electrode (**b**), some deposited electrons returning (**c**) (subfigures (**b**) and (**c**) are not to scale)

stage of the cycle has a current discharge that is relatively uniform due to the ease with which electrons can leave the electrode surface between 16 and 41 μs. During the second part of the cycle, when the voltage becomes increasingly positive, the deposited electrons are attracted back to the exposed electrode as shown in Fig. 4.3c. During this phase the electrons find it harder to leave the dielectric surface, this behaviour leads to a patchy discharge visible in the current trace between 58 and 83 μs.

Besides electrons other species such as negative oxygen ions exist also in the plasma, as a result of electron reattachment [31]. Since ions are larger in mass they can transfer more momentum to the neutral air compared to electrons. It has been shown that the majority of momentum coupling happens when the voltage is becoming more negative and the electrons and negative ions are attracted to the dielectric surface [32, 33].

4.3 Apparatus and Instrumentation

4.3.1 Wind Tunnel

The facility is an open-return low subsonic 'blower' with a 0.455 m square cross section by 1.4 m long test section. The ceiling and side walls are made of optical grade perspex to allow optical access for photography and visualisation. Turbulence in the test section is reduced by the presence of honeycombs located upstream of the test section. The turbulent intensity in the test section over the range of velocities used in the experiment is approximately 0.24%. The wind tunnel speed was monitored using a pitot-static tube placed upstream of the models mounted within the test section.

4.3.2 *Aerofoil*

The aerofoil used in the experiments was a NACA 0015 constructed from perspex with a chord length, c, of 0.24 m and span of 0.40 m. The characteristics of this aerofoil are well documented in the literature and it exhibits well-behaved leading edge separation at high angles of attack. For minimising the end effects, end plates are mounted on the sides of the aerofoil made from optical grade perspex to allow visual access for flow visualisation and measurements. The end plates were $0.685 \times 0.457 \times 0.01$ m (length × height × thickness). The leading and trailing ends of the end plates were machined to have a 5 mm radius. The maximum blockage of the aerofoil which occurs at the highest incidence was estimated to be 3.9%. Therefore, tunnel blockage effect on the aerodynamic coefficients considered minimal and no correction in the measurements taken were necessary.

4.3.3 *Cylinder*

At the working chord Reynolds number of 0.2×10^6, the frequency of the wake of the aerofoil at zero incidence is measured as 120 Hz and a Strouhal number, $St = fD/U_\infty$, of the cylinder estimated to be 0.2, where f is the frequency of the wake, D is the diameter of the cylinder and U_∞ is the free-stream velocity [34–36]. Therefore, a cylinder which can generate a frequency similar to that of the aerofoil should have a diameter of 0.02 m [20].

The vortex shedding of the cylinder depends strongly on its diameter and the distance between the cylinder and the aerofoil. This critical distance is found by Luk et al. [37] to be equal to the length of the vortex formation region of the circular cylinder measured without the presence of aerofoil, and is 1.5 times of the diameter of the cylinder behind it. The cylinder is placed between two end plates at 0.2 m upstream of the aerofoil.

4.3.4 *Plasma Actuator*

The configuration of plasma actuator examined in the current investigation is shown in the schematic of Fig. 4.4 [15]. The actuator is placed at the leading edge while the interface of the exposed electrode and the first covered electrode is located at $x/c = 0$. The encapsulated electrodes are aligned so that there is no offset between

	Width [mm]	Depth [mm]	Offset [mm]
Exposed electrode	5	0	0
A	5	0.24	0
B	40	0.48	5
C	5	0.18	45

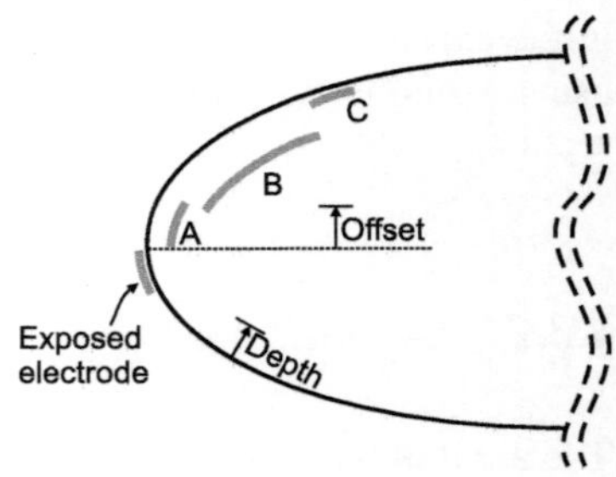

Fig. 4.4 Aerofoil actuator configuration at the leading edge (figure not to scale)

the edges of successive electrodes. All the electrodes are tinned copper foils, 74 μm thick and 200 mm in length, in the spanwise direction. Layered Kapton tape was used as a dielectric material with each layer having a 60 μm thickness. Dimensions and placement of the electrodes are also provided in Fig. 4.4. The offset listed in figure is the distance measured from the downstream edge of the exposed electrode. To have a uniform plasma along the span, a small amount of overlap is applied between the downstream edge of the exposed electrode and upstream edge of the first encapsulated one.

The actuator was controlled by a LabView program where the wave shape, driving frequency, modulation frequency and corresponding duty cycles are controlled. This signal was provided via National Instruments PCI-6713 to a control circuit board that was also connected to a Volkraft 3610 power supply capable of outputting 360 W. The powered control signal is then connected to a transformer cascade which provides the high voltage signal to drive the actuator. The transformer is capable of a maximum 40 kV_{p-p} at driving frequencies up to 30 kHz.

The voltage supplied to the exposed electrode is measured using a LeCroy PPE-20 kV high voltage probe while the current is monitored using a current probe attached to the transformer cascade output. The two probes are connected to a Picoscope 3206, 250 MHz oscilloscope connected to a PC so the output signals can be monitored and recorded.

The actuator spanned most of the width of the leading edge of aerofoil. A narrow gap was left at the centre line due to the presence of pressure taps along chordwise direction. However, both sides were electrically connected. A recess equal to the thickness of the actuator had been cut from the aerofoil in order to produce a flush surface after mounting the actuator without changing the profile shape of the aerofoil.

4.3.5 Pressure Measurements

A total of 23 surface pressure ports with an internal diameter of 1.5 mm were aligned in the flow direction at the half-span location of the aerofoil. The pressure tubes were passed through the side wall of the test section and connected to a scanning pressure valve that selectively connected each pressure port to a single pressure transducer.

A separate rake of pitot probes was mounted on a traversing mechanism and located downstream of the aerofoil at its spanwise centre line. Discrete points, with 2.54 mm distance between each probe, were sampled across the wake to determine the total pressure profile. The rake was placed one chord length downstream of the trailing edge in order to be far from any recirculation region that might exist in the separated flow.

For all pressure measurements, data was recorded for 4 seconds at a sampling rate of 10 kHz and averaged for both lift and drag measurements.

4.3.6 Particle Image Velocimetry (PIV)

Particle image velocimetry (PIV) uses the displacement of particles to determine various flow field parameters. A laser beam is manipulated into a thin sheet using an arrangement of lenses to illuminate the region of interest. Using two successive laser pulses separated by a known time, statistical analysis can be performed to measure the instantaneous velocity of the tracer particles captured with a high-speed camera. The PIV measurements were performed using a TSI system with an NDYAG 532 nm, Litron 120 mJ, pulsed laser. The laser was operated at a rate of 1 kHz. In these experiments the duration between pulses was set at 20 μs. This value is set based on the field of view size and expected induced velocity, of the order of 10 m/s. The laser was delivered using a laser arm mounted above the actuator to produce a laser sheet that ran along the centre-line of the aerofoil span with a bit of offset. The flow was seeded with light-scattering olive oil particles with a size of approximately 1 μm.

4.4 Results and Discussion

Figure 4.5 illustrates the comparison between mean velocity profiles 50 mm from the exposed electrode working in quiescent conditions of a standard DBD and optimised MEE-DBD actuator. The measurements were obtained from PIV experiments on actuators mounted on a flat surface. The actuators were operated at 15 kV_{p-p} with a driving frequency of 10 kHz using a sine wave input and 100% duty cycle. This driving frequency provides the actuator with a clean sinusoidal input signal not experienced at the other frequencies. The power consumption of the actuator was approximately 15 W. Identical settings are used on the aerofoil. As it is evident from the plot, the MEE-DBD actuator creates a faster induced jet with a greater jet thickness. This behaviour was also observed at different locations along the actuator.

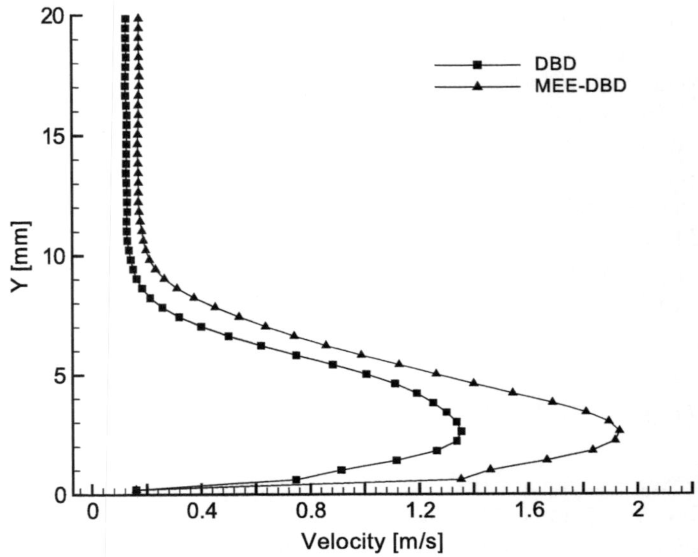

Fig. 4.5 Mean velocity profiles obtained with 15 kV_{p-p} and 10 kHz at x = 50 mm

4.4.1 Leading Edge Plasma Actuator

The results in this section document the ability of the leading edge MEE-DBD plasma actuator to reattach the flow at high angles of attack. Studies have shown that placing the actuator at the leading edge of an aerofoil, at the point of separation, has the best effect on the control of the leading edge separation [7, 38]. The presented results are taken at chord Reynolds number of 0.2×10^6 which enable us to compare with literature and in particular the results of Post and Corke [7], He et al. [6] and Rethmel et al. [10] which are taken at a similar Reynolds number. The effectiveness of MEE plasma actuator in separation flow control was evaluated on the basis of lift enhancement and drag reduction.

The pressure coefficient distribution and the momentum deficit in the wake of the aerofoil are used to assess the lift and drag coefficients for different angles of attack, with the plasma actuator off and actuator on.

Figure 4.6 shows the variation of lift and drag coefficients (C_l and C_d) at various incidences with the actuator off (baseline) and on. These values are calculated based on the procedure given by Erfani et al. [33, 39] At low angles of attack, the baseline values of C_l are in good agreement with thin aerofoil theory, namely $dC_l/d\alpha = 0.12$. The shape of the drag polar and C_l are also consistent with the results mentioned in literatures for this family of aerofoils [7, 40].

With the actuator off, a sudden reduction in lift coefficient after 14° and an increase in drag at $C_l = 1.2$ corresponds to the static stall angle which is accompanied by flow separation from the leading edge. The significant improvement that is produced

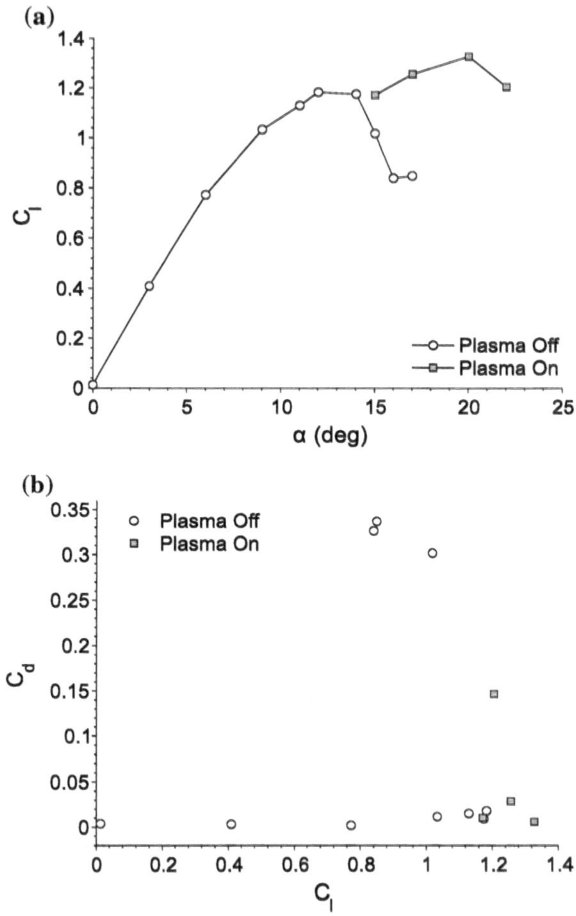

Fig. 4.6 Lift coefficient (**a**), and drag polar (**b**) of the aerofoil for plasma on and off cases

when the plasma actuator is switched on is the increase in stall angle from 14° to 22°, that is 8° past the natural stall angle of attack. By controlling the leading edge separation, the drag is also reduced, as can be seen in Fig. 4.6b. Comparing lift to drag ratios between the on and off cases shows a maximum improvement of 52 times for incidences between 15° and 22°. This value is greater than what has been presented in the literature for the standard DBD configuration [6, 7].

The pressure distributions for post-stall angles of 17° and 22° are shown in Fig. 4.7a, 17° is chosen since it corresponds to the largest lift to drag ratio and 22° is chosen to examine the drop in C_l observed at this incidence. When the actuator is off, the flow separates at the leading edge without reattaching, creating a large wake which extends beyond the trailing edge. This behaviour results in a constant magnitude of pressure coefficient, C_p, along the chord. When the actuator is switched on, the suction pressure is recovered and the separation is suppressed up to

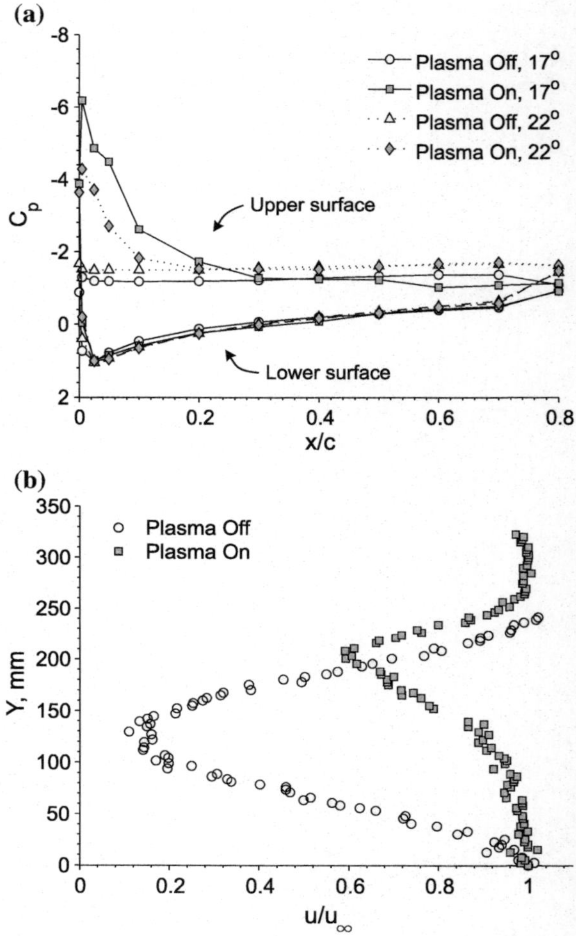

Fig. 4.7 Pressure distribution on the aerofoil surface (**a**), mean velocity profile of the aerofoil wake at $\alpha = 17°$ (**b**)

x/c = 0.6 and x/c = 0.3 for angles of 17° and 22°, respectively. The fact that the separation point is delayed only up to x/c = 0.3 at 22° incidence, justifies the drop in lift coefficient observed in Fig. 4.6a. Figure 4.7b shows mean velocity profile within the wake. It is observed that the wake region is narrower when the actuator is on which clearly illustrates the ability of plasma actuator in preventing or delaying the onset of separation. It is because of the introduction of high momentum jet to the surface at the point of separation which overcomes the adverse pressure gradient and therefore decreases the width of the separation region by approximately a third.

Figure 4.8 compares the PIV data obtained for incidences of 17° and 22° with actuators off and on. A white PIV mask is applied to the images to remove any bad vectors close to the aerofoil surface. These bad vectors result from the reflection of the laser beam from the surface. Due to the large area under investigation and the

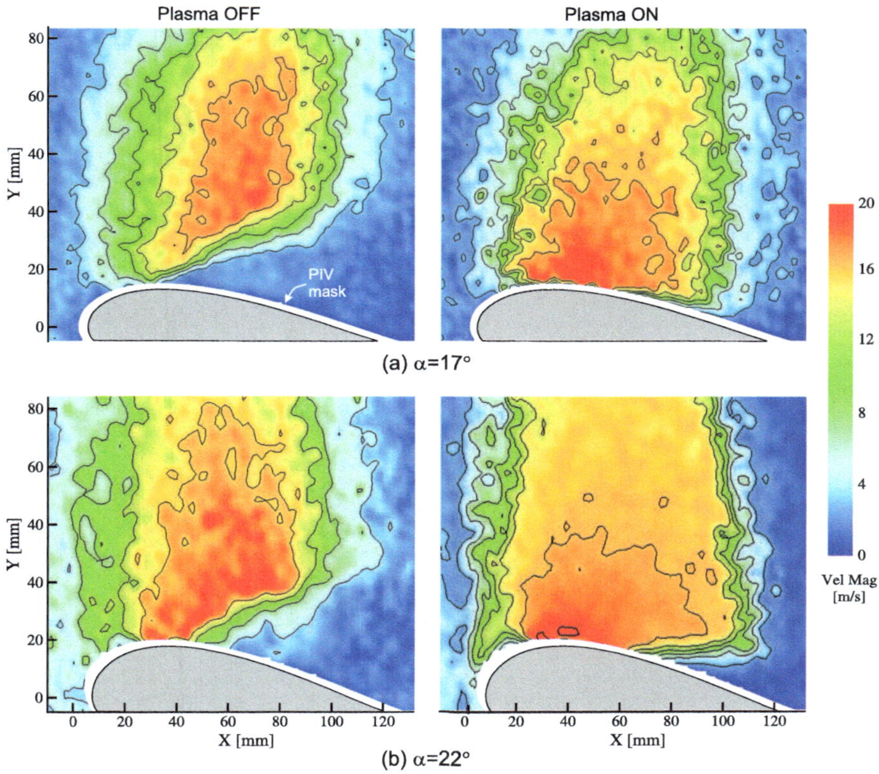

Fig. 4.8 Velocity contours on the leading edge portion of the upper surface of the aerofoil at $\alpha =$, **a** 17°, **b** 22°

lack of illumination size of laser beam, only the region in the immediate vicinity of the separation zone had adequate illumination. For the actuator off case the measurements verify that the drop in lift at higher angles of attack is due to the leading edge flow separation. The flow separation starts from the leading edge and covers the entire suction surface of the aerofoil. For the plasma on case, at incidence of 22° the separation point corresponds to approximately x/c = 0.3. However, due to the limited illumination size it was not possible to concur that the rear part of the reattachment zone for 17° occurs at x/c = 0.6.

4.4.2 Aerofoil-Wake Interaction

The results in this section examine the effects of the leading edge MEE-DBD plasma actuator on the aerodynamic performance of the aerofoil at chord Reynolds number of 0.2×10^6, whilst in the wake of a circular cylinder.

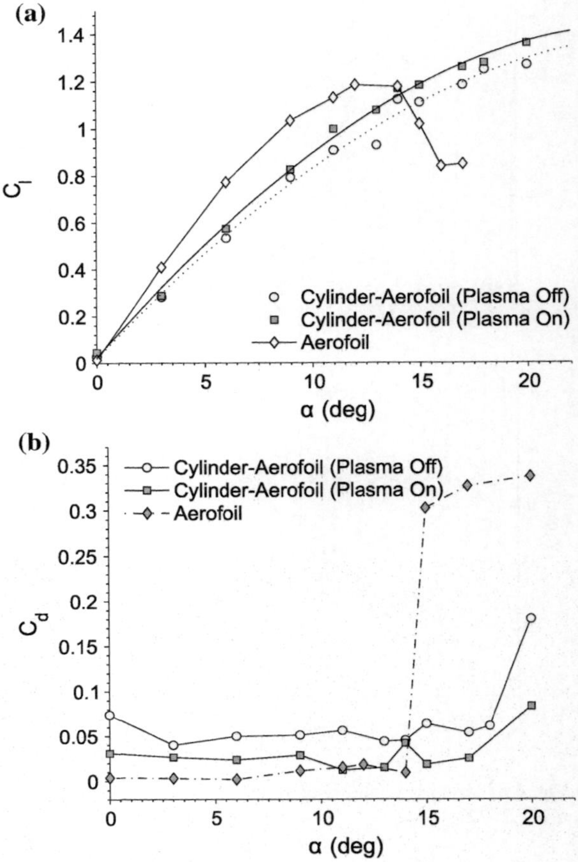

Fig. 4.9 Lift coefficient (**a**), and drag coefficient (**b**)

The lift and drag coefficients of the aerofoil at various angles of attack facing the cylinder wake are presented in Figs. 4.9a, b, respectively. The lift and drag coefficients of the plane aerofoil are also shown in the figure as the baseline case for comparative purposes. With the actuator in off mode, it is observed that up to the natural stall angle of the aerofoil $\alpha = 14°$, the presence of the cylinder wake has an adverse effect on the performance of the aerofoil by reducing the lift and increasing the drag. However, at higher incidences the presence of the cylinder wake suppresses the stall and reduced the drag.

Figure 4.9 also includes the results for plasma on case. In all angles of attack, the actuator on case shows an increase in C_l compared to the actuator off case. Up to $\alpha =$ 14° the drag coefficient does not change with increasing incidence for all cases, with the actuator on leading to a reduced level of drag compared to the actuator off case. After $\alpha = 14°$ a significant rise in drag coefficient is observed for the baseline case whilst the presence of the cylinder wake suppresses the maximum drag coefficient.

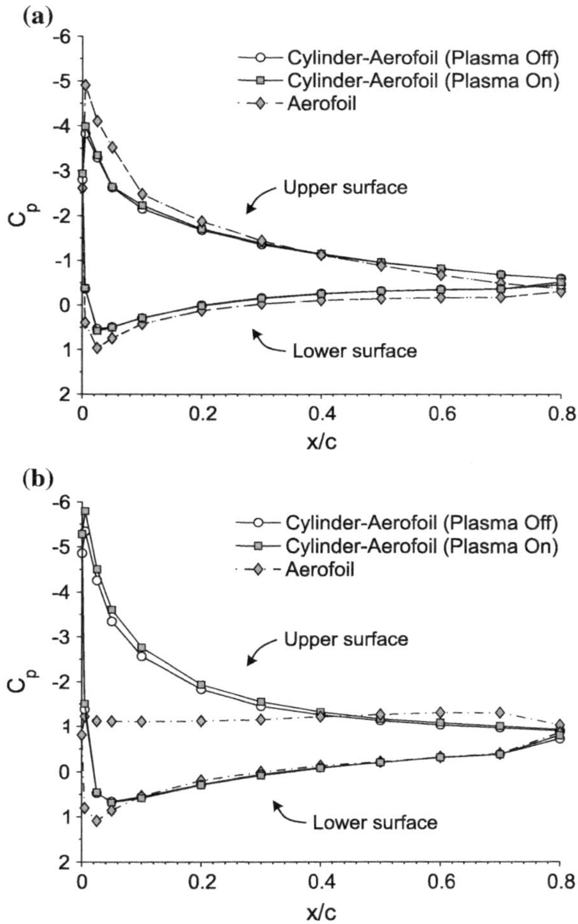

Fig. 4.10 Pressure distribution on aerofoil surface for $\alpha = 11°$ **a** (pre-stall), and **b** 17° (post stall)

The drag coefficient begins to increase after 17° for the cylinder wake case, but still does not approach the results without the cylinder case. This increase is believed to be due to the flow separation and formation of separation region at the trailing edge.

The pressure distribution on the aerofoil surface for a pre-stall angle of 11°, and a post stall angle of 17° is shown in Figs. 4.10a, b, respectively. At 11° incidence the pressure distribution shows a reduction in pressure on the lower surface and increased pressure on the upper surface of the aerofoil compared to the baseline measurements of the aerofoil. No formation of a separated region was detected from the C_p curves in this range of pre-stall angles of attack. This justifies the constant values of C_d observed in Fig. 4.9b for different angles. For the 17° post stall case, the pressure distribution on the upper surface reveals a large decrease in pressure with the presence of the cylinder wake. This indicates an attached flow pattern on the surface due to the velocity fluctuations of the cylinder wake and its interaction with the boundary layer over the aerofoil. The separation of the boundary layer on the suction side is removed and a turbulent boundary layer forms over the surface. Flow over the lower surface of the aerofoil is also promoted to a turbulent state due to the influence of the cylinder wake which shows itself as a reduction in pressure values in Fig. 4.10.

4.5 Conclusions and Future Work

The influence of MEE-DBD plasma actuator proposed by Erfani et al. [15] on the aerodynamic performance of a NACA 0015 aerofoil has been investigated in two test cases: leading edge separation control, and aerofoil-wake interaction. Both cases were studied in a chord Reynolds number of 0.2×10^6. Overall, the new plasma actuator configuration produced an improvement in aerodynamic coefficients in both test cases by increasing the lift and reducing the drag.

The stall angle was increased by 8° past the α_{stall} angle of 14° by delaying the leading edge separation. Using the conventional DBD plasma actuator and working at a similar Reynolds number with a NACA 0015 aerofoil, He et al. [6] was able to increase the stall angle up to 18° which was 4° past the α_{stall} angle. Using a NACA 66_3-018 aerofoil and chord Reynolds number of 0.15×10^6, Post and Corke [7] were able to only achieve an increase of 6° past the α_{stall} angle of 14°.

At high angles of attack, due to the wake excitation of the cylinder wake, the separation region at the suction side of the aerofoil was restrained, leading to the suppression of the stall angle. However, at all incidences the MEE actuator was able to increase the magnitude of the lift coefficient and decrease the level of drag coefficient.

Since a plasma actuator is designed to produce a steady two-dimensional wall jet in the flow direction on the upper surface, the pressure distribution on the lower surface remains unaffected. It would be an interesting study to examine the effect of the plasma actuators placed on the lower surface of the aerofoil.

References

1. Velkoff H, Ketcham J (1968) Effect of an electrostatic field on boundary layer transition. AIAA J 6(7):1381–1383
2. Roth DM, Sherman JR and Wilkinson SP (1998) Boundary layer flow control with a one atmosphere uniform glow discharge surface plasma. In: 36th Aerospace sciences meeting & exhibit, Reno
3. Opaits DF, Roupassov DV, Starikovskaia SM, Starikovskii AY, Zavialov IN, Saddoughi SG (2005) Plasma control of boundary layer using low-temperature non-equilibrium plasma of gas discharge. AIAA J 1180(43):10–13

4. Corke TC, Cavalieri DA, Matlis E (2002) Boundary-layer instability on sharp cone at Mach 3.5 with controlled input. AIAA J 40:1015–1018
5. Grundmann S, Tropea C (2008) Delay of boundary-layer transition using plasma actuators. In: 46th AIAA aerospace sciences meeting and exhibit, paper number AIAA-2008-1369
6. He C, Corke TC, Patel MP (2009) Plasma flaps and slats: an application of weakly ionized plasma actuators. J Aircr 46(3):864–873
7. Post ML, Corke TC (2004) Separation control on high angle of attack airfoil using plasma actuators. AIAA J 42(11):2177–2184
8. Huang J, Corke TC, Thomas FO (2006) Plasma actuators for separation control of low-pressure turbine blades. AIAA J 44(1):51–57
9. Ruisi R, Zare-Behtash H, Kontis K, Erfani R (2016) Active flow control over a backward-facing step using plasma actuation. Acta Astronaut 126:354–363
10. Rethmel C, Little J, Takashima K, Sinha A, Adamovich I, Samimy M (2011) Flow separation control over an airfoil with nanosecond pulse driven dbd plasma actuators. In: 49th AIAA aerospace sciences meeting including the New Horizons forum and aerospace exposition, Orlando, Florida, paper number AIAA-2011-487
11. Erfani R, Zare-Behtash H, Kontis K (2012) Plasma actuator: influence of dielectric surface temperature. Exp Therm Fluid Sci 42:258–264
12. Hale C, Erfani R, Kontis K (2010) Plasma actuators with multiple encapsulated electrodes to influence the induced velocity. In: 48th AIAA aerospace sciences meeting including the New Horizons forum and aerospace exposition, AIAA-2010-1223
13. Hale C, Erfani R, Kontis K (2010) Plasma actuators with multiple encapsulated electrodes to influence the induced velocity : further configurations. In: 40th Fluid dynamics conference and exhibit, AIAA-2010-5106, number 2010–5106
14. Erfani R, Hale C, Kontis K (2011) The influence of electrode configuration and dielectric temperature on plasma actuator performance. In: 49th AIAA aerospace sciences meeting including the New Horizons forum and aerospace exposition, Orlando, AIAA-2011-955
15. Erfani R, Erfani T, Utyuzhnikov SV, Kontis K (2013) Optimisation of multiple encapsulated electrode plasma actuator. Aerosp Sci Technol 26(1):120–127
16. Erfani R, Erfani T, Hale C, Kontis K, Utyuzhnikov SV (2011) Optimization of induced velocity for plasma actuator with multiple encapsulated electrodes using response surface methodology. In: 49th AIAA aerospace sciences meeting including the New Horizons forum and aerospace exposition, Orlando, AIAA-2011-1206
17. Squire LC (1989) Interactions between wakes and boundary-layers. Prog Aerosp Sci 26(3):261–288
18. Doligalski TL, Walker JDA (1984) The boundary layer induced by a convected two-dimensional vortex. J Fluid Mech 139(1):1–28
19. Gartshore IS, Durbin PA, Hunt JCR (1983) The production of turbulent stress in a Shear flow by irrotational fluctuations. J Fluid Mech 137(1):307–329
20. Takagi Y, Fujisawa N, Nakano T, Nashimoto A (2006) Cylinder wake influence on the tonal noise and aerodynamic characteristics of a NACA0018 airfoil. J Sound Vib 297(3–5):563–577
21. Pfeil H, Herbst R, Schroder T (1982) Investigation of the laminar-turbulent transition of boundary layers disturbed by wakes. In: American society of mechanical engineers, 27th international gas turbine conference and exhibit, London, England
22. Liu X, Rodi W (2006) Experiments on transitional boundary layers with wake-induced unsteadiness. J Fluid Mech Digital Archive 231:229–256
23. Kyriakides NK, Kastrinakis EG, Nychas SG, Goulas A (1999) Aspects of flow structure during a cylinder wake-induced laminar/turbulent transition. AIAA J 37(10):1197–1205
24. Mailach R, Vogeler K (2002) Wake-induced boundary layer transition in a low-speed axial compressor. Flow Turbul Combust 69(3):271–294
25. Bloy AW, West MG, Lea K (1993) Lateral aerodynamics interference between tanker and receiver in air-to-air refueling. J Aircr 30(5):705–710
26. Iversen JD, Bernstein S (1974) Trailing vortex effects on following aircraft. J Aircr 11:60

27. Kornilov VI, Pailhas G, Aupoix B (2002) Airfoil-boundary layer subjected to a two-dimensional asymmetrical turbulent wake. AIAA J 40(8):1549–1558
28. Enloe CL, McLaughlin TE, VanDyken RD, Kachner KD, Jumper EJ, Corke TC, Post M, Haddad O (2004) Mechanisms and responses of a single dielectric barrier plasma actuator: geometric effects. AIAA J 42(3):595–604
29. Erfani R, Zare-Behtash H, Kontis K (2012) Influence of shock wave propagation on dielectric barrier discharge plasma actuator performance. J Phys D Appl Phys 45:225201
30. Enloe CL, McLaughlin TE, VanDyken RD, Kachner KD, Jumper EJ, Corke TC (2004) Mechanisms and responses of a single dielectric barrier plasma actuator: plasma morphology. AIAA J 42(3):589–594
31. Fang Z, Lin J, Xie X, Qiu Y, Kuffel E (2009) Experimental study on the transition of the discharge modes in air dielectric barrier discharge. J Phys D Appl Phys 42:085203
32. Enloe CL, McHarg MG, McLaughlin TE (2008) Time-correlated force production measurements of the dielectric barrier discharge plasma aerodynamic actuator. J Appl Phys 103:073302
33. Erfani R, Zare-Behtash H, Hale C, Kontis K (2015) Development of dbd plasma actuators: the double encapsulated electrode. Acta Astronaut 109:132–143
34. Zdravkovich MM (1997) Flow around circular cylinders: fundamentals. Oxford science publications, Oxford University Press, Oxford
35. Nakagawa T (1986) A formation mechanism of alternating vortices behind a circular cylinder at high reynolds number. J Wind Eng Ind Aerodyn 25(1):113–129
36. Roshko A (1961) Experiments on the flow past a circular cylinder at very high reynolds number. J Fluid Mech 10(03):345–356
37. Luk KF, So RMC, Kot SC, Lau YL, Leung RCK (2002) Airfoil vibration due to upstream alternating vortices generated by a circular cylinder. ASME Appl Mech Div Publ-AMD 253(A):79–88
38. Roth JR, Dai X (2006) Optimization of the aerodynamic plasma actuator as an electrohydrodynamic (EHD) electrical device. In: 44th AIAA aerospace sciences meeting and exhibit, Reno, paper number AIAA 2006-1203 pp 9–12,
39. Erfani R, Hale C, Kontis K (2012) Flow control of a NACA 0015 airfoil in a turbulent wake using plasma actuators. In: 50th AIAA aerospace sciences meeting including the New Horizons forum and aerospace exposition, AIAA-2012-187
40. Abbott IH, Doenhoff AEV (1959) Theory of wing sections, including a summary of airfoil data. Dover books on physics and chemistry, Dover Publications

Chapter 5
Plasma Models in Hybrid RANS-LES Simulation for Backward Facing Step Flow Control

Palma Gonzalez and Ning Qin

Abstract This paper presents a study on the effects of a single discharge barrier dielectric barrier device on the periodic components of the turbulent shear layers and the Reynolds stresses. 3D simulations using improved delayed detached eddy simulation, a hybrid RANS/LES technique, has been used for the study. The geometry for the study is taken from the experimental configurations for this case. The case comprises a turbulent flow over a backward facing step (BFS), where separation is induced after the step edge. The results from the simulations are compared to the experimental data with and without control. The active flow control device is a single dielectric barrier discharge, DBD, plasma actuator located upstream of the step. The effects of quasi-steady and unsteady—or pulsated-plasma actuation using two different phenomenological models are studied. The resulting turbulent structures, Reynolds stresses, and velocity profiles are analysed applying two different models to simulate the plasma actuation. The results for quasi-steady plasma mode show good agreement with the available experimental data and a reduction of the reattachment length. Regarding modulated actuation of the DBD plasma device, three dimensional simulations are carried out and the results also show excellent agreement of the overall behaviour flow when compared to the experimental data.

Keywords Plasma models · Detached Eddy Simulation · Separation control

5.1 Introduction

The problem of a turbulent flow over a backward facing step, BFS, has been widely investigated both experimentally and numerically [1, 2, 7, 9, 11, 16, 18]. One of the most comprehensive BFS cases is by Driver and Seegmiller, case which was tested in order to validate our in-house code by Xia, Wang [34].

A quasi-steady case was firstly investigated and compared with the experimental data. Finally, an unsteady forcing study was carried out comparing two different

P. Gonzalez · N. Qin (✉)
Department of Mechanical Engineering, University of Sheffield, Sheffield S1 3JD, England, UK
e-mail: n.qin@sheffield.ac.uk

N. Qin et al. (eds.), *Advances in Effective Flow Separation Control for Aircraft Drag Reduction*, Computational Methods in Applied Sciences 52,
https://doi.org/10.1007/978-3-030-29688-9_5

phenomenological plasma models as it has been experimentally observed that this plasma operation mode showed to be more effective regarding the reduction of the reattachment area of the flow [31].

5.2 DBD Plasma Models and Their Implementation

In the current section, a dielectric barrier discharge, DBD, plasma actuator will be described. In order to perform appropriate simulations using this control device, two different models were implemented in the solver: Shyy's model, [32] and the model by Singh and Roy [30].

Plasma actuators have shown several advantages in aerospace applications when compared to other flow control devices actuating at a wide range of speed regimes [Low speed: 10, 24], High speed: 17, 20]. For instance and unlike other active flow control devices, plasma actuators have no mechanical components and the required power to operate them is moderate; furthermore, the actuation has a quick response and the generated force can be significant.

Several studies have been carried out in the past to investigate how the plasma force is generated when using a DBD plasma actuator and its effects when applied to different cases such a BFS or other geometries such as aerofoils under different initial configurations, [3–6, 8, 13, 19, 21, 25, 26]. A single DBD plasma actuator consists of two electrodes asymmetrically placed at different sides of a flat dielectric plate, [22]; however, different configuration of the electrodes and the dielectric material can be configured [8, 14, 29]. The chosen configuration in this thesis is shown in Fig. 5.2 where a brief diagram of how plasma is generated is depicted: two electrodes are located on both sides of a dielectric material of certain thickness (in the order of micro/millimetres). A high voltage is applied between the electrodes by means of a transformer and the air in the indicated regions is ionised and turned into plasma, Fig. 5.2. Plasma actuation consists in generating an electrical discharge at the wall within the boundary layer. The objective is to add momentum with the discharge-induced electro-hydro-dynamic force and electric wind (Fig. 5.1).

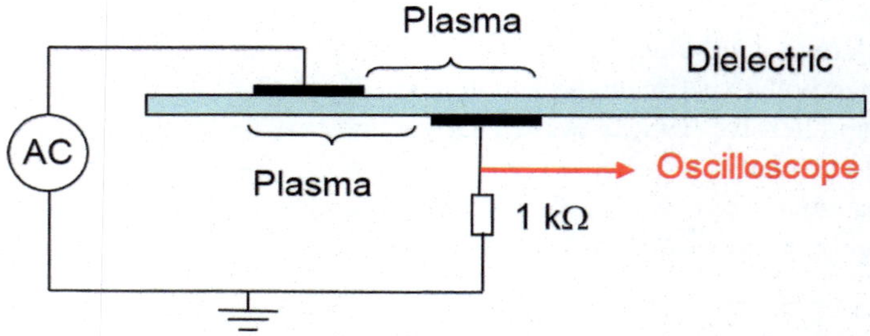

Fig. 5.1 Basic principle of an axisymmetric DBD plasma actuator

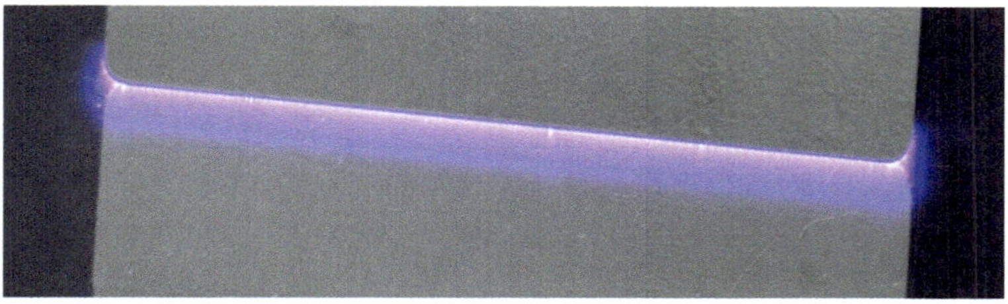

Fig. 5.2 Plasma visualisation along the spanwise of wind tunnel at University of Poitiers

In the work carried out by the experimental partner, the electrode downstream of the flow is grounded and insulated, so the plasma will be generated only on the upper surface of the dielectric surface. Due to the distance between the electrodes along the stream wise direction and due to the difference of pressure produced by the application of high voltage, a discharge barrier is created in the adjacent region to the exposed electrode and it ionises the air surrounding this region and an optical glow appears as shown in Fig. 5.2. This plasma is highly collisional: due to the added energy, the air dissociates because of the collisions of particles whose molecular binding energy is lower than the thermal kinetic energy. This effect together with the electric body force induces a momentum transfer from the electric field into the flow field, influencing the near-wall boundary layer. In another arrangement, a pulsation frequency–much lower than the AC frequency-can be applied. In this case, the excitation induces an unsteady force which will influence the natural large-scale flow structures of the flow rather than the near-wall boundary layer.

The generated force will be dependent on the applied voltage—applied waveform and voltage -, the applied frequency–normally in the range of radio frequency wave lengths- and the geometry and material characteristics of the device, i.e., distance and thickness of electrodes and thickness and permittivity of dielectric material.

Two different models were implemented in DGDES: Shyy's model and Singh and Roy's model. A complete explanation and description of these models is presented below.

5.2.1 *Plasma Models in DGDES*

In this section an explanation of two different models implemented in the solver is going to be given. In both cases, the force generated by the plasma device is considered as a body force and it will be added as a source term in the Navier–Stokes governing equations of the flow.

5.2.1.1 Shyy's Model and Its Implementation into DGDES

Shyy et al. modelled the fluid dynamics and heat transfer induced by an asymmetric dielectric barrier discharge plasma actuator. In their work, a single DBD plasma actuator was installed on a flat plate and the effects of the generated plasma were investigated. A formulation for the body force and its addition into the governing equations was obtained.

The DBD actuator operated at atmospheric pressure and plasma was generated by applying a certain frequency and voltage. Regarding the order of frequency and voltage, the frequency was in the range of kilohertzs and the voltage was also in the range of a few kilovolts.

Since the electric field lines emerge from the exposed electrode to the grounded electrode, in Shyy's model plasma is considered to interact with air inside a triangular region defined by the position of the electrodes, Fig. 5.3. The electric field lines are assumed to be linear and moving from A to B so the electric field, hence the induced force is strongest near the edge of the exposed electrode and weakens as moving beyond the embedded electrode and beyond line A-B.

According to these assumptions, an equation for the electric field variation can be written as,

$$|E| = E_0 - k_1 x - k_2 y \tag{5.1}$$

where E_0 is the electric field in the region between the left edge of the exposed electrode and the right edge of the left electrode and it is approximated to be,

$$E_0 = \frac{V}{d} \tag{5.2}$$

where d is the distance between electrodes in the x-direction. k_1 and k_2 are constants whose values were obtained by means of the condition that "the field strength is the breakdown value at the plasma-fluid boundary". Regarding the electric field, $\vec{E}$, it is restricted to be two dimensional in this approximation, their components were the

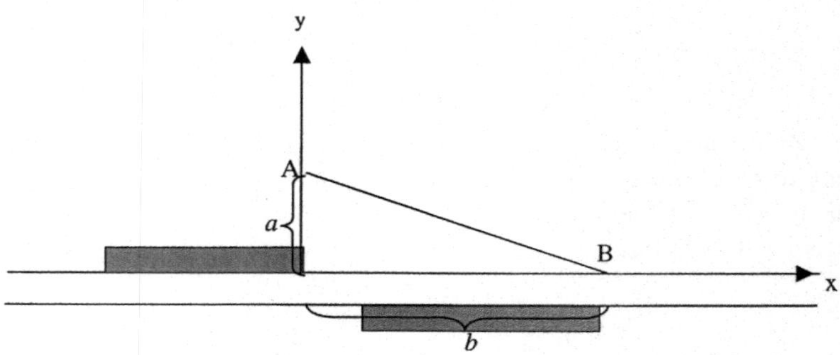

Fig. 5.3 Triangular area of actuation of plasma in Shyy's model

x-direction component given by,

$$E_x = \frac{Ek_2}{\sqrt{k_1^2 + k_2^2}} \tag{5.3}$$

and the y component was,

$$E_y = \frac{Ek_1}{\sqrt{k_1^2 + k_2^2}} \tag{5.4}$$

The force components were obtained by multiplying each electric field component by the electron charge, e (C) and the electron number density, ρ_c (m^{-3}),

$$f_x = E_x \rho_c e \tag{5.5}$$

$$f_y = E_y \rho_c e \tag{5.6}$$

However, this force only influences the area where the plasma is present, i.e., the triangular region therefore a blending function δ is needed to make sure plasma acts in the correct region in the whole domain,

$$\begin{cases} \delta = 1 for E < E_{crit} \\ \delta = 0 for E \geq E_{crit} \end{cases} \tag{5.7}$$

E_{crit} is the aforementioned breakdown electric field strength.

In addition to this, as plasma is a highly collisional phenomenon, a constant named α was also introduced to account for the collision efficiency. Taking into account the blending function and the efficiency assumptions, the final effective force components can be easily obtained,

$$f_{effx} = \alpha f_x \delta \tag{5.8}$$

$$f_{effy} = \alpha f_y \delta \tag{5.9}$$

Finally, this body force is not acting constantly but when plasma is formed during Δt time, which is half cycle of the AC voltage as explained in Shyy's paper. However, as the frequency of the applied voltage is high, the acting force is considered to be steady, hence a time averaged force was obtained,

$$F_{tavex} = \vartheta f_{effx} \Delta t \tag{5.10}$$

$$F_{tavey} = \vartheta f_{effy} \Delta t \tag{5.11}$$

Note that ϑ is the frequency of the applied AC voltage.

These two components were added then as a source term in the momentum equations and the effects of plasma were analysed.

At high frequencies, plasma shows to dramatically increase the stream wise component of the velocity.

5.2.1.2 Singh and Roy's Model and Its Implementation into DGDES

Singh and Roy [27] proposed another simple and straightforward model for obtaining the electro-hydro-dynamic force generated by a simple DBD actuator.

In the model proposed by Singh and Roy, several assumptions were made: a sinusoidal voltage was applied to the exposed electrode and in this case the embedded electrode was grounded as well as in Shyy's model. The current flows perpendicular to the dielectric surface and no slip condition was assumed for this surface. The generated force was assumed to be two dimensional, therefore it will only be defined by two components in the x and y direction, respectively and the air was considered to be a quiescent gas flow. The formulation does not include the frequency of the AC voltage applied so if in reality the plasma actuation is quasi-steady–radiofrequency waves—this model considers the plasma actuation as steady.

With regards to the area where plasma has got influence in the flow field, unlike in Shyy's model where the plasma region is limited to a triangle defined by the geometry of the device, in Singh and Roy's model plasma is considered to have influence in the whole domain. Furthermore, the value of each EHD force component will be dependent on the distance and position with respect to the DBD actuator location.

According to these ideas and assuming the time averaged force interacts with air, the body force was approximated by a fourth order polynomial equation dependent on the applied voltage, ϕ_0 and the DBD geometry,

$$\begin{aligned}\vec{F} = {} & F_{x0}\phi_0^4 exp\left(-\left\{\frac{[x - x_0 - (y - y_0)]}{y}\right\}^2 - \beta_x(y - y_0)^2\right)\vec{\mathrm{i}} \\ & + F_{y0}\phi_0^4 exp\left\{-\left[\frac{(x - x_0)}{y}\right]^2 - \beta_y(y - y_0)^2\right\}\vec{\mathrm{j}}\end{aligned} \tag{5.12}$$

where the values for the following constants: F_{x0} and F_{y0} were taken from the paper and their values are obtained by averaging the solution of the body force obtained when solving air-plasma coupled equations. These values were obtained after performing a study of the plasma actuator under nine conditions explained in the original paper therefore the model will require further adjustment when the air is moving/flowing. Besides, β_x and β_y are constants that depend on the dielectric material characteristics and their values were initially and for test purposes assumed to be equal as those provided in Singh and Roy's paper. However, and because of taking such values for the above mentioned constants, when Singh and Roy's model was implemented in the flow solver, several problems due to large generated plasma force values were found in the first CFD simulations. The introduction of a constant solved this problem, [28]. Finally, x_0 and y_0 are the midpoint between the electrodes and the thickness of the dielectric material, respectively, Fig. 5.7.

It can be seen from the force equation that its value over the domain of actuation is spatial dependent: the closer to the actuator, the more intense the force will be. This is one of the main differences with the formulation provided by Shyy in which the plasma only operates in a triangular area and it does not have any effect beyond the limits of such triangle.

With regards to its implementation into DGDES, the model assumes the generated body force follows the Eq. (5.12). In this study, the only investigated configuration of the two electrodes considered is the one depicted back in Fig. 5.4.

As already discussed, in the original plasma model formulation several conditions were assumed and tested in order to obtain all the coefficients needed to produce the values of the plasma force which matched the experimental measurements. This means that the force given by the model is sensitive to the initial conditions. In this approach, the initial conditions of the analysed experimental case are completely different to the original paper formulation. Because of that, a constant needs to be applied to the plasma equation as it will be seen later on. Apart from that reason, according to experimental results the efficiency of the plasma actuator is not 100% and both components of the force would have to be multiplied by a constant in order to match the experimental results. In addition to all this, due to the original Singh and Roy's plasma model formulation, in which the magnitude of the force increases with the fourth power of the maximum applied voltage, when high voltages such as the experimental case carried out in the MARS project framework are applied (which can reach up to 25 kV), the resulting EHD force is a huge number which leads to not physically reasonable values for the force hence an assumption was made: the plasma body force term had to be reduced by the application of another

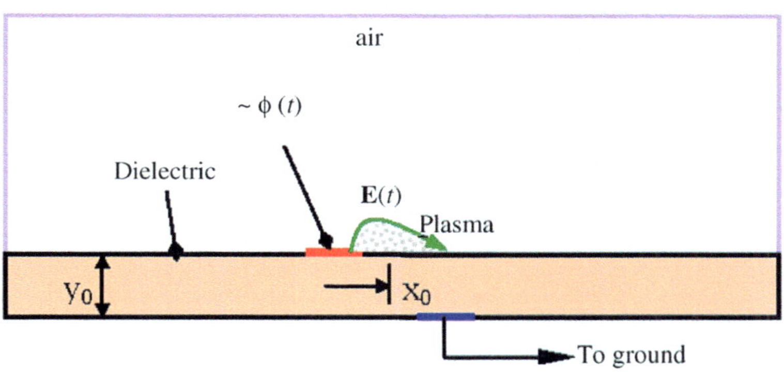

Fig. 5.4 Single DBD plasma actuator diagram according to Singh and Roy's model

constant which would multiply both components of the force. Another important assumption as already mentioned before in this section, the plasma actuation when it is really quasi-steady—in the order of kilohertzs—Singh and Roy's model consider the actuation as purely steady and constant plasma actuation.

With regards to the introduction of a constant into the force components using the original Singh and Roy's model formulation, the application of such constant was later on found in the paper "Bulk flow modification with horseshoe and serpentines plasma actuators" by Roy and Wang on [28]. Supported by that reason, a re-formulation of the original equation was performed by introducing a constant, C_{ROY}, which multiplies both force components,

$$\vec{F} = C_{ROY} F_{x0} \phi_0^4 exp\left(-\left\{ \frac{[x - x_0 - (y - y_0)]}{y} \right\}^2 - \beta_x (y - y_0)^2 \right) \vec{\mathrm{i}} + C_{ROY} F_{y0} \phi_0^4 exp\left\{ -\left[\frac{(x - x_0)}{y} \right]^2 - \beta_y (y - y_0)^2 \right\} \vec{\mathrm{j}} \tag{5.13}$$

Regarding the implementation of Singh and Roy's model into DGDES, it is clear that the model is a electromagnetical phenomenological model and therefore the generated force was included as a source term in the Navier–Stokes equations, which means, when plasma is switched on in Eq. (3.9),

$$\frac{\partial}{\partial t} \iiint_V \mathbf{W} \, \partial V + \frac{\partial}{\partial \tau} \iiint_V \mathbf{W} \partial V + \iint_{\partial V} (\boldsymbol{F} - \boldsymbol{G}) \cdot \boldsymbol{n} dA = \int_V \boldsymbol{S} dV \tag{5.14}$$

The term $\int_V \boldsymbol{S} dV$ is the summation of the all the body forces such as electric fields, magnetic fields, gravity, etc. will be equal to the force generated by the plasma actuator. When the plasma device is deactivated, $\int_V \boldsymbol{S} dV = 0$.

5.3 Geometry and Computational Domain

Experiments performed at the wind-tunnel in Poitiers facilities, Fig. 5.5, were characterised by the following configuration: the step height, H, was 0.03 m with a span wise length of 0.3 m. The wind tunnel test section is 0.3 × 0.3x1. Measurements were taken at the middle section covering x/H = {−2, 7} and y/H = {−1.5, 2}.

Experiments were carried out for a free stream velocity of 15 m/s, therefore the Reynolds number based on step height is 30,000. The boundary layer thickness was calculated by means of velocity profiles measurement at various locations upstream of the step and it was determined to be approximately 13 mm at 1H before the step edge.

The actuator was mounted as shown in Fig. 5.6,

Fig. 5.5 BFS model in wind-tunnel (CNRS PPRIME Poitiers)

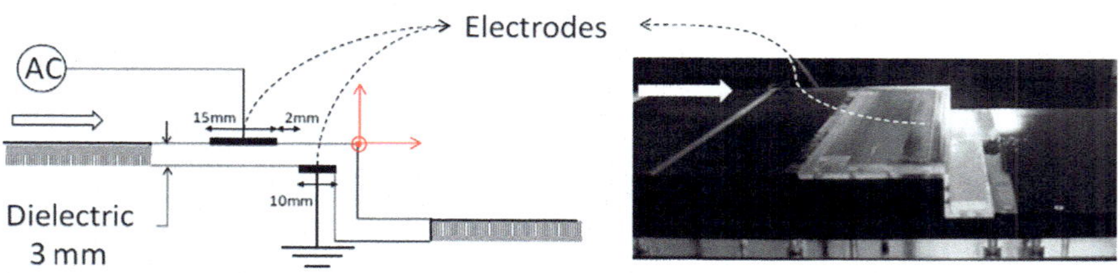

Fig. 5.6 DBD plasma actuator configuration

As it can be seen in Fig. 5.6, the exposed electrode is 0.015 m long. The embedded electrode, on the other hand, is located 0.002 m downwards the exposed and it is 0.01 m long. The dielectric thickness in this case is 0.003 mm.

The plasma actuation consists in generating an electrical discharge at the wall within the boundary layer of the air flow, adding momentum with the discharge-induced electro-hydro-dynamic force and the electric wind. Additionally, DBD plasma actuators can be operated in two different modes; the first one known as quasi-steady or steady forcing—depending on the AC frequency—modifies the near-wall flow imparting flow perturbations in the boundary layer at the frequency of the applied electrical signal. In first case, the frequency of the alternating current was 1 kHz so it is quasi-steady [26] actuation of the device; however, in the simulations with Singh and Roy's model, the plasma actuation will be considered purely steady as this model does only consider the applied voltage and not the AC frequency as stated in the previous chapter. For such reason and for comfort, steady actuation will be the term used when referring to steady or quasi-steady actuation of the plasma in the simulations; nonetheless this assumption needs to be borne in mind.

The second mode of actuation of plasma is the modulated forcing which was also performed. Experimentally, pulsating plasma produced a smaller recirculation region and a larger reduction of Reynolds stress therefore a series of CFD simulations will be performed and compared with the experimental database expecting the same or very similar effects. Visbal et al., [33] and Riherd et al., [23], carried out numerical investigation of pulsated plasma, showing a better separation control than and better effectiveness than when a steady forcing is applied. These studies also showed that the application of modulated plasma have a quick transient respond producing a quicker reattachment of the flow when compared to the steady force. The configuration of the case was therefore taken from the work by Bernard et al. [3]. Unsteady forcing of plasma influences in the large turbulent structures of the flow, thus the expected effects are different from those caused by steady plasma. The most promising results were experimentally obtained at a frequency of the pulsation of 125 Hz with a duty cycle of the 50%, which means in every pulse, plasma is on the 50% of the period of that cycle and off the rest 50% of time and so on. The experimental results showed a reduction of a 20% of the reattachment length applying modulated plasma with that frequency.

Regarding the mesh, a 8 million cells provided with very reliable results. Figure 5.7 shows the computational mesh and the location of the DBD actuator.

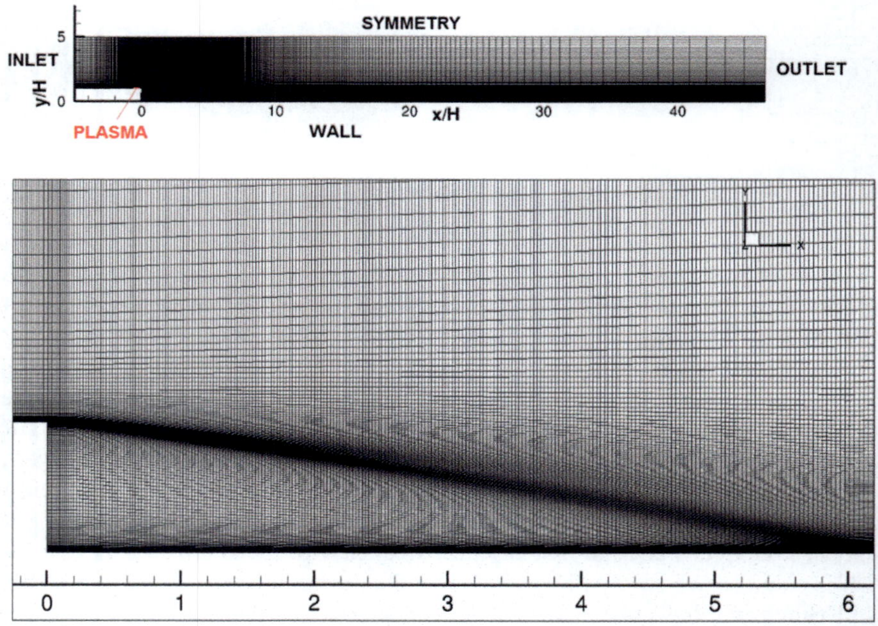

Fig. 5.7 x-y plane of 8M computational mesh (top) and detail of step region (bottom)

5.4 Results and Discussion: Comparison with Experimental Database

5.4.1 Quasi-steady Plasma

5.4.1.1 Analysis of Coherent Structures

Q-criterion [15] was used in this study to identify the turbulent structures in the flow field.

Figure 5.8 displays a comparison Q-criterion at (a) 100,000 and (c) 200,000 for the baseline flow and the steady plasma case.

Before the step for the baseline case, quasi-two dimensional vortex tubes are formed and then shed from the step edge. Then, these tubes interact with the structures in the recirculation region, break down and hairpin structures and Falco eddies [12] are observed further downstream. It is also observed than the lower parts of these larger structures produce the shear layer in the recirculation and break into smaller structures towards the wall for both uncontrolled and controlled simulations. Overall, in the plasma actuation cases, the aforementioned quasi-two dimensional vortex tubes are flattered due to the acceleration of the flow in the near-wall boundary layer that steady plasma induces. Also, the hairpin-like structures are formed also after the step and are predominant over the typical eddies. Finally, steady forcing influences the flow after the reattachment area and larger structures appear after the primary air bubble when compared to the baseline case.

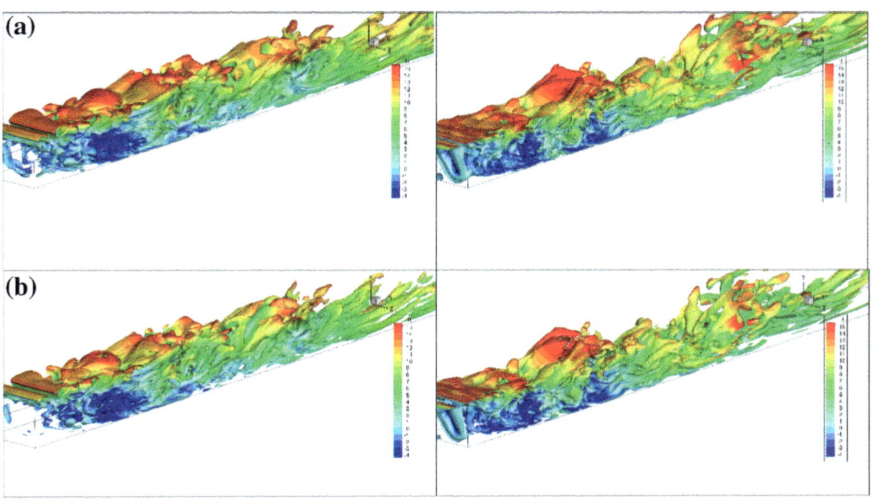

Fig. 5.8 Q-criterion of baseline case–left column- and the steady plasma actuation-right column-at **a** 100,000 and **b** 200,000. Iso-surfaces coloured by stream wise velocity

5.4.1.2 Velocity Profiles

Unfortunately, there was no available data for the steady forcing experimental velocity profiles only the baseline case without control. Velocity profiles are depicted at nine different x/H locations, where the experimental data for the baseline flow was measured (Fig. 5.9).

At-1H, velocity profiles of the three set of data are practically identical. The data at the inlet was provided by Poitiers and it shows that the development of the flow before the step edge is very accurate, the boundary thickness from experiments was 13 mm and it can be seen all simulations match it perfectly. At this location, the flow

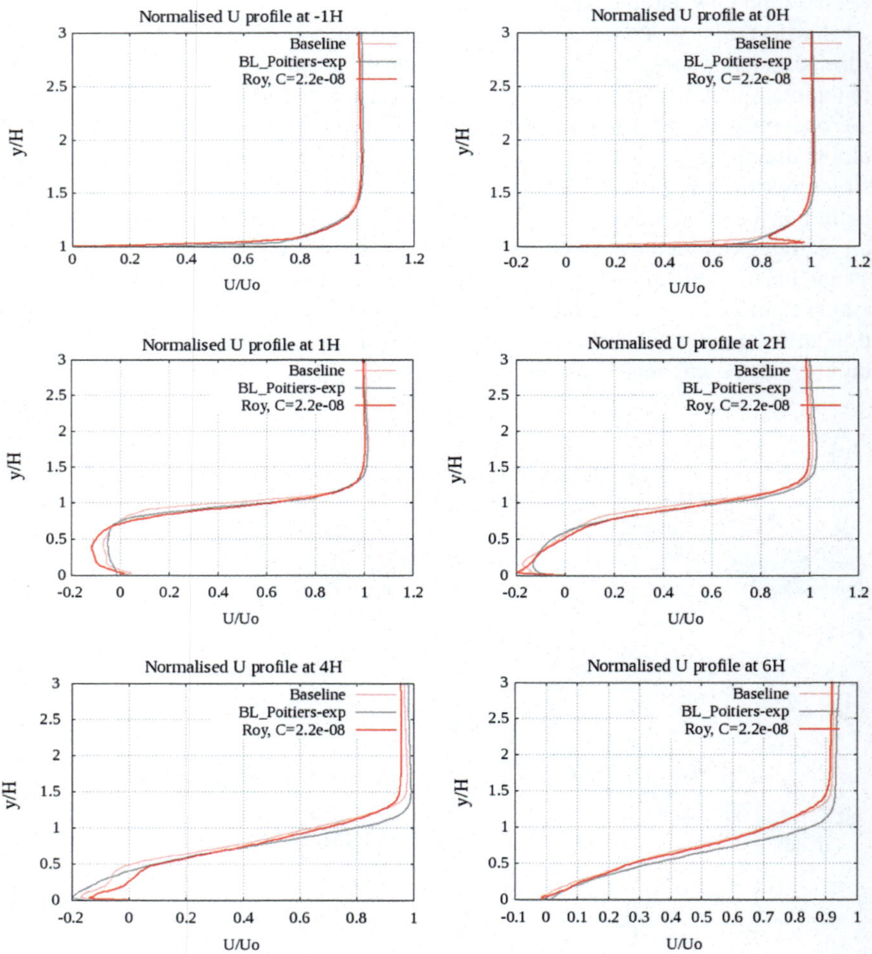

Fig. 5.9 Comparison of velocity profiles of baseline cases–simulation and experiments- and steady plasma force—simulation and experiment-at-1H, 0H, 1H, 2H, 4H and 6H

has not reached the plasma actuator therefore it is obvious that the velocity profile is coincident with the baseline case profile.

At step, the effects of plasma are clearly seen: the flow near the wall suffers a dramatically acceleration and stream wise velocity increases up to a 95% of the free-stream velocity.

Once the flow moves downstream of the step, at 1H the baseline simulation matches very well the experimental profile and the baseline secondary bubble appears to be larger than the plasma secondary region. This will be clearly seen later on when the flow streamlines is depicted in section next section. At the location of 2H, simulation shows good agreement with experimental data and plasma shows this time higher negative values of velocity near to the wall and the shear layer is located closer to the wall than the baseline cases due to the effects of plasma.

Closer to the reattachment point, at the location 4H, the velocity of the plasma case in the near wall region and within the circulation bubble increases. This implies that the flow will be attached to the wall before the baseline cases. Further downstream at 6H, simulations show a small negative value, which indicates the presence of small embedded eddies near the wall. The velocity profile from experiments, on the other hand, shows positive values at all y/H locations indicating the flow is fully attached to the wall is such location.

5.4.1.3 Reattachment Region

This section comprises the analysis of the reattachment area for the baseline and steady plasma forcing cases and compares the simulation results with the experimental database. Besides, an exploration on the skin friction coefficient was carried out and analysed.

In Fig. 5.10 the streamlines of both experiments and simulations are shown. The baseline flow reattaches to the wall at 5.8H in the experiments and at 6H in the simulation. This over-prediction was acceptable and the reason was explained in the previous section in the paper. With regards to plasma, the experimental reattachment length occurs at 5.6H and at 5.7H at simulation. This slight difference is caused by the grid high-stretching ratio in the shear layer slowing down the flow velocity and hence increasing the length of the reattachment. The secondary bubble is well predicted in both uncontrolled and controlled cases: for the baseline simulation it is located at ~1.6H which is equal to the experimental baseline; and for plasma in the experiments a smaller secondary bubble appears due to the effects on plasma and it reattaches at 1H as well as in simulation results.

Despite of the slight differences in the primary reattachment length, very good and satisfactory agreement was found between simulations and experiments.

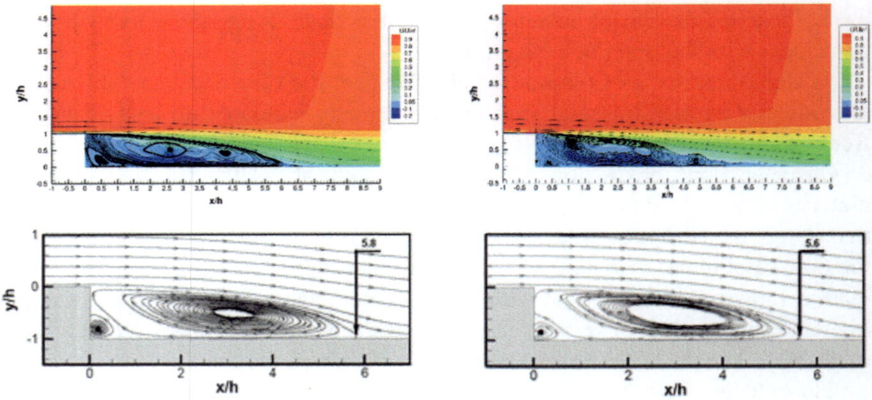

Fig. 5.10 Streamlines of flow field: baseline-left column-and steady plasma actuation-right column-comparison. Top row: simulations; bottom: experimental data

5.4.2 *Unsteady Plasma*

5.4.2.1 Analysis of Coherent Structures

The same analysis structure of the results carried out for steady plasma will be followed for analysis of the modulation of plasma results. Firstly the turbulent coherent structures of the two different—Singh and Roy's and Shyy's model—cases is analysed by means of the Q-criterion at 1000, Fig. 5.11 and at 100,000, Fig. 5.12.

In both figures, Fig. 5.13 and Fig. 5.14, in Shyy's model case the small scale structures within the circulation zone are destroyed and do not even appear; on the other hand, much smaller structures appear in Singh and Roy's case. This is mainly due to the mesh density difference: the 3M mesh used for Shyy's model—this simulation was inherited from previous studies for the MARS project—is not able to capture such small turbulent structures whereas the 8M mesh—used for Singh and Roy—does. It is also observed in the figures, especially at Q-criterion at 1000, how the pulsation of the plasma produces large scale structures at the free shear layer. Their wavy shape is related to the duty cycle of the device: a pulsated injection of

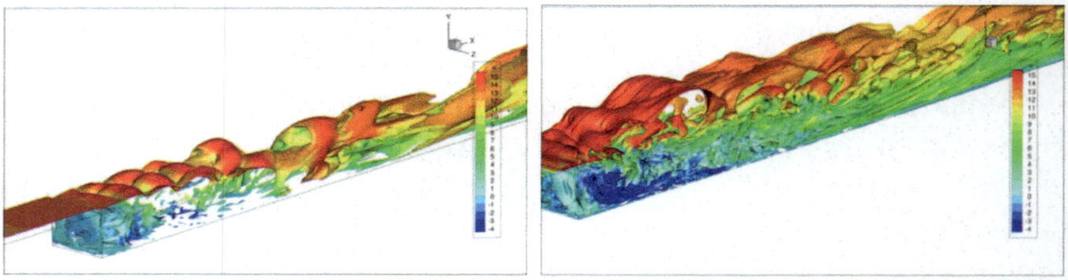

Fig. 5.11 Q-criterion at 1000 comparison: Shyy's model (left) versus Singh and Roy's model (right)

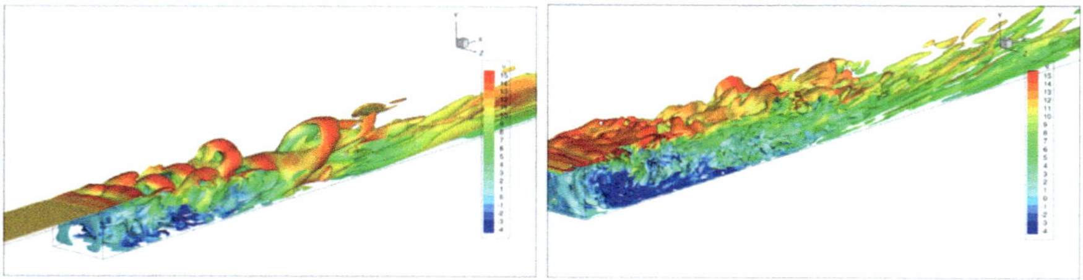

Fig. 5.12 Q-criterion at 100,000 comparison: Shyy's model (left) versus Singh and Roy's model (right)

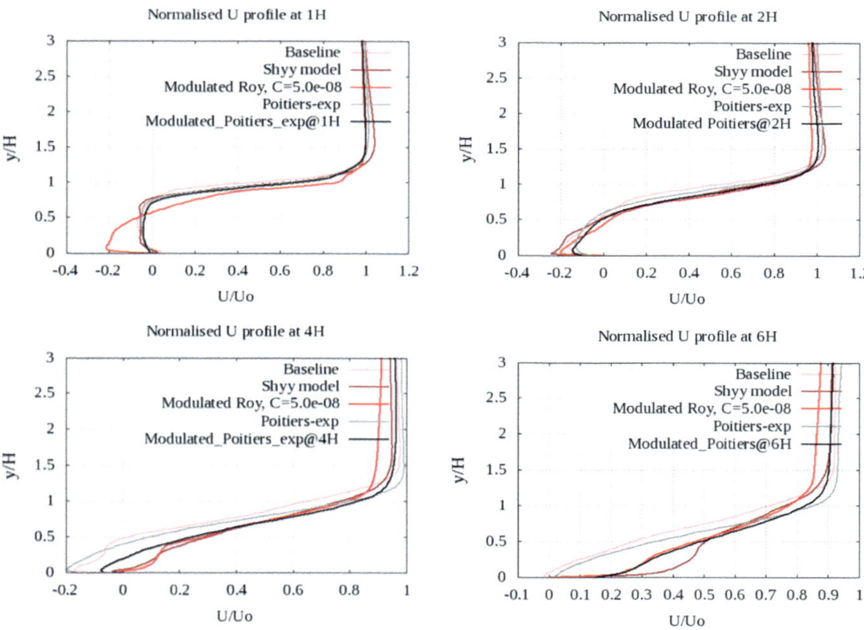

Fig. 5.13 Normalised stream wise velocity at 1H, 2H, 4H and 6H

Fig. 5.14 Experimental flow streamlines

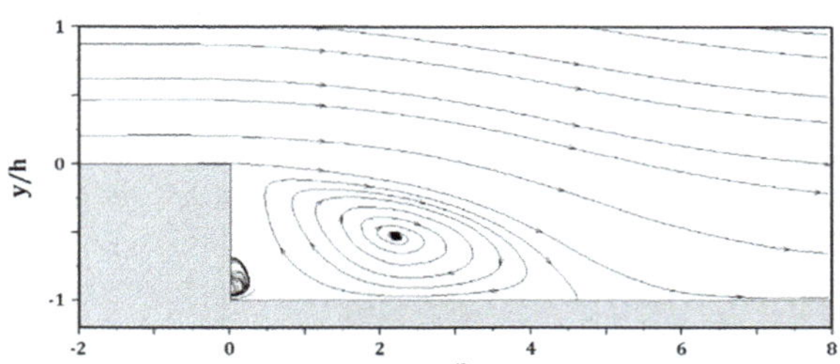

plasma produces wave-like turbulent structures downstream of the actuator whereas the turbulent structures for a steady forcing of plasma showed large structures but really long and flatter—not as wavy-. These large structures however are rapidly destroyed downstream, showing a good recovery of the flow after the reattachment area. This fact will be again mentioned then the velocity profiles at different stream wise locations are shown. At that point, it will be seen that Singh and Roy's model has got a better recovery of the flow after the circulation area.

5.4.2.2 Velocity Profiles

In this section velocity profiles are depicted at four crucial x/H locations: 1H, 2H, 4H and 6H where the experimental baseline and the experimental pulsated plasma measurements were taken, Fig. 5.13. At each location, both experimental and simulation baseline cases, Shyy's model case, Singh and Roy's model case and the experimental profile with modulated plasma actuation is depicted.

At location 1H, the plasma effects on the flow are not strong as all the profiles have got a similar shape, almost equal. Simulation and experimental data are very close for all the cases except for the Singh and Roy's model. This is due to the strong effects of plasma in the region near the step, causing a smaller secondary recirculation region on the corner right after it. The secondary bubble is hence "smashed". Moving downstream, at 2H, Singh and Roy's model starts to be closer to the experimental velocities, especially in the near wall region. Shyy's model produces smaller velocities in such region. For the baseline cases, both experimental and simulation data are in very good agreement. At 4H, near the reattachment point, the baseline cases are again very close to each other and it is seen that for both experimental and computational results, modulated plasma effects cause higher velocities hence a smaller circulation bubble after the step region as expected. In this case, Shyy and Singh and Roy's model have a very similar distribution near the wall and Shyy's model is closer to the experimental profile in the free stream region. Finally at 6H, a location where all the cases have already passed the reattachment region and the flow is again developing, Singh and Roy's model shows an exceptional agreement in the recovery of the boundary layer—it can be observed in the figure that Singh and Roy's model follows the same distribution of the experiment. Again the computational profile show very good agreement with the experimental velocity distribution.

As a conclusion, Shyy's model predicts the behaviour of the flow better within the circulation region, in particular near the step, producing a more realistic prediction of the flow behaviour as it is been shown. On the other hand, the effects of the plasma produced by Singh and Roy's model are stronger near the edge of the step resulting in a smaller secondary bubble in the corner right after the step edge—this can be clearly seen when the velocity streamlines are depicted in the next section. These results are related to the way the plasma is modelled: Shyy's model restricts the plasma effects to a triangular region, whereas Singh and Roy's model plasma actuation affects the whole domain and its effects diminish proportionally as the flow moves further from the actuator location. The plasma effects are also related to the way the force of the

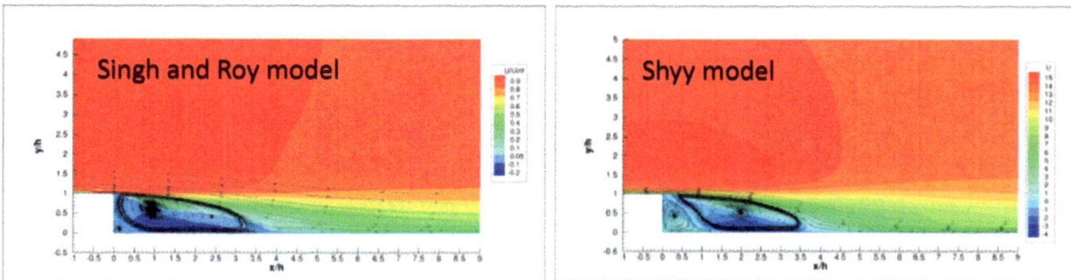

Fig. 5.15 Streamlines of flow field and Normalised U contours; top–Singh and Roy's model using two different constants; bottom; Shyy's model

plasma is calculated as it was shown in Chap. 3 where the models were explained: Singh and Roy's model force is related to the fourth of the applied voltage and the aforementioned distance to the actuator; Shyy's model only depends on the voltage and frequency of the device and it is constraint by the triangle geometry.

As the flow moves downstream towards the outlet, Singh and Roy's shows a better prediction of the flow behaviour and it matches the experimental boundary layer development after the reattachment of the flow to the bottom wall after the step. This effect makes perfect sense as Shyy's plasma effects weaken earlier due to the restricted area of actuation of the plasma force.

5.4.2.3 Reattachment Region

The resulting reattachment length matched the experimental length, Fig. 5.14. The normalised velocity streamlines for the experiment without control and the 3D CFD simulation were shown in Fig. 5.15.

Singh and Roy's model shows a smaller secondary bubble right after the step than Shyy's model, which means the momentum added due to the plasma is higher in Singh and Roy's model causing a larger acceleration after the device's location and hence smashing the bigger bubble and reducing the size of the smaller in the corner. Plasma effects in Singh and Roy's model are visible at the device location even after performing the time and space averaging whereas when Shyy's model is used, the effects are not visible.

5.5 Conclusion

An investigation of the effects on a flow over a backward facing step of a single DBD plasma device located upstream of the edge of the step was carried out.

Simulations were carried out and afterwards compared according to the experimental partner in the MARS project, the University of Poitiers (France), for this particular case. In order to simulate the actuation of plasma in the flow field, a phenomenological model was implemented in the CFD in-house code utilised to run all calculations in this study. Roy's model was the model selected to simulate the plasma actuator.

Singh and Roy's model provided reasonable agreement and reliability when it was compared to the available experimental database, reducing the reattachment length a 5% compared to the baseline; value which was acceptable as in experiments the reduction was a 3.4% when applying steady plasma. In the simulations of steady plasma, the actuation of plasma in the experiment had a frequency of 1 kHz which is considered quasi-steady. Singh and Roy's model does not include the frequency of the applied voltage but it considers the plasma actuation to be purely steady yet it showed very good agreement with the experimental database.

Finally, a modulation of plasma with 125 Hz pulsation frequency was investigated testing as well two initial different Singh and Roy's constants but the model showed to over-predict the reattachment length when plasma was applied in unsteady mode using low values of the constant of the model. A further investigation hence was performed, as the effects of plasma using Singh and Roy's model were not strong enough resulting in highly unsteady flow. A set of two larger constants was investigated finding finally an appropriate value which showed reasonable agreement with the experimental results from the University of Poitiers.

The difference between models lies on the region of actuation of the plasma in the model and also on the mathematical formulation. In Shyy's model, the region of actuation of plasma is restricted to a triangular region and a blending function controls whether the plasma is present or not. In Singh and Roy's model, the plasma actuation is more realistic as it affects the whole domain and it is a function of the distance of certain point in the domain to the actuator location. Because of this, the effects of plasma affect more the whole mean flow when Singh and Roy's model is used than when Shyy's model is used.

In Shyy's model the created plasma force is strongest near the edge of the exposed electro and it weakens as it moves beyond the embedded electrode. On the other hand, according to the mathematical model of Singh and Roy's model, the plasma EHD force will be maximum in the middle of the distance of the two electrodes on top of the dielectric and will vary all over the domain depending on the position within it. Another difference is the parameters on which each model depends on: Shyy's model depends on the electric field and on the applied AC frequency: in its formulation, the plasma force is linearly related to these two parameters. On the other hand, Singh and Roy's model is a function of the dielectric material properties, two empirical constants adjusted for a quiescent environment and it is related to the applied critical applied voltage to the fourth power, making the model extremely sensitive to the applied voltage and requiring the introduction of a constant to adjust the effects of the induced body force. This model does not consider the AC frequency.

As final most conclusions, it has been shown that experimental and numerically a single DBD plasma actuator induces a wall jet flow with steady or periodic fluctuations that clearly influences the flow fields on a backward facing step. The reattachment point can be reduced by 20% when applying modulated plasma due to the strong influence of the discharge on the shear layer development and resulting

Reynolds stresses: higher velocities before the step enhance the flow mixture so that the energy is conveyed from the mean flow to the recirculation area. The shear stress thus increases in the shear layer region causing a reduction of the reattachment length and a reduction of the skin friction coefficient. With regards to the analysis of steady plasma using Singh and Roy's model, it has been shown that despite the model does not take into account the frequency of the applied voltage, the model produces some reliable results. This is a clearly advantage not only with regards to the reliability of the prediction of the flow behaviour but also computationally speaking because there is less computational expenses as frequency is not included in the formulation of the model. The application of modulated plasma produces better control than a steady application; the real advantage of this is that only a portion of the energy—which will depend on the duty cycle of the pulsation—applied to the steady plasma is the energy needed for the pulsation of the plasma, leading to energy saving and potential use of this device in the pulsating mode for real application.

Acknowledgements This research has been partially funded by the European Commission (EC), though the Framework Programme 7 (FP7) Project #266326 entitled: "Manipulation of Reynolds Stress 382 for Separation Control and Drag Reduction" (MARS). The authors would also like to thank Dr W Wang for her help in running the cases.

References

1. Armaly BF et al (1983) Experimental and theoretical investigation of backward facing step flow. J Fluid Mech 127(473):20
2. Barri M et al (2010) DNS of backward-facing step flow with fully turbulent inflow. Int J Numer Meth Fluids 64:777–792
3. Benard N, Sujar-Garrido P, Bayoda KD, Bonnet JP, Moreau E (2014) Pulsed dielectric barrier discharge for manipulation of turbulent flow downstream a backward facing step. AIAA paper 2014-1127
4. Benard N, Braud P, Jolibois J, Moreau E (2008) Airflow reattachment along a NACA 0015 airfoil by a surface dielectric barrier discharge actuator—time-resolved particle image velocimetry investigation. AIAA paper 2008-4202
5. Benard N, Bonnet JP, Touchard G, Moreau E (2008) Flow control by dielectric barrier discharge actuators–jet mixing enhancement. AIAA J 46(9):2293–2305
6. Benard, N., Braud, P., Pons, J., Touchard, G., Moreau, E (2007) Quasi-steady and unsteady actuation by surface non-thermal plasma discharge for control of a turbulent round air jet. J Turbulence, Vol. 49
7. Chiang TP, Sheu TWH (1998) A numerical revisit of backward-facing step flow problem. Phys Fluids 11(4):862–874
8. Corke TC, Enloe CL, Wilkinson SP (2010) Dielectric barrier discharge plasma actuators for flow control. Ann Rev Fluid Mech 2010(42):505–529
9. Driver DM, Seegmiller HL (1985) Features of a reattaching turbulent shear layer in divergent channel flow. AIAA J 23:163–171
10. Enloe CL, McLaughin TE, VanDyken RD, Kachner KD, Jumper EJ, Corke TC (2004) Mechanisms and responses of a single dielectric barrier plasma actuator: plasma morphology. AIAA J 42(3):589–594
11. Fadai-Ghotbi A et al (2008) Revisiting URANS computations of the backward-facing step flow using second moment closures

12. Falco RE (1979) A review of the current state of knowledge of turbulent boundary structure. Summary of the AFOSR/MSU research specialists workshop on coherent structure in turbulent boundary layers, AFOSR-TR-80-0290, 1979
13. Gregory JW (2007) Force production mechanisms of a dielectric-barrier discharge plasma actuator. AIAA paper 2007-185, 2007
14. Hoskinson, AR, Hershkowitz, N (2008) Flow measurements and plasma simulations of double and single DBD plasma actuators in quiescent air. In: 46th AIAA aerospace sciences meeting and exhibit, pp 1370, Reno, NV, 2008
15. Hunt JCR, Wray AA, Moin P (1988) Eddies, stream, and convergence zones in turbulent flows. Center for Turbulence Research Report CTR-S88, pp. 193–208
16. Le H, Moin P, Kim J (1997) Direct numerical simulation of turbulent flow over a backward facing step. J Fluid Mech 330:349–374
17. Leonov SB, Yarantsev DA (2007) Quasi-DC discharge in high-speed flow for combustion enhancement. ICPIG 15–20, 2007
18. Lien FS, Leschziner MA (1994) Assessment of the turbulence-transport models including non-linear RNG eddy viscosity formulation and second-moment closure for flow over a BFS
19. Mertz BE, Corke TC (2011) Single-dielectric barrier discharge plasma actuator modelling and validation. J Fluid Mech 2011(669):557–583
20. Miles RB (2000) Flow control by energy addition into high-speed air. AIAA 2000-2324, 2000
21. Opaits DF (2012) Dielectric barrier discharge plasma actuator for flow control. NASA/CR—2012-217655, 2012
22. Pons J, Moreau E, Touchard G (2005) Asymmetric surface dielectric barrier discharge in air at atmospheric pressure: electrical properties and induced airflow characteristics. J Phys D Appl Phys 38:3635
23. Riherd M, Roy S, Rizzetta D, Visbal M (2011) Study of transient and unsteady effects of plasma actuation in transitional flow over an SD7003 airfoil. AIAA Paper 2011-1075, 49th AIAA aerospace meeting including the new horizons forum and aerospace exposition, Orlando, Florida, 4–7 Jan 2011
24. Roth JR (2003) Aerodynamic flow acceleration using paraelectric and peristaltic electrohydrodynamic effects on a one atmosphere uniform glow discharge plasma. Phys Plasmas 10:2117
25. Roy S, Gaitonde DV (2006) Force interaction of high pressure glow discharge with fluid flow for active separation control. Phys Plasmas 13:023503
26. Roy S, Singh KP, Gaitonde DV (2007) Air plasma actuators for effective flow control. AIAA paper, 2007-184, 2007
27. Roy S, Singh KP (2007) Modeling plasma actuators with air chemistry for effective flow control. J Appl Phys 101:123308
28. Roy S, Wang CC (2009) Bulk flow modification with horseshoe and serpentine plasma actuators. J Phys D: Appl Phys, vol 42
29. Santhanakrishnan A, Jacob JD (2006) Flow control using plasma actuators and linear/annular plasma synthetic jet actuators. In: 3rd AIAA flow control conference, San Francisco, CA, 2006, p 3033
30. Singh KP, Roy S (2008) Force approximation for a plasma actuator operating in atmospheric air. J Appl Phys 103:013305
31. Sujar-Garrido P, Bernard N, Laurentie JC, Bonnet JP, Moreau E (2012) Modifications du tenseur de Reynolds turbulent en aval d'une marche descendante par actionneur plasma. 13ième Congrès Francophone de Techniques Laser, CFTL 2012—ROUEN, 18–21 Sept 2012
32. Shyy W, Jayaraman B, Andersson A (2002) Modelling of glow discharge-induced fluid dynamics. J Appl Phys 92:6434
33. Visbal MR, Gaitonde DV, Roy S (2006) Control of transitional and turbulent flows using plasma-based actuators. AIAA Paper 2006-3230, San Francisco, CA, 5–8 June 2006
34. Wang W (2013) Passive and active flow control studies using hybrid RANS/LES simulations, Thesis (Ph.D.), The University of Sheffield, 2013

Chapter 6
The Effects of Plasma-Based Body Force on Flow Separation Suppression

Junji Huang, Baopeng Hu, Zexiang Li, Jinbai Zhang, Zhansen Qian and Shilong Lan

Abstract The effects of dielectric barrier discharge (DBD) plasma-based body force on Reynolds stresses and flow structures are studied numerically on two basic test cases, which are NACA0015 airfoil flow and backward facing step flow. The function of the plasma actuator is considered as the body force, based on a phenomenological model, which is suggested by W. Shyy. A detached eddy simulation (DES) methods are applied in both benchmark flows. Two effects of the continuous force are identified in present study, one is acceleration of flow in the vicinity of the actuator or the forcing region, another is reduction of Reynolds stresses. As for pulsed force, Reynolds stresses will be increased, which is different from that of the continuous force cases. The underlying mechanisms are proposed in present paper.

Keywords Plasma · Body force · Separated flow · Detached eddy simulation

6.1 Introduction

On moving vehicles it is common to find the separated flows, which generally result in more drag. Effective control devices reduce separation region, or pressure drag, thus a large amount of energy can be saved. Many engineers and fluid dynamicists have engaged in such devices. Some control devices, such as fluctuating wall, oscillating or moving wall, synthetic jet, pulsed jet, plasma actuator, etc., are already available and extensively studied. Among them plasma actuator appears relatively late, but gains promising expectation. However, the relation between the Reynolds stresses

---

J. Huang · B. Hu · Z. Li · J. Zhang · S. Lan (✉)
Ministry of Education Key Laboratory of Fluid Mechanics, Beihang University, Beijing 100191, People's Republic of China
e-mail: lanshilong@126.com

J. Huang
Department of Mechanical and Aerospace Engineering, Missouri University of Science and Technology, Rolla, MO 65409, USA

Z. Qian
AVIC Aerodynamics Research Institute, Beijing, People's Republic of China

N. Qin et al. (eds.), *Advances in Effective Flow Separation Control for Aircraft Drag Reduction*, Computational Methods in Applied Sciences 52,
https://doi.org/10.1007/978-3-030-29688-9_6

and plasma actuator is not very clear yet. As we know, Reynolds stress tensor is the key quantity, underlying the separated flow or the attached flow. This paper aims to investigate how the body force induced by the dielectric barrier discharge (DBD) plasma affects the Reynolds stresses and flow separation.

As it is mentioned in MARS acronym, one of the main objectives of the project is to manipulate Reynolds stresses for separation control by devices of different type. As the member of one of the working package of MARS, we focused on numerical study of flow control by DBD plasma actuator. There are two benchmark cases in MARS, one is the flow around NACA0015 airfoil [1] and another is the flow across the backward facing step [2, 3]. For NACA0015 case, the experiment by Siauw [1] showed that a stall or separated flow occurred near the trailing edge when angle of attack was 11°. The onset of separation near the trailing edge depends on the structures of the boundary layer before the separation point. Therefore, the Reynolds stresses in the boundary layer must be computed with sufficient accuracy to predict the separation point. For the second case, although the separated flow always initiate at the corner of the step, the evolution of the free shear layer is still greatly influenced by the upstream boundary layer. In all, the resolution of the Reynolds stresses in the boundary layer, further the manipulation of them may play very important role on the flow control.

In the present study a phenomenological model suggested by Shyy [4] was used to account for the action of the DBD plasma on neutral air molecules. The Shyy's model [4] is simple and easy to be implemented in the existing code. It is widely used as the plasma model in the flow control [5–7]. A description about this model is given in Sect. 6.2. DES and numerical methods used in our study are also introduced in Sect. 6.2. The results of the NACA0015 flow manipulated by continuous forcing and backward facing step flow manipulated by pulsed forcing are given in Sect. 6.3 and the conclusions are drawn in Sect. 6.4.

6.2 Modellings and Numerical Methods

6.2.1 The Plasma-Based Body Force Modelling

In the last two decades, a very promising flow control device based on alternating current dielectric barrier discharge(AC-DBD) has been intensively studied. The experiments [8, 9] on AC-DBD plasma actuators showed that the air flows could be induced by glow discharge. The role of plasma on the neutral air is commonly considered as the body force [1, 8, 10]. There are a number of body force models of different approximation levels, for example, some are based on first principles [11–13], some are simplified phenomenological models [4, 14–16]. The Shyy's model can roughly capture the spatial distribution of the force acted by plasma. In addition, this model can be conveniently implemented in the flow governing equations. Since the Shyy's model has a good balance between the accuracy and efficiency, we used this model in our study.

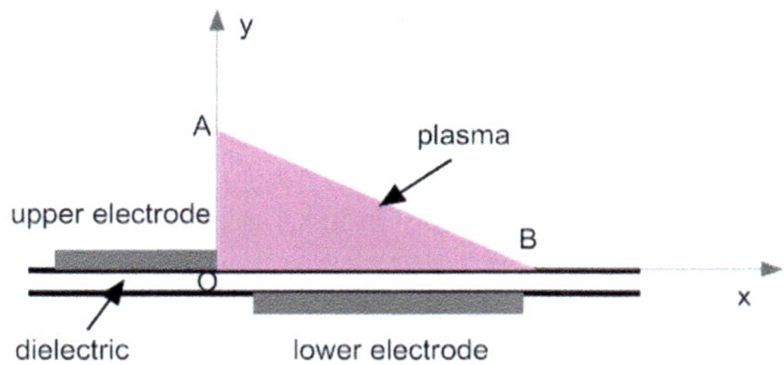

Fig. 6.1 The schematic of the dielectric barrier discharge actuator and plasma region

As proposed by Shyy [4], the plasma region or the body force region is assumed as a triangle for a cross-section, and the field strength is linearly distributed inside the triangle, the function of the field strength is given by

$$\left|\vec{E}\right| = E_0 - k_1 x - k_2 y, \tag{6.1}$$

where $k_1 = \frac{E_0 - E_b}{b}$, $k_2 = \frac{E_0 - E_b}{a}$. E_b is the breakdown electric field strength (30 kV/cm), $E_0 = \frac{V_0}{d}$, V_0 is the peak of applied voltage. The plasma is assumed to exist in a right triangle ABO(see Fig. 6.1), and in general, a is the height of the plasma, $a = \text{OA} = 1.5$ mm, and b is the width of the plasma, $b = \text{OB} = 2$ mm.

The field strength is written by

$$\vec{E} = \frac{\left|\vec{E}\right|}{\sqrt{k_1^2 + k_2^2}}(k_2, k_1). \tag{6.2}$$

The effective force on the neutral molecules is given by

$$\vec{F} = \alpha \vartheta T_c \rho_c e_c \vec{E}, \tag{6.3}$$

where ϑ is the frequency of the applied voltage, T_c is the discharge time, α is the factor to account for collision efficiency and can be assumed to be one in view of the disparity between the electron and neutral particle number densities.

The dimensionless strength parameter for the body force is expressed by Dc, which represents the ratio of electric field force to inertial force, and following the definition in Gaitonde [5], it can be written by

$$\text{Dc} = \frac{\rho_c e_c E_0 c}{\rho_\infty U_\infty^2}, \tag{6.4}$$

where ρ_c is charge density of electrons, $1.0 \times 10^{11}/\text{cm}^3$(approximated by Shyy [4], e_c is elementary charge, 1.602×10^{-19}, ρ_∞ is air density, 1.225 kg/m^3. E_0 is the strongest electric field.

The dimensionless momentum equation is

$$\frac{\partial \vec{u}}{\partial t} + \vec{u} \cdot \nabla \vec{u} = -\nabla p + \frac{\Delta \vec{u}}{Re_m} + \vec{F} \tag{6.5}$$

The third term in the right hand side of Eq. (6.5) is the plasma-based body force, $\vec{F}$ is defined by

$$\vec{F} = \vartheta T_c \, \mathrm{Dc} \vec{E}, \tag{6.6}$$

where ϑ is frequency of applied voltage, $\vec{E}$ is dimensionless electric field, which is same as the definition by Shyy[4]. If $\vartheta = 3$ kHz, $T_c = 67$ μs are fixed in this study, thus the magnitude of the plasma-based body force in Eq. (6.5) is uniquely determined by the strength parameter Dc.

For example, if the applied voltage is 4 kV(rms), the peak is 5.6568 kV, and d is the thickness of the dielectric barrier, 0.25 mm, corresponding E_0 is 226.26 kV/cm, then Dc = 64.7. And if $\vartheta = 3000$ kHz, $T_c = 67$ μs, it can be derived that the maximum dimensionless force induced by plasma is almost 13.

6.2.2 *Turbulence Modelling and Numerical Methods*

For the low wind speeds in basic test cases, we assumed the flows incompressible. The incompressible Navier-Stokes equations were computed with the pseudo-compressibility method. The 2nd order Roe scheme was used for inviscid fluxes, and the 2nd order central difference scheme was used for viscous fluxes. Dual time step method was used for time marching, that is, the LU-SGS method for pseudo time, and the 2nd order backward difference for physical time. The detailed see references [17–19]. Since the DES approach proposed by Spalart [20], a great improvement has been made on the turbulence modelling to resolve the separation flow. In present study a kind of hybrid LES and RANS method [21] was used, in which Menter k-ω SST [22] was used for RANS. This model is briefly introduced here.

The turbulent kinetic energy (k) equation is

$$\frac{D\rho k}{Dt} = \tau_{ij}\frac{\partial u_i}{\partial x_j} - \beta^* \rho k \omega + \frac{\partial}{\partial x_j}\left[(\mu + \sigma_k \mu_t)\frac{\partial k}{\partial x_j}\right]. \tag{6.7}$$

The dissipation rate (ω) equation is

$$\frac{D\rho\omega}{Dt} = \frac{\gamma}{\upsilon_t}\tau_{ij}\frac{\partial u_i}{\partial x_j} - \rho\beta\omega^2 + \frac{\partial}{\partial x_j}\left[(\mu + \sigma_\omega \mu_t)\frac{\partial \omega}{\partial x_j}\right] + 2(1 - F_1)\rho\sigma_{\omega 2}\frac{1}{\omega}\frac{\partial k}{\partial x_j}\frac{\partial \omega}{\partial x_j}, \tag{6.8}$$

where the eddy viscosity $\upsilon_t = \min\left[\frac{k}{\omega}, \frac{a_1 k}{\Omega F_2}\right]$, For concise, the constants and parameters are not listed here, see Menter's paper [22]. Next, a length scale $l_{k-\omega} = \frac{k^{1/2}}{\beta^* \omega}$ is defined [21] by, and the k equation is rewritten by

$$\frac{D\rho k}{Dt} = \tau_{ij}\frac{\partial u_i}{\partial x_j} - \frac{\rho k^{3/2}}{l_{k-\omega}} + \frac{\partial}{\partial x_j}\left[(\mu + \sigma_k \mu_t)\frac{\partial k}{\partial x_j}\right]. \tag{6.9}$$

For the hybrid LES and RANS model [21], a new length scale $\tilde{l}$ is introduced as

$$\tilde{l} = \min(l_{k-\omega}, C_{DES}\Delta), \tag{6.10}$$

where Δ is the filter length, and the coefficient $C_{DES} = (1 - F_1)C_{DES}^{\kappa-\varepsilon} + F_1 C_{DES}^{\kappa-\omega}$, where $C_{DES}^{\kappa-\varepsilon} = 0.61$, $C_{DES}^{\kappa-w} = 0.78$. Finally, the k equation in the hybrid LES and RANS model is written by

$$\frac{D\rho k}{Dt} = \tau_{ij}\frac{\partial u_i}{\partial x_j} - \frac{\rho k^{3/2}}{\tilde{l}} + \frac{\partial}{\partial x_j}\left[(\mu + \sigma_k \mu_t)\frac{\partial k}{\partial x_j}\right]. \tag{6.11}$$

The advection terms in this k equation and the ω equation are discretized with the first order upwind difference scheme.

6.3 Results and Discussions

6.3.1 NACA0015 Flow

A C-type structured grid is generated for NACA0015 airfoil, and extended $0.2c$ in spanwise direction, as shown in Fig. 6.2. There are 225 grid points wrapping the airfoil surface, and 159 points in normal direction. From the trailing edge to the downstream border, 150 points are in streamwise direction and 317 points are in y direction. There are 40 points in span. Total grid is of 3.41 million points. The distance between the second grid line and airfoil surface is $4 \times 10^{-6}c$,most of the grid points cluster within one chord from the airfoil surface. The outer border is $50c$ away. Boundary conditions are specified as, no-slip condition on the airfoil surface, freestream at inlet, and periodicity in span. The non-dimensional time step is 0.006, which is sufficient to resolve the large eddies shedding in the wake.

Considering the great computational costs, only two grids were used for grid-independent study. One coarse grid is of 3.41 million points, the other fine grid is of 13.6 million points.

Let's examine the time history of the lift coefficient of the two grids (Fig. 6.3). For coarse grid, the dimensionless time step is 0.006. The computation initiated from the uniform flow and it seems approach statistically steady state after time is 40. The lift

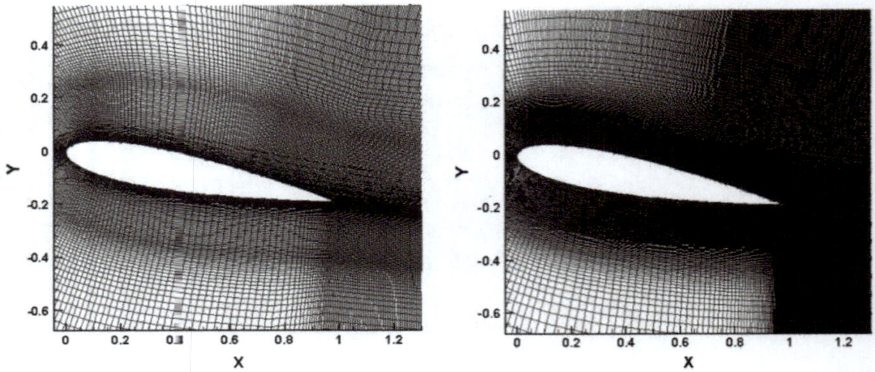

Fig. 6.2 The coarse(left) and fine grids(right) of NACA0015 airfoil

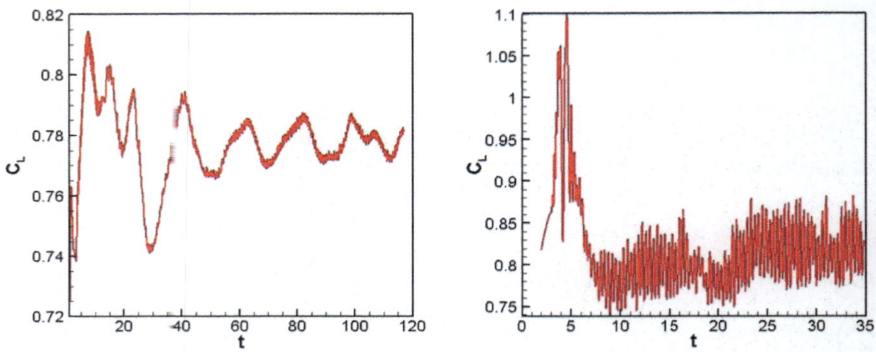

Fig. 6.3 The time history of the lift coefficients of the coarse(left) and fine(right) grids

time history shows the apparent oscillations with relatively low frequency, accompanied by much smaller and higher frequency oscillations. The mean lift coefficient is 0.77, and much less than the measured data 0.90 [1].

For finer grid, the dimensionless time step is 0.002, only one third of the time step of the coarse grid case. For huge computational work, only 17500 time step is run for the fine grid. It can be seen that the amplitude of oscillations are increased by a factor of 5, and more oscillations are invoked (Fig. 6.3). The mean lift coefficient computed from time 10 to 35 is around 0.81, much closer to the measured data.

Tsinghua University was involved in the same work package in MARS. Prof. Fu's group also used the NACA0015 case as the baseline. They studied the same case with LLR-DDES and a mesh of 33 million points. A blind comparison was made between their results and our data. The u-profile at x/c = 0.51 and x/c = 1.98 are shown in Fig. 6.4, the x-component velocities in our study are less than that of THU, and our fine grid results are close to the THU's data. For Reynolds shear stress –<u′v′>, there are remarkable differences between the coarse grid cases and the fine grid cases

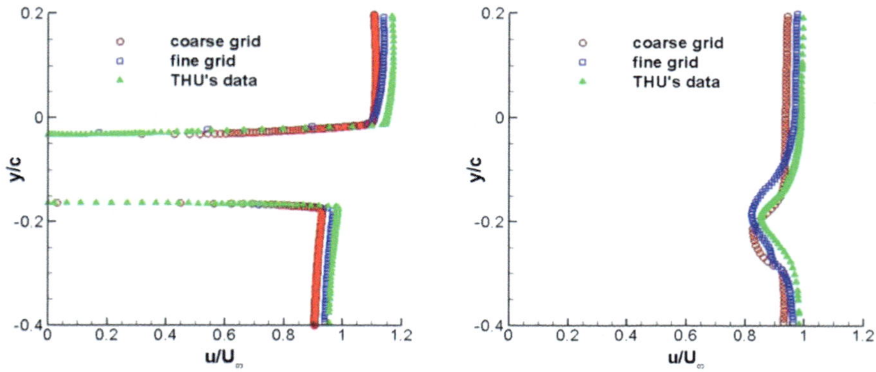

Fig. 6.4 The u-profiles of our data and THU's data(obtained from http://www.cimne.com/mars/) at x/c = 0.51(left) and x/c = 1.98(right)

(Fig. 6.5). Yet, our results of fine grid case agree well with THU's data. THU used even more grid points, almost 3 times that of our fine grid case, more eddies could be captured in the wake, which means higher level of $-\langle u'v' \rangle$ there.

The u profiles at locations x/c = 0.3, 0.5, normalized by the local velocities at the edge of the boundary layer, are shown in Fig. 6.6. The experimental data are contributed by another group led by Prof. Liu in BUAA, and the data also obtained from http://www.cimne.com/mars/. Our data are computed from both the coarse and the fine grid. The u profiles of coarse grid case show better agreement with experiments than those of the fine grid case. This is because the fine grid, in coupled with the aforementioned DES, over predicts the separation flow on the airfoil. In Fig. 6.6, at x/c = 0.5, x-component velocity gradient in y direction is much less than those of the other two cases, which implies it close to the separation point.

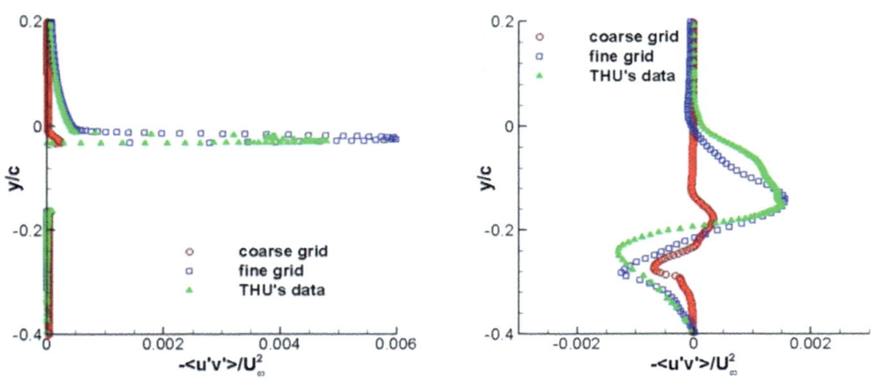

Fig. 6.5 The Reynolds shear stress $-\langle u'v' \rangle$ of our data and THU's data(obtained from http://www.cimne.com/mars/) at x/c = 0.51(left) and x/c = 1.98(right)

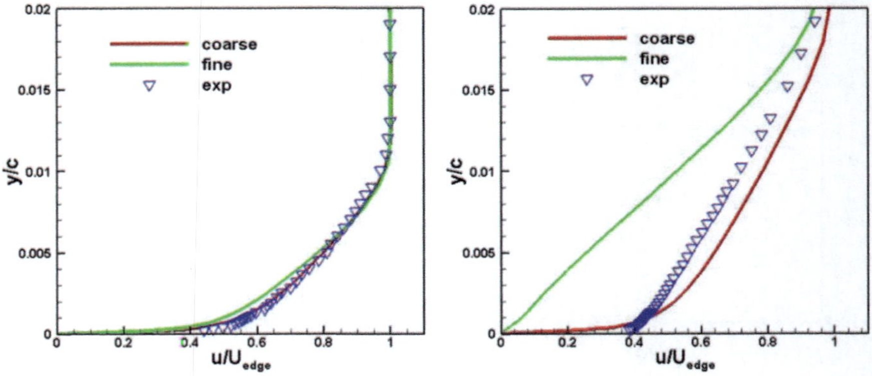

Fig. 6.6 The u-profiles of our data and the experiments by Prof. Liu(obtained from http://www.cimne.com/mars/) at x/c = 0.3(left) and x/c = 0.5(right)

Some improvements, such as DDES [23] or IDDES [24], may be used to cope with such problem when the fine grid used. It is not easy to simulate this mild trailing edge separation [25], both the turbulence models and the grids should be carefully selected. Based on the preceding results, it can be drawn that the coarse grid and DES used here can give reasonable results. Particularly, much less computer capacity is needed for the coarse grid, so it was used for cases studies here.

For the NACA0015 case, a continuous body force was used to manipulate the separation flow. According to the results of baseline flow, the separation point is around 75% position from the leading edge, thus the actuator is placed near the separation point. The body force region covers 0.024c in the streamwise direction, shown in Fig. 6.7. Three cases of different magnitude of body force are considered

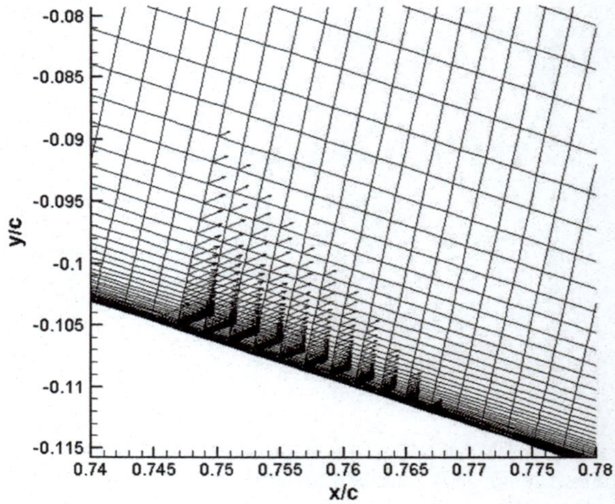

Fig. 6.7 The DBD plasma-induced forcing region on the surface of the airfoil

in this paper. For controlling cases, the frequency of actuator is fixed at 3000 Hz. In case II, the voltage is 2 kV and Dc = 32.4; In case III, the voltage doubled, i.e., 4 kV, and Dc = 64.7; In case IV, the voltage quadrupled, i.e., 8 kV, and Dc = 129.4, see Table 6.1. Corresponding maximum dimensionless body force $F_{\max}$ is 6.47, 12.94 and 25.88 respectively.

The mean streamline fields coloured by mean pressure coefficients in the four cases are shown in Fig. 6.8. Compared to baseline, the separation region is reduced by the body force in all three control cases. And the flow in case IV even attaches to the airfoil surface, no separation at all.

The pressure coefficients on the airfoil surface are shown in the Fig. 6.9. It can be seen that the higher suck peak in controlled cases compared to baseline case. The upper surface pressure in front of the forcing region becomes lower when force becomes larger, and quickly recovers during the forcing region.

Figure 6.10 shows that the lift coefficients increase obviously as soon as the force is exerted in all three control cases. The maximum increment of 26.1% is achieved in case IV.

Figure 6.11 shows the instantaneous vortex structures, which identified with $\lambda_2 = -10$. And the iso-surface is coloured by streamwise velocity. Compared to the baseline flow, it is very clear that the flow structures approach two-dimensionality in case III and IV. As for case II, the forcing is much smaller than those of case III

Table 6.1 The parameters in the four continuous forcing cases

Cases	Case I	Case II	Case III	Case IV
Dc	0	32.4	64.7	129.4
F_{max}	0	6.47	12.94	25.88

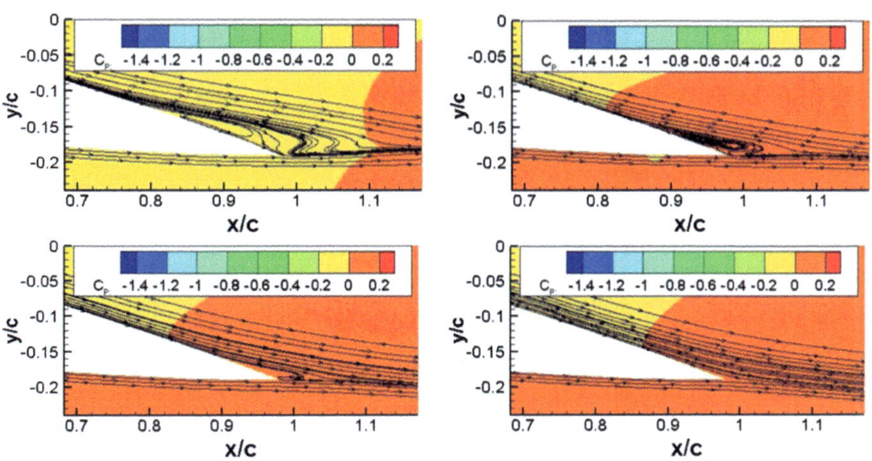

Fig. 6.8 Mean streamline fields and mean pressure distributions for case I, case II, case III and case IV(from the top left to bottom right)

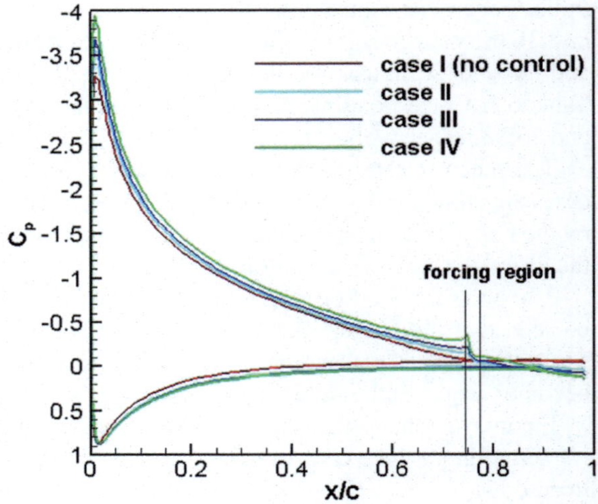

Fig. 6.9 The pressure coefficient on the airfoil surface

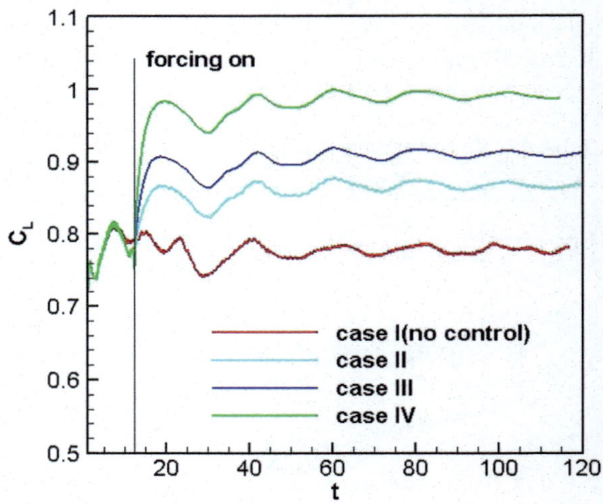

Fig. 6.10 The lift coefficients C_L versus dimensionless time

and IV, but the effects of the forcing on the vortices still can be seen, which is less developed in spanwise direction in comparison with the no-control flow.

Figure 6.12 shows the mean x-component velocities at four different locations. It can be seen that the induced velocities are not obvious at x/c = 70%, in front of the forcing region, and increase rapidly inside the forcing region, as shown at x/c = 80%. This increment of velocity is like a co-flow jet beneath the boundary layer. For example, at x/c = 90%, y/c = 0.01, the increments of velocities are about 20 m/s in case IV, 13 m/s in case III, and 6 m/s in case II, respectively. As the distance away from the wall increasing, the increments of velocities due to body force gradually decrease. It is just the co-flow jet induced by the continuous forcing that suppresses

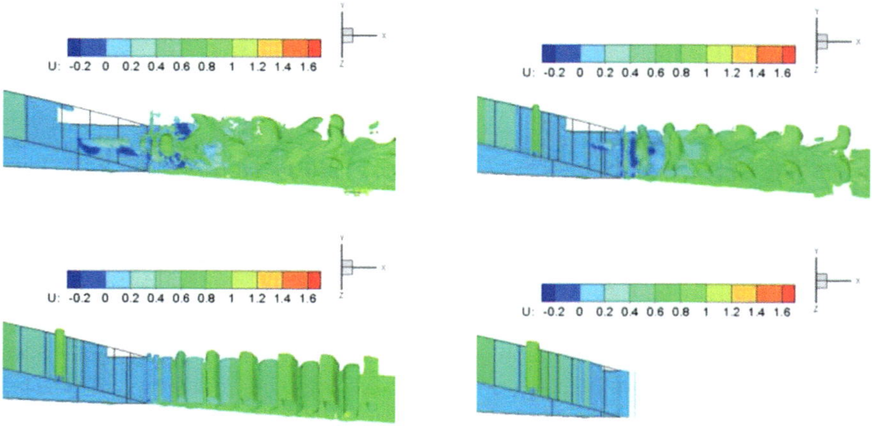

Fig. 6.11 The instantaneous vortex structures identified by $\lambda_2 = -10$ for case I, case II, case III and case IV(from the top left to bottom right)

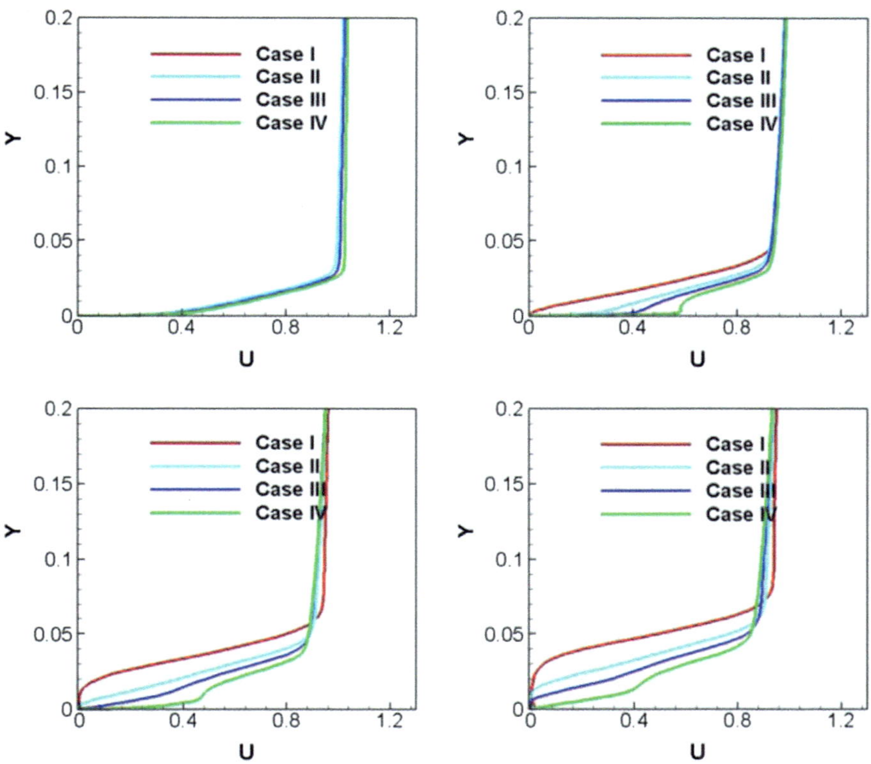

Fig. 6.12 The mean velocity profiles near the wall, at x/c = 70%, 80%, 90% and 96%(from the top left to bottom right)

the flow separation. Keep in mind that the magnitude of the forcing used in case IV almost approaches the practicable maximum, the induced velocity near the surface is around 10–20 m/s, which may not work well for reducing the massive separation of the high Re flow.

Figure 6.13 shows the Reynolds shear stresses –<u′v′> at four different locations. It can be seen that the Reynolds shear stresses are greatly reduced in all three control cases compared to the baseline flow, even at x/c = 70%, just in front of the forcing region, the levels of –<u′v′> are remarkably lowered. The results of three control cases are almost coincidental, which may be due to the under resolution by the RANS model. Compared to the separation flow in baseline, the velocity fluctuations or Reynolds shear stresses are definitely suppressed due to the reattached boundary layer. In another aspect, we note that the forcing has a regular spatial distribution, which the forcing gets maximum value on the surface and gradually decreases away from the surface. This kind of spatial distribution of the body force results in vorticity with opposite sign compared to the head of the hairpin-like vortex. As pointed by the same authors (2013, 2015), it may suppress the formation of the hairpin-like vortex, the source of the Reynolds stresses. It can be found that in Fig. 6.14, the

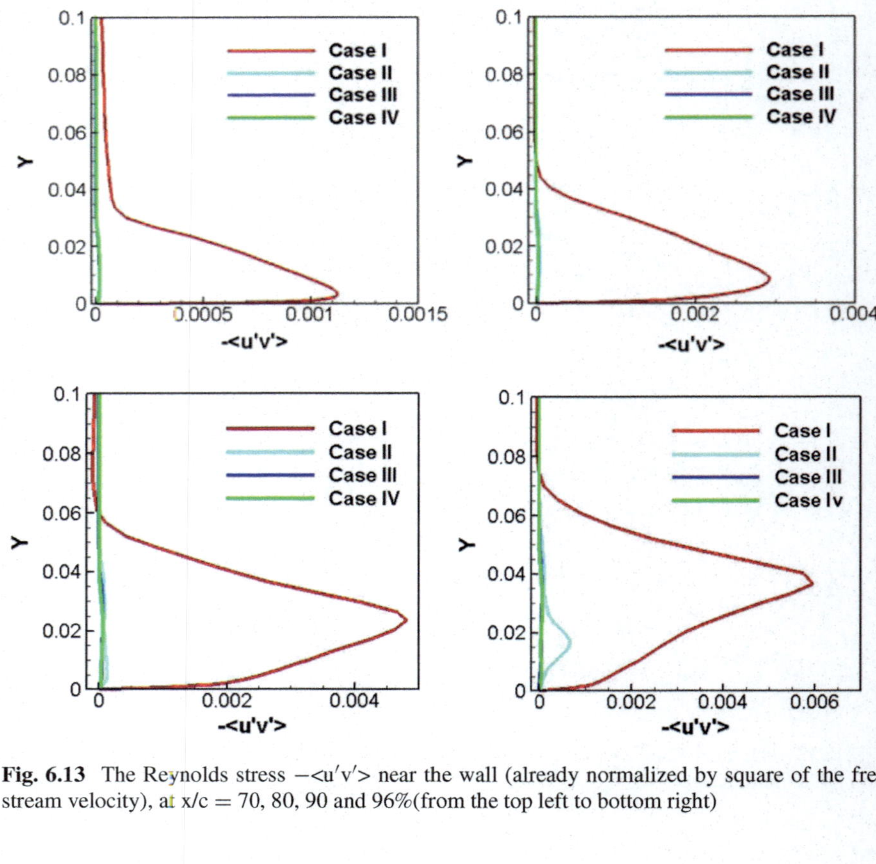

Fig. 6.13 The Reynolds stress $-\langle u'v' \rangle$ near the wall (already normalized by square of the free stream velocity), at x/c = 70, 80, 90 and 96%(from the top left to bottom right)

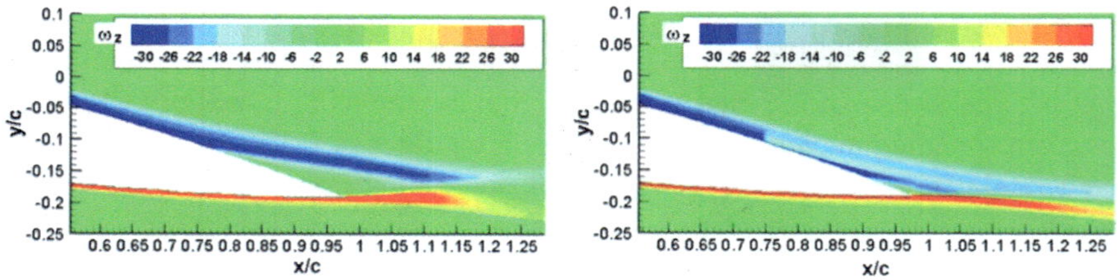

Fig. 6.14 Mean vorticity in z direction for baseline(left) and case III(right)

z-component vorticity in the body forcing region is apparently decreased, and this tendency continues downstream the forcing region. In all, there are two factors to reduce the Reynolds shear stress, one is the boundary layer reattachment by the co-flow jet induced by the forcing, another is the vorticity produced by the forcing.

6.3.2 *Backward Facing Step Flow*

Another test case in MARS is backward facing step flow. We followed the case definition by Prof. Qin. The step height h = 30 mm and the free stream velocity is 15 m/s, and Reynolds number (based on h) is Re = 30000. In present paper, length scales are non-dimensioned by step height, and velocity scales by free stream velocity. The computational domain and mesh are shown in Fig. 6.15. The origin of the frame oxy is just at the sharp corner of the step. The domain from inlet to the step corner is 10 h × 9 h, and covered with mesh points of 73 × 93, the domain from step to outlet is 35 h × 10 h, and with mesh points of 298 × 152. The spanwise length is 1.4 h, with a resolution of 80 points. Total number of mesh points is 4,218,885. A velocity profile measured from LDV, which given by Prof. Qin, is used as velocity inlet. Pressure outlet is applied to the exit. The periodic condition is applied in span. Both the bottom and top surfaces are no slip walls.

In the NACA0015 cases, we used continuous forcing. But in the back-facing step flow, we used pulsed forcing. As suggested by Prof. Qin, the pulsed forcing is

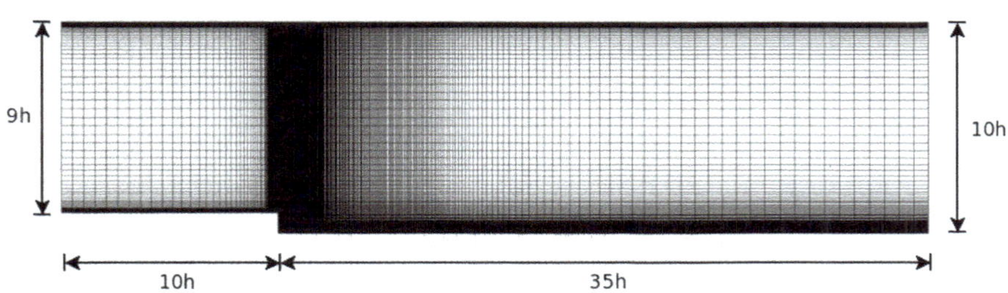

Fig. 6.15 Computational domain and mesh (according to Qin's case description)

achieved by a modulation frequency of 125 Hz and 50% duty. The location of the actuator or the forcing region is illustrated in Fig. 6.16. The parameters of the actuator are listed in Table 6.2. For this case, we used the strength of plasma Dc = 29.6, and the corresponding maximum dimensionless body force is 1.98. The modelling and numerical methods used here are same as the preceding cases.

The mean streamlines of the baseline flow and the controlled flow are shown in Fig. 6.17. It can be seen that the reattachment length is 6.1 for the baseline flow and 4.2 for the controlled flow, and a reduction of 31% in recirculation length is achieved.

Figure 6.18 shows the contour of Reynolds shear stress (–<u′v′>) of the baseline flow and the controlled flow. It is clearly seen that the level of Reynolds shear stress in the recirculation zone is increased when the pulsed forcing on. The frequency of the pulsed forcing here is approximately same as the frequency of the dominant eddies shedding(according to the PSD analysis provided by Qin). The fluctuations or eddies are intensified by the pulsed forcing, which accounts for the increasing of Reynolds stresses in the recirculation zone. The results show that the pulsed forcing increases the Reynolds stresses. It can be noticed that the continuous forcing in the NACA0015 case weakens the Reynolds stresses.

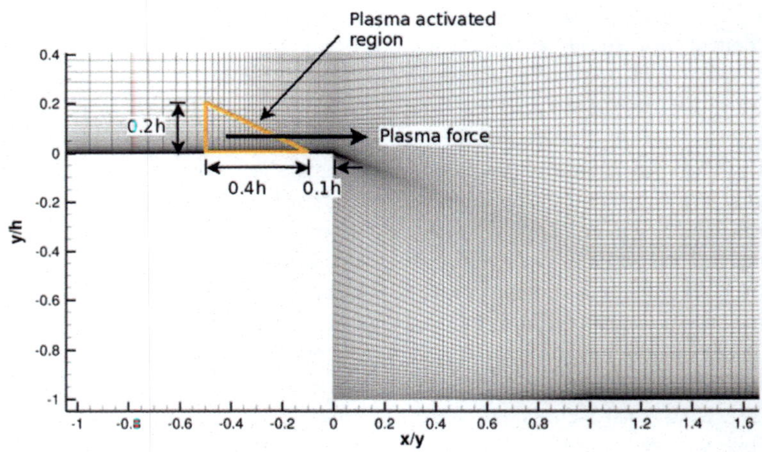

Fig. 6.16 Plasma activated region or the body force region(according to Qin's case description)

Table 6.2 The DBD plasma operating parameters

Parameter	Value
Voltage (peak to peak)	24 kV
AC frequency	1 kHz
Modulation frequency	125 Hz
Duty cycle	50%
Dc	29.6
F_{max}	1.98

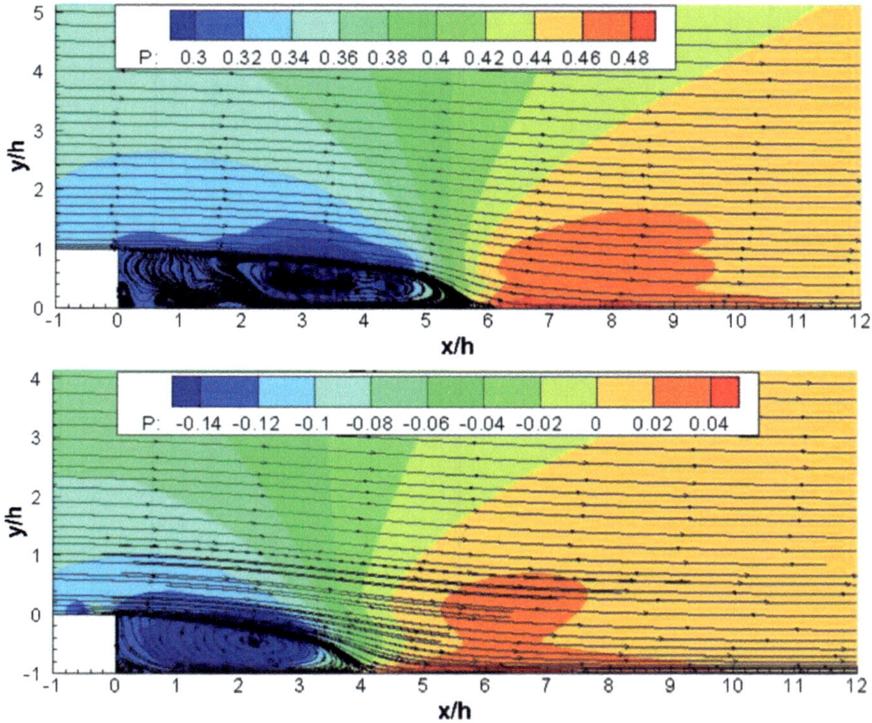

Fig. 6.17 Time averaged streamlines and pressure contour for the baseline(top) and the pulsed forcing flow(bottom)

6.4 Conclusions

In this paper two baseline flows are investigated, one is the mild separation flow on the NACA0015 with $Re = 10^6$, another is the massive separation flow across the backward facing step with medium Reynolds number of the order of 10^4. To suppress the separation flow of the NACA0015 case, the continuous forcing was used. The force obtained from the available single DBD actuators now can apparently reduce the separation zone. The underlying physics is the co-flow jet induced by the continuous forcing, which is well-known to the researchers. However, this may work only for moderate separation flow at such high Re number. It is also found that the continuous forcing can greatly reduce the Reynolds shear stress, which is unfavourable for flow mixing. This is rarely mentioned in the public literatures. In same authors' DNS work [26, 27] on channel flow, they showed that the continuous forcing can suppress the Reynolds stresses too. For the backward facing step flow, the separation line fixes on the sharp step corner, the continuous forcing may not work well. Additionally, it lowers the Reynolds shear stress. So the pulsed forcing was used for this case. The

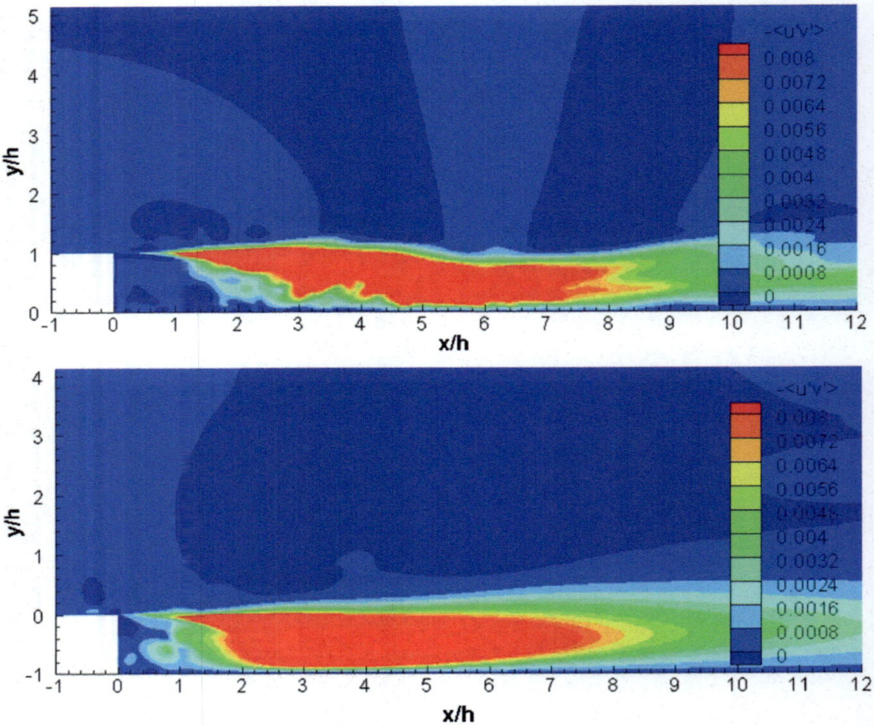

Fig. 6.18 The contour of Reynolds shear stress($-\langle u'v' \rangle$, normalized by square of the free stream velocity), for the baseline(top) and the pulsed forcing flow(bottom)

pulsed forcing can increase the Reynolds shear stress, subsequently the flow mixing, thus the reattachment length is decreased.

Acknowledgements This work was co-funded by PRC-MIIT and EC-DG RTD project "Manipulation of Reynolds Stress for Separation Control and Drag Reduction". National laboratory for computational fluid dynamics provided most of the computational resources.

References

1. Siauw WL et al (2009) Physics of separated flow over a NACA 0015 airfoil and detection of flow separation. AIAA Paper 2009-144
2. Driver DM, Seegmiller HL (1985) Features of a reattaching turbulent shear layer in divergent channel flow. AIAA J 23(2):163–171
3. Driver DM, Seegmiller HL, Marvin JG (1987) Time-dependent behavior of a reattaching shear layer. AIAA J 25(7):914–919
4. Shyy W, Jayaraman B, Andersson A (2002) Modeling of glow discharge-induced fluid dynamics. J Appl Phys 92(11):6434–6443

5. Gaitonde DV (2005) Control of flow past a wing section with plasma-based body forces. AIAA Paper 2005-5302
6. Asada et al (2009) Airfoil flow experiment on the duty cycle of DBD plasma actuator. AIAA Paper 2009-0531
7. Nonomura et al (2013) Control mechanism of plasma actuator for separated flow around NACA0015 at reynolds number 63000. AIAA Paper 2013-0853
8. Roth J, Sherman D, Wilkinson S (1998) Boundary layer flow control with one atmosphere uniform glow discharge surface plasma. AIAA Paper 1998-0328
9. Corke TC, Jumper EJ, Post ML, Orlov D, Mclaughlin TE (2002) Application of weakly-ionized plasmas as wing flow-control devices. AIAA Paper 2002-0350
10. Enloe CL et al (2004) Mechanisms and responses of a single dielectric barrier plasma actuator: geometric effects. AIAA J 42(3):595–604
11. Singh KP, Roy S, Gaitonde DV (2006) Modeling of dielectric barrier discharge plasma actuator with atmospheric air chemistry. AIAA Paper 2006-3381
12. Singh KP, Roy S (2007) Modeling plasma actuators with air chemistry for effective flow control. J Appl Phys 101:122308
13. Likhanskii AV, Shneider MN, Opaits DF, Miles RB, Macheret SO (2007) Numerical modeling of DBD plasma actuators and the induced air flow. AIAA Paper 2007-4533
14. Suzen Y, Huang P, Jacob J, Ashpis D (2005) Numerical simulations of plasma based flow control applications. AIAA Paper 2005-4633
15. Hall KD, Jumper EJ, Corke TC, Mclaughlin TE (2005) Potential flow model of a plasma actuator as a lift enhancement device. AIAA Paper 2005-0783
16. Orlov DM, Corke TC (2005) Numerical simulation of aerodynamic plasma actuators. AIAA Paper 2005-1083
17. Qian Z, Zhang J, Li C (2010) Preconditioned pseudo-compressibility methods for incompressible Navier-Stokes equations. Sci China Phys Mech Astron 53(11):2090–2102
18. Qian Z, Zhang J (2012) Implicit preconditioned high order compact scheme for the simulation of the three dimensional incompressible Navier–Stokes equations with pseudo-compressibility method. Int J Numer Meth Fluids 69(7):1165–1185
19. Zhang J, Qian Z (2012) Implicit eighth-order central compact scheme for the numerical simulation of steady and unsteady incompressible Navier–Stokes equations. Int J Comput Fluid Dyn 26(4):247–261
20. Spalart PR, Jou WH, Strelets MK, Allmaras SR (1997) Comments on the feasibility of LES for wings, and on a hybrid RANS/LES Approach. In: Ruston LA advances in DES/LES(1st AFOESR int conf on DNS/LES) Colunbus, OH Greyden Press
21. Travin AK, Shur ML, Strelets MKh, Spalart PR (2000) Detached-eddy simulations past a circular cylinder. Flow Turbul Combust 63:293–313
22. Menter FR (1993) Zonal two equation k-w turbulence models for aerodynamic flows. AIAA Paper 1993-2906
23. Spalart PR, Deck S, Shur ML, Squires KD, Strelets MK, Travin AK (2006) A new version of detached-eddy simulation, resistant to ambiguous grid densities. Theor Comput Fluid Dyn 2006(20):181–195
24. Shur ML, Spalart PR, Strelets MKh, Travin AK (2008) A hybrid RANS-LES approach with delayed-DES and wall-modelled LES capabilities. Int J Heat Fluid Flow 29:1638–1649
25. Durrani N, Qin N (2011) Behavior of detached-eddy simulations for mild airfoil trailing-edge separation. J Aircr 48(1):193–202
26. Lan S, Huang J, Zhang J (2013) Turbulence statistics and coherent structures of channel flow with the dielectrics barrier discharge plasma actuator. In: Proceedings of the 14th Asian Congress of Fluid Mechanics(14th ACFM). Publishing House for Science and Technology, pp 702–706
27. Li Z, Hu B, Lan S, Zhang J, Huang J (2015) Control of turbulent channel flow using a plasma-based body force. Comput Fluids 119:26–36

Chapter 7
Effects of DBD Plasma Frequency, Voltage, and Modulation on BFS Flow Reattachment

Spiridon Siouris and Ning Qin

Abstract The work in this paper presents the results from numerical investigation of Dielectric Barrier Discharge plasma actuators, applied on a backward facing step geometry. 2D and 3D simulations were carried out indicating that optimisation of plasma operating parameters is only reliable when dealing with 3D simulations. It was found that there is a parameter space in which reduction in reattachment length takes maximum values, up to 22% reduction, and beyond this, the reduction in reattachment length decreases.

Keywords Hybrid RANS-LES simulation · Active flow control · Dielectric barrier discharge plasma · Actuator

7.1 Introduction

Current aircraft wing designs are highly optimised in shape with the primary aim of reducing drag and control separation. Any further advances in increasing an aircraft's efficiency through wing shape optimisation is becoming increasingly difficult and for this reason there has been a significant focus on active flow control technologies. The aim of these is to manipulate the flow field around a wing in an active manner by energy transfer from the device to the flow field [1]. One type of active flow control device is the Dielectric Barrier Discharge (DBD) plasma actuator [2] that comprises of two electrodes in which high voltage difference is applied between them, as illustrated in Fig. 7.1.

When the voltage difference between the two electrodes is high enough, then a discharge current flows from one electrode to the other due to dielectric (air) ionisation. The movement of electrons results in momentum transfer to the air particles and therefore accelerating the flow near the plasma region. This phenomenon can be considered as the fundamental principle of plasma active flow control, and can be used

S. Siouris · N. Qin (✉)
Department of Mechanical Engineering, The University of Sheffield,
Mappin Street, Sheffield S1 3JD, England, UK
e-mail: n.qin@sheffield.ac.uk

N. Qin et al. (eds.), *Advances in Effective Flow Separation Control for Aircraft Drag Reduction*, Computational Methods in Applied Sciences 52,
https://doi.org/10.1007/978-3-030-29688-9_7

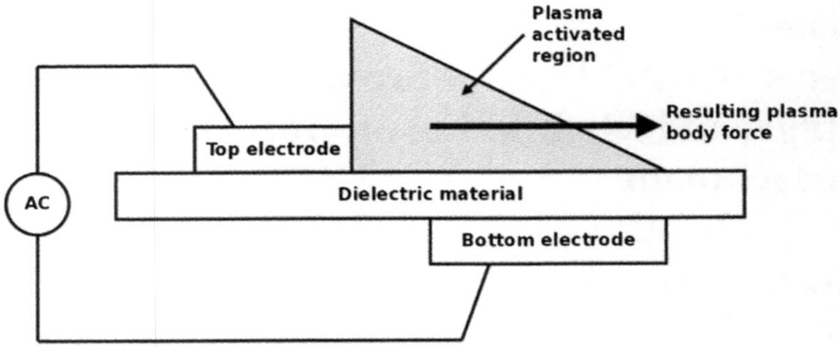

Fig. 7.1 Schematic of a typical DBD plasma actuator, and its electric field and body force according to Shyy et al. [3]

to control flow separation and ultimately drag reduction. The advantages of this type of active flow control are apparatus simplicity, robustness due to no moving parts, ability of fast response at high frequencies, and low power consumption. The operating parameters for generating the sinusoidal voltage signal, such as frequency and amplitude, contribute greatly to the resulting effect of accelerating the flow around the plasma region. In addition to this, by turning on and off the application of voltage, also known as modulation, at specific frequency and time duration in which plasma is on or off (duty cycle), a more effective flow control can be achieved with the appropriate parameters [4]. The work in this paper presents a study on the influence of the plasma controlling parameters in a backward facing step configuration on the properties of the flow such as reattachment point and Reynolds stresses.

7.2 Numerical Setup

7.2.1 Flow Solver Implementation

The in-house CFD solver "Dynamic Grid-Detached Eddy Simulation" (DG-DES) [5, 6] used in this study is an unstructured, density based, cell-centered, finite volume method code that solves the Navier-Stokes equations, as shown below.

$$\partial_t \int_{\Omega} \mathbf{W} d\Omega + \int_{S} (\mathbf{F}^c - \mathbf{F}^{\upsilon}) dS = 0 \tag{7.1}$$

In Eq. (7.1) $\boldsymbol{W} = (\rho, \rho \boldsymbol{u}, E)^{\mathrm{T}}$ is the vector of conservative variables; ρ is the fluid density; $\boldsymbol{u} = (u, v, w)^{\mathrm{T}}$ is the velocity vector of the flow field; $E = \rho(e + 0.5\boldsymbol{u}^2)$ is the total energy $\mathbf{F}^c$ is the convective flux, and $\mathbf{F}^u$ is the viscous flux.

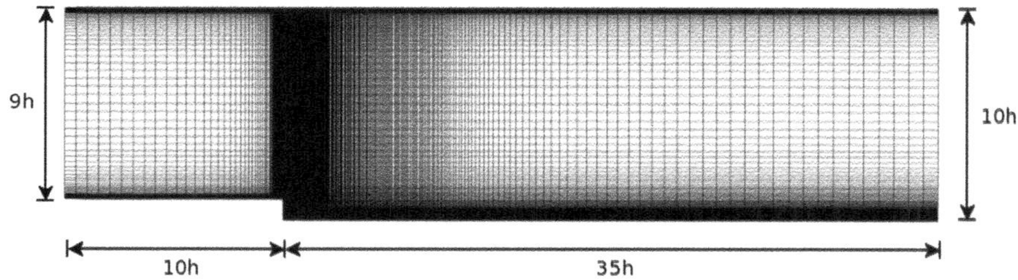

Fig. 7.2 The BFS geometry based on Sújar-Garrido et al. [12], and fully structured grid

In order to solve Eq. (7.1), the dual-time stepping technique is applied [7]. The physical time term is discretized with the implicit second order Euler's backward difference scheme, and third order Runge-Kutta method is used to iterate the pseudo time term. The Roe scheme [8] is used to discretize the convective flux terms, and the central difference scheme for the viscous flux. Details on the flux definition and discretization can be found in Qin and Xia [5]. This solver implements S-A based DES models to simulate complex flows with separation and reattachment such as DES [9], DDES [10], and IDDES [11].

7.2.2 Backward Facing Step

The backward facing step geometry with DBD plasma actuator used in this study is modelled after the configuration in Sujar-Garrido et al. [12]. The BFS geometry and fully structured grid can be seen in Fig. 7.2. In this case, the top and bottom walls are included in the grid with a no-slip boundary condition. The inlet velocity profile applied at −10 h has been based according to experimental LDV measurements, and the velocity profile can be seen in Fig. 7.3. The 2D model has been extended in the spanwise direction by 1.4 h, with a grid resolution of 50 cells, resulting in a total cell count of approximately 4.5 million cells.

7.3 Results

7.3.1 Plasma Modelling Validation

The DBD plasma implementation in the DG-DES CFD solver is based on the phenomenological model by Shyy et al. [3]. Here, a plasma DBD actuator is placed on a flat plate in which the flow is accelerated through particle ionization. Velocity profiles upstream and downstream of the electrode of the published data and current implementation are compared and this can be seen in Fig. 7.4.

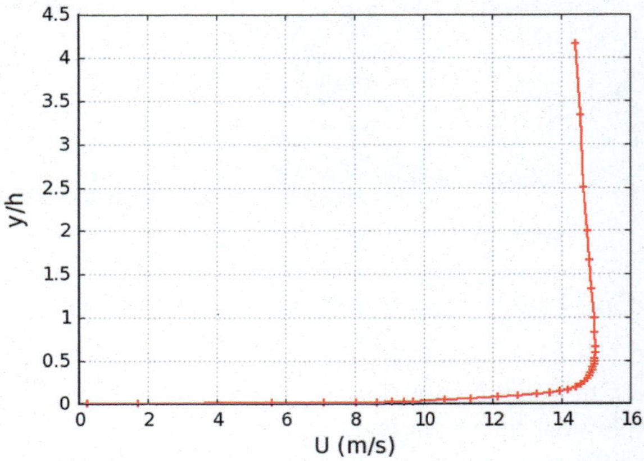

Fig. 7.3 Velocity profile at −10 h according to experimental LDV measurements by Sújar-Garrido et al. [12]

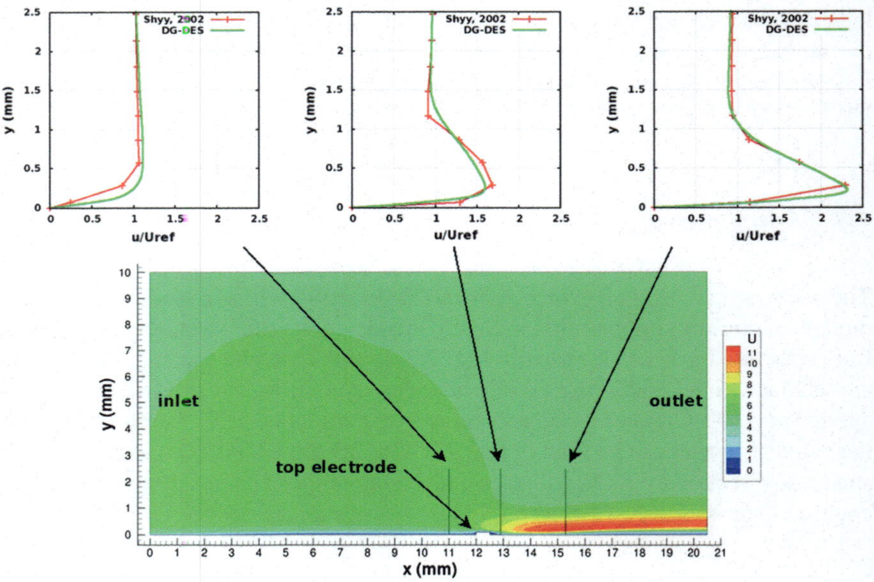

Fig. 7.4 Validation of the plasma induced velocity field in a given cross flow based on the body force model from Shyy et al. [3]

It can be seen that the first profile at 11 mm downstream of the inlet displays a typical boundary layer growth that is unaffected by the plasma region. The published data and computational results match very well. Further downstream, at 13 mm, it can be seen that the plasma force is already affecting the velocity profile of the flow. The acceleration of the fluid at this point is up to 50% of its original value, and affecting a region of up to 1.5 mm away from the wall. At 15.2 mm the flow reaches its maximum velocity which is 140% of the original value had there not been a plasma

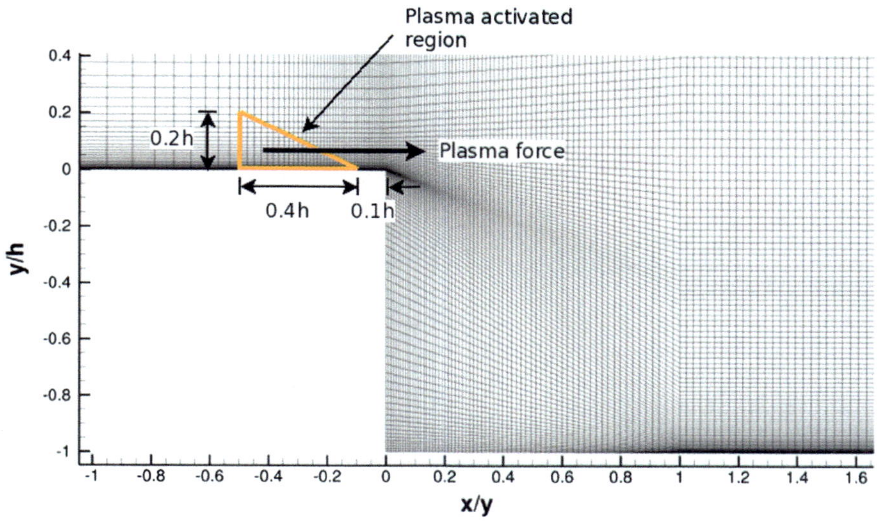

Fig. 7.5 Plasma activated region for the BFS case

Table 7.1 DBD plasma operating parameters

Parameter	Value
Voltage (peak to peak)	24 kV
AC frequency	1 kHz
Modulation frequency	125 Hz
Duty cycle	50%

actuator. It can be seen that the computed results are in very good agreement with the published data from the plasma model's authors and the implementation can then be considered as accurate.

Having ensured that the plasma implementation is correct, it is subsequently applied on the BFS geometry according to Sújar-Garrido et al. [12], and the extent of the plasma region can be seen in Fig. 7.5. The operating parameters for the DBD actuator are as in and are listed in Table 7.1.

7.3.2 2D Simulations

Initial 2D simulations were carried out for the calibration of the amount of plasma forcing via the efficiency factor of Shyy's 2002 model. This factor accounts for the collision efficiency (and therefore momentum transfer) between the ionised and non-ionised particles. Based on the experimental observations for the reattachment point,

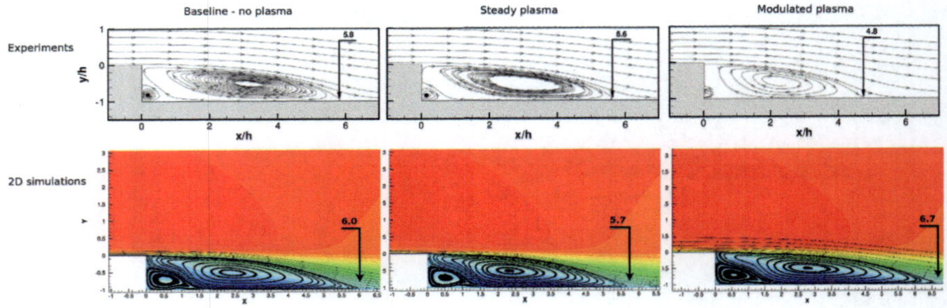

Fig. 7.6 Comparison between experimental and numerical simulations for the BFS baseline and actuated cases with steady and modulated plasma

it was found that an efficiency factor of 30% would provide the correct amount of forcing to match the experimental conditions, and this is seen in Fig. 7.6.

From the figure, the reattachment point of the baseline case is over predicted by 0.2 h, whereas the steady plasma case is over predicted by 0.1 h. This suggests that the inclusion of the 3D turbulent phenomena may assist in achieving modelling results that are in better agreement with experimental data due to the inclusion is turbulent dissipation in the 3D simulations. The 2D simulations for the modulated plasma case do not agree with experimental data as these increase the reattachment length rather than decrease. For this reason it can be stated that 2D modulated plasma simulations can lead to unreliable results and should therefore be avoided.

7.3.3 2D Optimisation of Plasma Operating Parameters

Having assessed the performance of the 2D plasma case, several simulations where then run with varying plasma operating parameters in terms of AC frequency, and voltage differential across the DBD electrodes. The AC frequency range was varied between 0.5 and 3 kHz, whereas the applied voltage difference varied between 5 and 25 kV, and this resulted to 15 2D simulations in total. The results from all simulations are presented in Fig. 7.7, for example, at operating parameters of 1.5 kHz and 15 kV, it can be seen that the resulting percent reduction on reattachment length is approximately 5%. The highest percent reduction in reattachment length was noted at 2.5 kHz and 25 kV with a value of 11%. It is interesting to note that the landscape of this parametric study with two variables does not suggest that the higher the AC frequency and voltage differential will produce higher amount of reattachment length. Beyond certain combination of these two parameters the reduction in reattachment length may decrease suggesting that there is an optimal region in which plasma actuation has a maximum effect.

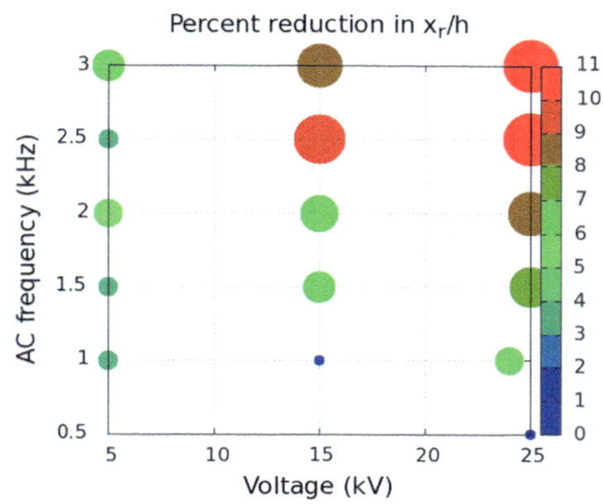

Fig. 7.7 2D steady state plasma operating parameter optimisation

7.3.4 *3D Simulations*

Having found a value for the efficiency factor by 2D simulations, 3D unsteady simulations were then carried out for a number of AC frequencies between 200 and 1800 Hz for steady and modulated plasma actuation. The results can be seen in Fig. 7.8 where it is evident that there is an optimal region for plasma operating conditions that are suitable for maximum reduction in reattachment length.

It can be seen that the optimal AC frequency for the steady and modulated cases are 270 and 890 Hz respectively. The optimal range for AC frequency is also much narrower for steady actuation compared to the modulated case. At low AC frequency, the steady actuation behaves similarly to the modulated case, however beyond the optimal point of AC frequency for the steady plasma, there is a decrease in reduction

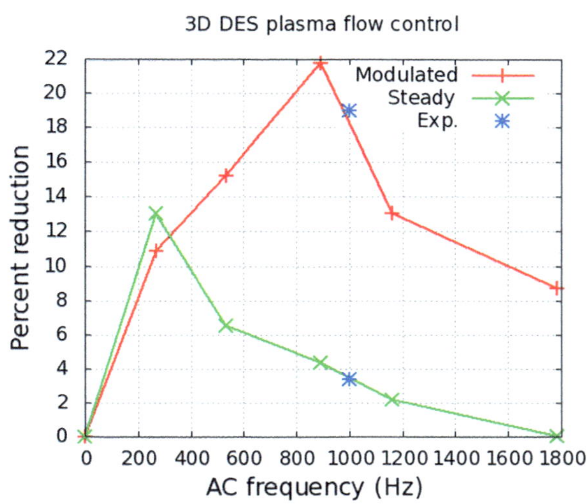

Fig. 7.8 Percent reduction in reattachment length with varying AC frequency (3D cases)

of reattachment length. For the modulated case, the optimal AC frequency is three times higher than the optimal value for the steady case, and the reduction in reattachment length is approximately twice as large (22%) compared to the maximum reduction with the steady case (13%).

The effect of plasma on the velocity field can be better investigated by plotting the changes in velocity between the baseline and actuated cases for a number of AC frequencies, and this can be seen in Fig. 7.9. Here, the increase in velocity in the free shear layer due to plasma actuation can be seen, along with reduction in velocity in the primary circulation zone. This is most evident in the extreme case of steady actuation at 1785 Hz AC frequency. However, such modification of the velocity flow field does not translate in a similar reduction of reattachment length as shown in Fig. 7.9.

The Reynolds stresses for the baseline and actuated cases for the 890 Hz AC frequency can be seen in Fig. 7.10. For the steady state plasma, an overall reduction of $u'u'$, $u'v'$ stresses can be seen, and also an increase for the $v'v'$ and $w'w'$ stresses. The modulated plasma has greater overall reduction of $u'u'$ and $u'v'$, and greater overall increase of the $v'v'$ and $w'w'$ stresses than steady plasma.

As with Fig. 7.9, the changes in Reynolds stresses due to plasma actuation can provide a clearer insight into the effects of plasma, and this is shown in Fig. 7.11. For the normal stresses $u'u'$, there is a reduction downstream of the free shear layer, and an increase in the circulation region. The case with the most reduction in reattachment length (modulated at 890 Hz) indicates that an increase in the $u'u'$ Reynolds stresses in the circulation zone is more important than a reduction downstream from the free shear layer such as in the steady case at 1785 Hz where there is no change in reattachment length. Similarly for the $u'v'$ stresses, the important aspect is their reduction in the circulation region for effective control. For the $v'v'$ and $w'w'$ stresses, again it is observed that for the maximum reduction in reattachment length there has to be a stress increase within the circulation zone.

The three dimensional turbulent structures can be seen through visualization of the Q-criteria for the baseline, steady and modulated cases in Fig. 7.12. For the steady state plasma, the small scale structures within the circulation zone are reduced, and

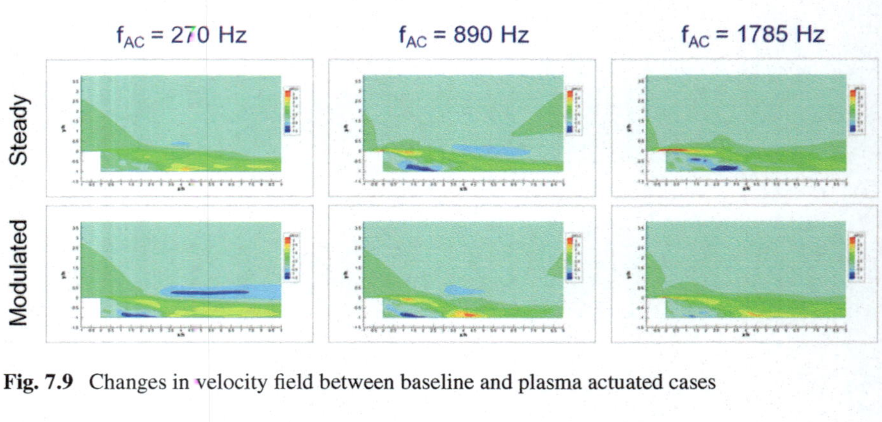

Fig. 7.9 Changes in velocity field between baseline and plasma actuated cases

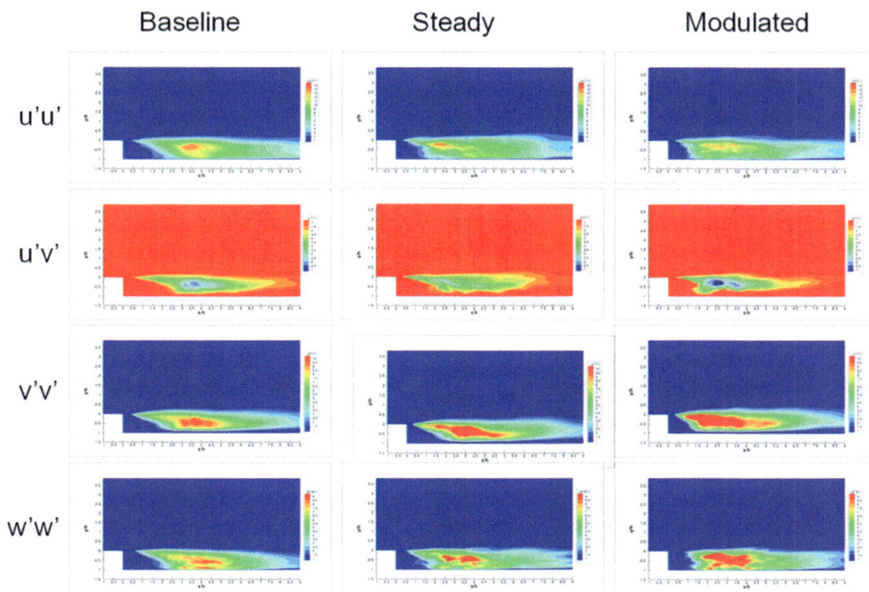

Fig. 7.10 Reynolds stresses for baseline, steady and modulated cases for 890 Hz AC frequency

larger structures downstream are more evident compared to the baseline case. For modulated plasma, the small scale structures are destroyed even more, and large scale structures are present at the shear layer due to the plasma pulsation, but these are destroyed quickly further downstream.

Finally, the modulation frequency was also investigated for values of 75, 125 and 175 Hz, with a 50% duty cycle for all cases. The results for the reattachment length can be seen in Fig. 7.13, where the numerical and experimental results suggest a narrow band for optimal operation of the modulated plasma. In addition to this, Fig. 7.14 also presents the velocity contours, changes in velocity field, changes in Reynolds stresses, and vorticity and Q-criteria for visualization of the turbulent structures.

7.4 Conclusions

The work in this study dealt with 2D and 3D numerical simulations of DBD plasma flow control in a BFS geometry. For the 2D cases, it was found that 2D plasma actuation can be reliably match experimental data through calibration of plasma force. However, 2D plasma optimisation cannot be carried out due to lack of resolving 3D turbulent structures.

3D baseline simulations did provide better agreement between experimental data, but proved to be too computationally expensive to optimise operating parameters

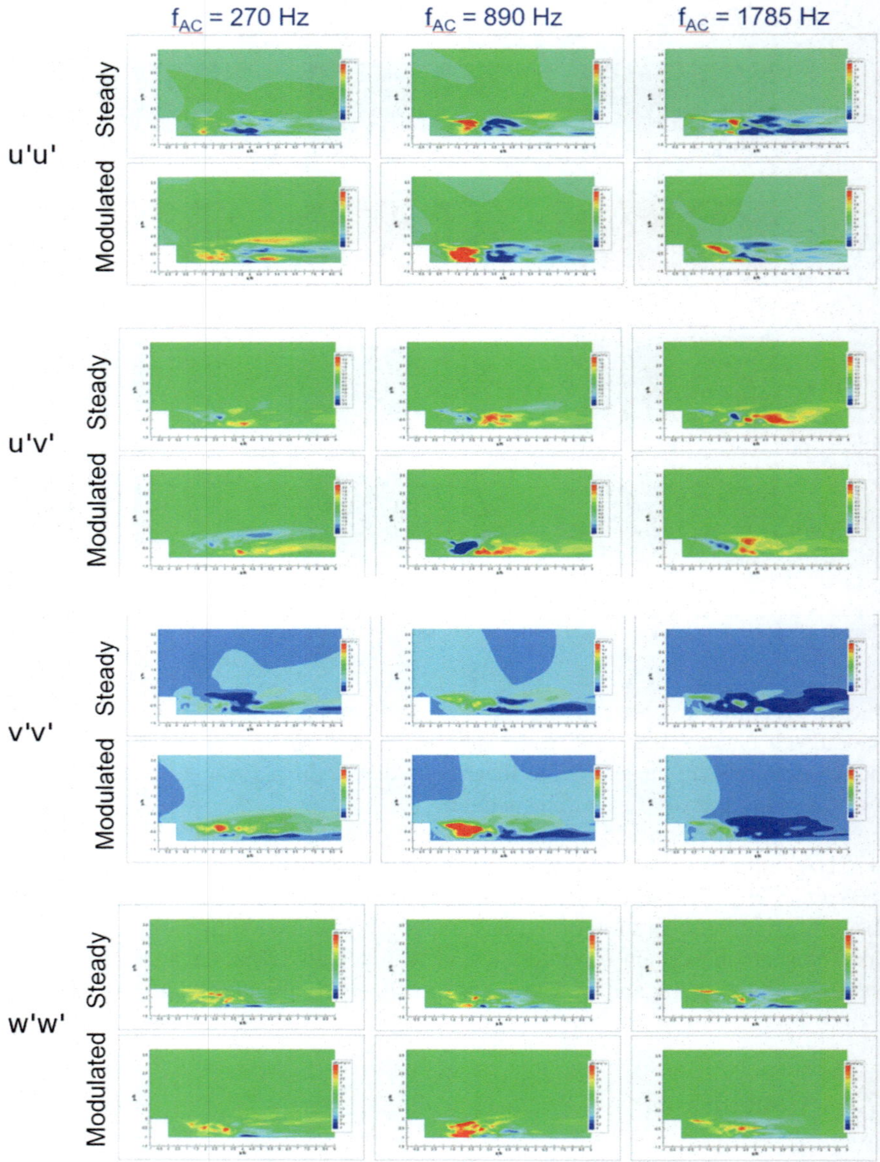

Fig. 7.11 Changes in Reynolds stresses between baseline and actuated cases for 270, 890, and 1785 Hz

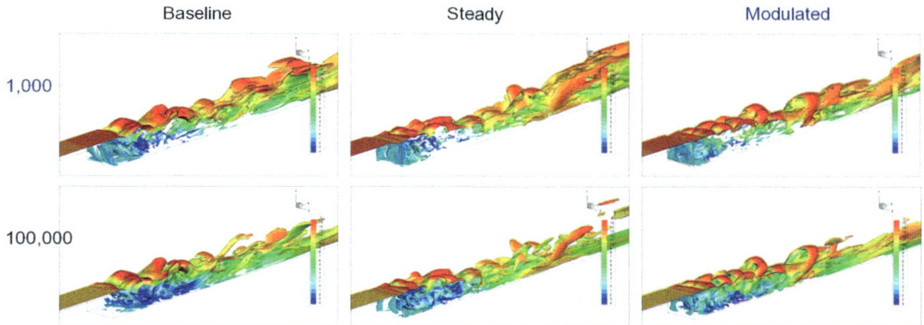

Fig. 7.12 Iso surfaces of Q-criteria at 1000 and 100,000 for the baseline, steady and modulated cases at 890 Hz, coloured by streamwise velocity

Fig. 7.13 Reattachment length versus modulation frequency

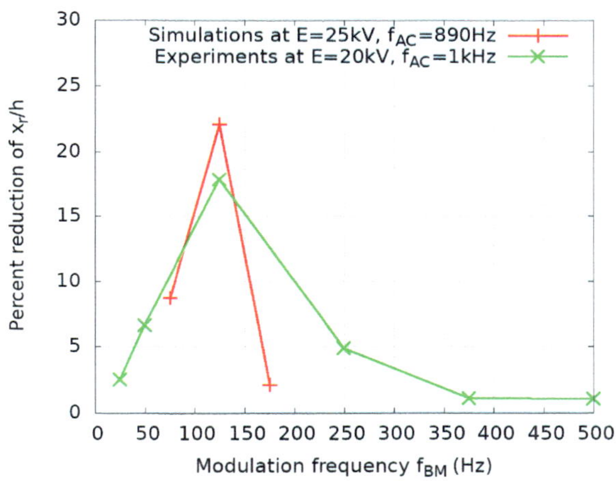

due to requirement of large mesh (cell count 8M). For this reason, 3D optimisation was carried out on smaller grid (4M cells) and the effect of plasma was studied and matched with experimental data. The results of this showed that steady plasma at low AC frequencies has similar effect as modulated plasma at the same AC frequencies. However, steady plasma has a much narrower band of effective operation in terms of f_{AC} compared to modulated. Modulated plasma can be as twice as effective as steady plasma, and by using half the energy input (at 50% duty cycle).

Reduction in reattachment length due to plasma comes from the destruction of small scale eddies within the primary circulation zone, increase in eddy size at free shear layer, and destruction of large scales downstream from the reattachment. For effective plasma actuation, the $u'u'$, $v'v'$, $w'w'$ Reynolds stresses are shifted closer to the step, whereas the $u'v'$ need to be shifted away from the step. Preliminary results for the effects of modulation frequency show that as f_{AC} is reduced, there is a shorter band where f_{BM} can result to large reduction in reattachment length.

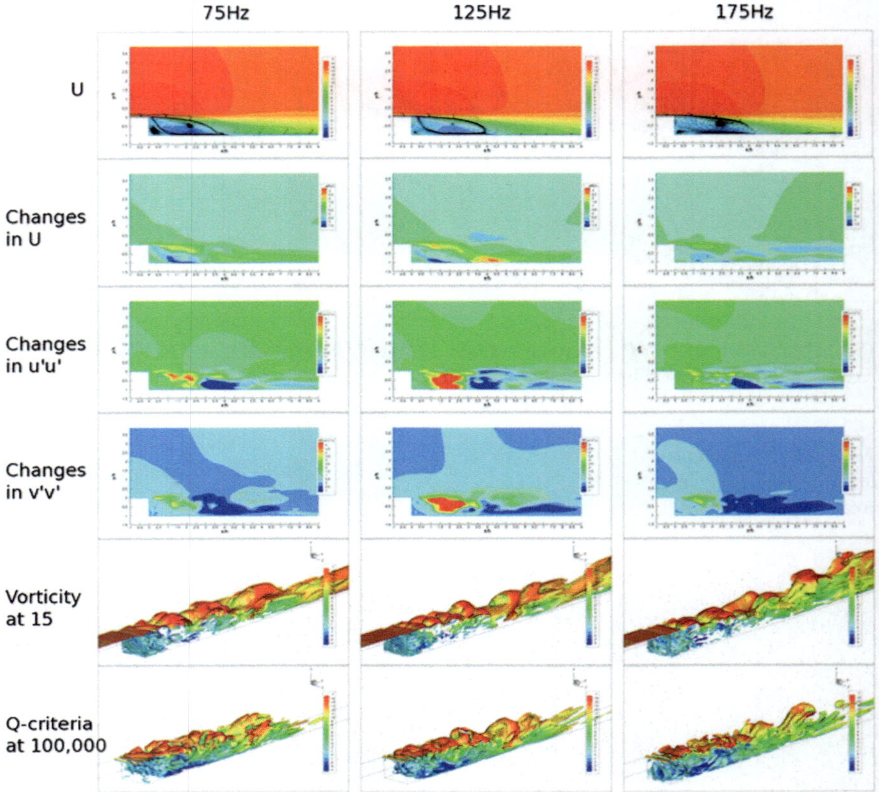

Fig. 7.14 Effects of change in modulation frequency

Acknowledgements This work was supported by European Union FP7/2010–2013, MARS (grant agreement n°266326).

References

1. Cattafesta LN, Sheplak M (2011) Actuators for active flow control. Annu Rev Fluid Mech 43:247–272
2. El-Khabiry S, Colver GM (1997) Drag reduction by dc corona discharge along an electrically conductive flat plate for small Reynolds number flow. Phys Fluids 9(3):587–599
3. Shyy W, Jayaraman B, Andersson A (2002) Modeling of glow discharge-induced fluid dynamics. J Appl Phys 92(11):6434–6443
4. Benard N, Moreau E (2010) Capabilities of the dielectric barrier discharge plasma actuator for multi-frequency excitations. J Phys D Appl Phys 43:145201–145215
5. Qin N, Xia H (2008) Detached eddy simulation of a synthetic jet for flow control. Proc Inst Mech Eng Part I J Syst Control Eng 222(5):373–380

6. Wang W, Siouris S, Qin N (2014) Hybrid RANS/LES for active flow control. Aircr Eng Aerosp Technol 86(3):179–187
7. Weis JM, Smith MA (1995) Preconditioning applied to variable and constant density flows. AIAA J 33(11):2050–2057
8. Roe PL (1981) Approximate Riemann solvers, parameters vectors and difference schemes. J Comput Phys 43:357–372
9. Spalart PR, Jou W-H, Strelets MK, Allmaras SR (1997) Comments on the feasibility of LES for wings and on a hybrid RANS/LES approach. Presented in Advances in DNS/LES, 1st AFOSR international conference on DNS/LES, Greyden Press, Columbus Ohio, 4–8 Aug 1997
10. Spalart PR, Deck S, Shur ML, Squires KD, Strelets MK, Travin A (2006) A new version of detached-eddy simulation, resistant to ambiguous grid densities. Theor Comput Fluid Dyn 20:181–195
11. Shur M, Spalart P, Strelets M, Travin A (2008) A hybrid RANS-LES approach with delayed-DES and wall-modelled LES capabilities. Int J Heat Fluid Flow 29:1638–1649
12. Sujar-Garrido P, Benard N, Moreau E, Bonnet JP (2012) Modifications of the shear layer downstream a backward facing step by dielectric barrier discharge plasma actuator. In: International symposium on turbulence and shear flow phenomena (TSFP-8), 28–30 Aug 2012

Chapter 8
The Optimization of Reynolds Stress Flow Control for the Backward-Facing Step with Synthetic Jet Based on Numerical Simulation and Test

Zhenghong Gao, Dong Li, Fang Xu, Xiao Ming, Lu Xia, Yuhang Wang and Rui Ma

Abstract Active flow control, which has great application prospects in aerodynamic design, can restrain flow separation and reduce drag. This paper provides a review on our works relating to the China–EU MARS research project. Backward-facing step (BFS) flow, with sudden expansion of cross-section and complex vortices between the free stream and the main vortex, is a typical kind of separated flow. Synthetic jet, which transports momentum with zero net mass flux and affects structure of the flow field with length scales higher than its characteristic length scales, is applied to control BFS flow field. The optimization processes for the backward-facing step flow control with synthetic jet based on both numerical simulation and experimental data are introduced. In the numerical simulation aspect, experimental and computational data are synthesized and a computational model is proposed based on physical characteristics of wind tunnel. An adaptive double stage surrogate model based optimization framework is constructed and applied to optimize parameters of the synthetic jet. Two optimal schemes with similar control effects are obtained through design optimization based on computational data, revealing two inherently different flow control mechanisms. Pressure coefficient, skin friction and Reynolds stress of the two schemes are compared and analyzed. The flow control scheme for one problem based on Reynolds stress may not be definite. In the experimental aspect, the synthetic jet device has non-linear oscillation of reciprocating piston actuator into a pipe and can eject/suck mass flux through a jet slot. An in-looped design optimization system based on experimental data adopting hybrid searching algorithm is constructed and applied to optimize parameters of the synthetic jet. Then, power consumed in

Z. Gao (✉) · D. Li · F. Xu · L. Xia · Y. Wang · R. Ma
National Key Laboratory of Aerodynamic Design and Research, Northwestern Polytechnical University, Xi'an 710072, People's Republic of China
e-mail: zgao@nwpu.edu.cn

R. Ma
e-mail: marui9125@163.com

X. Ming
College of Aeronautics and Astronautics, Nanjing University of Aeronautics and Astronautics, Nanjing 210016, People's Republic of China

N. Qin et al. (eds.), *Advances in Effective Flow Separation Control for Aircraft Drag Reduction*, Computational Methods in Applied Sciences 52,
https://doi.org/10.1007/978-3-030-29688-9_8

driving reciprocator is considered to derive a multi-objective optimum scheme. Flow control mechanism of synthetic jet device is preliminarily revealed, with theoretical and experimental data analysis of velocity profiles and Reynolds stress distributions. The optimization process and the analysis of optimum state provide guidance to the design of active flow control devices.

Keywords Flow control · Synthetic jet · Reynolds stress · Design optimization · Backward-facing step · Wind tunnel · Numerical simulation

8.1 Introduction

Though traditional aerodynamic design tends to avoid separation on aircraft surface, separated flows still commonly occur in engineering practices. Separation leads to dramatic increase in drag and changes in moment. The complex flow structure of separation induces non-linear variation of aerodynamic forces, which has negative effects on maneuverability and stability of the aircraft. Thus, it is of engineering significance to take measures to restrain flow separation. In the aspect of modern aircraft design, active flow control technique introduces inputs which corresponding to instability of separated flow and obtains distinct effects with less energy consumption, thus restrains flow separation in broader flight envelope with acceptable of energy input [1].

Backward-facing step flow is a typical kind of separated flow, representing the flow around a blunt body with sudden expansion of cross-section. Shear layer structure is observed between the free stream and the main vortex, with complex vortices and pressure loss. The length of recirculation zone and the wall shear stress are important parameters for analysis of the flow field. Extensive experimental efforts on flow field structure have been conducted by Troutt et al. [2], Driver et al. [3], Heenan and Morrison [4], among others. Armaly et al. [5] reported the influence of Reynolds number on separation length and structure of flow regimes and revealed inability to predict the flow field over relatively high Reynolds numbers.

Synthetic jet (zero mass jet) works similarly to pulse jet, but its working medium is the fluid of external flow field and its momentum transportation is conducted with zero net mass flux, as reported by Smith and Glezer [6]. Interaction with the external flow field by synthetic jet can distort the local streamlines, resulting in changes of the wall boundary shape, and affecting structure of the flow field with length scales one to two orders of magnitude higher than the characteristic length scale of synthetic jet, according to investigation by Glezer and Amitay [7]. It should be highlighted that Ming [8] independently discovered the zero mass jet and clarified the formation mechanism as rectifying effect in later researches [9]. Currently, flow control mechanism of synthetic jet driven by piston vibration remains unsolved and needs further investigation, as compared to mechanism of spanwise instability and interaction of successive vortex pairs of other synthetic jet devices [6].

Separation control cases of backward-facing step using synthetic jet with various parameters are extensively reported. Researches on separated flow control of backward-facing step mainly focus on experimental works or sophisticated numerical simulations separately. However, since angle, velocity and frequency are coupled due to inherent non-linear characteristics of synthetic jet, selection of the optimum control parameters is often tough and time-consuming. To date, literature considering synthetic jet flow control of backward-facing step in the experimental aspect utilizing optimization method is not yet available, and research on Reynolds stress flow control in the computational aspect utilizing optimization method is limited.

The tasks of MARS project are firstly to understand the mechanisms of separated flow with Reynolds stress from synthesis of experimental and computational data for backward-facing step (BFS); secondly to develop optimization methods based on the wind tunnel test and CFD simulation respectively to find more suitable parameters for the synthetic jet flow control devices.

In this paper, an adaptive double stage surrogate model based optimization framework is constructed and applied to optimize parameters of the synthetic jet. Two optimal schemes with similar control effects are obtained through design optimization based on computational data, revealing two inherently different flow control mechanisms. Hybrid particle swarm pattern search algorithm based on in-looped update surrogate model is developed. Due to the fast convergence and accurate prediction characteristics of this algorithm, design optimization and experimental work can be effectively combined. This paper introduces the newly developed synthetic jet apparatus and measuring techniques, and describes design optimization system based on in-looped update surrogate model. A multi-objective optimum scheme is derived considering power consumption in driving reciprocator.

8.2 Design Optimization System

8.2.1 Summarization of Optimization System

The optimization method is composed of three parts: construction of optimization model, analyzing method and searching algorithm. Hierarchy diagram of design optimization system is shown in Fig. 8.1. Numerical simulation and wind tunnel test are the two methods adopted in analyzing function value of optimization objective. Due to characteristics of low cost, high efficiency and low equipment requirement (only digital computers), numerical simulation is commonly chosen as the analyzing method. But for design problems without in-depth understanding of the working mechanism, like newly developed synthetic jet device, numerical simulation is not convincing and cannot be adapted in the design optimization process. Wind tunnel test, in contrast, provides fast and reliable features in evaluating objective function and does not require theoretical understanding of mechanism in advance.

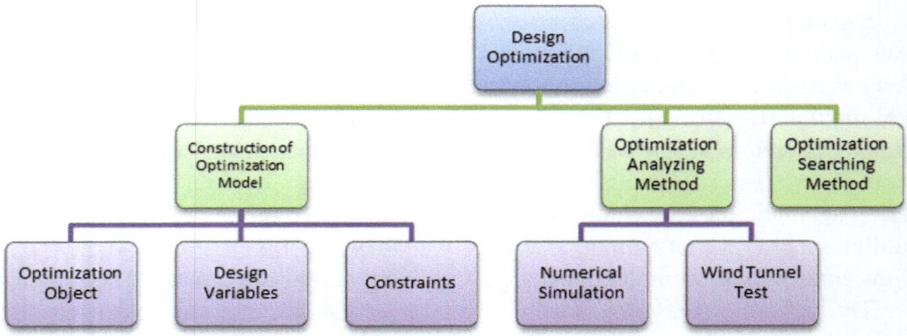

Fig. 8.1 Structure of optimization system

8.2.2 Hybrid Searching Algorithm

With one or more algorithms properly embedded in a major optimization algorithm, hybrid optimization algorithm can effectively improve the robustness and convergence rate.

Particle swarm pattern search algorithm is developed by Vaz and Vicente [10], which combines pattern search algorithm and PSO algorithm. With the introduction of intelligent searching method, the ultimate hybrid optimization algorithm is equipped with global searching ability. Therefore, hybrid particle swarm pattern search algorithm is taken as the optimization searching method. Flow chart of the algorithm is shown in Fig. 8.2.

8.2.3 Surrogate Model

For complex systems, optimization methods require time-consuming calculations of the objective function and constraints. In order to improve the predictive accuracy, double-stage surrogate model (DSM) based on the principles and characteristics of the interpolation-based and regression-based surrogate models are introduced. It takes advantage of the numerical noise filtering capability of the regression-based surrogate model and the points fitting ability of interpolation-based surrogate model to improve overall performance. In the optimization iteration process, validation of surrogate model is needed to ensure the accuracy of the model for every optimization iteration step. If the accuracy of surrogate model does not meet the requirements, then new samples are added and surrogate model is updated.

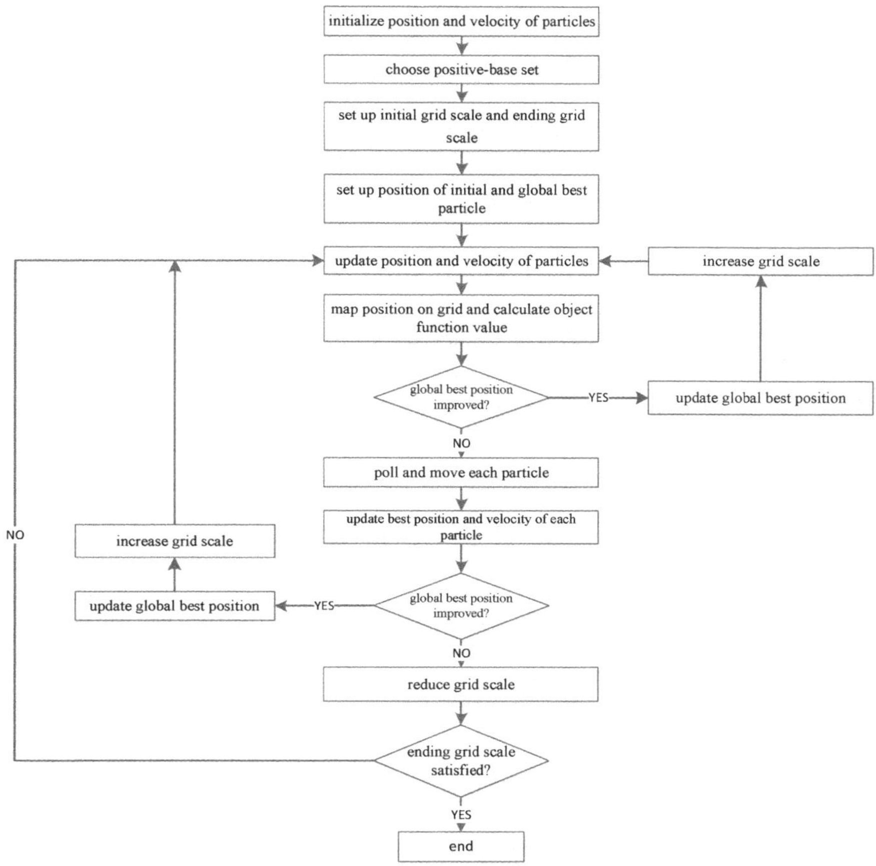

Fig. 8.2 Flow chart of hybrid particle swarm pattern search algorithm

8.3 Design Optimization Based on Numerical Simulation

8.3.1 BFS Experimental and Computational Data Synthesis

The wind tunnel experiment is conducted by National Key Laboratory of Aerodynamic Design and Research at Northwestern Polytechnical University. Parameters of the wind tunnel are: 2000 mm from entrance to test section, 2500 mm from test section to the exit. Cross section of test section is 240 mm $(H) \times 720$ mm (W). The height of backward-facing step is 40 mm with expansion ratio of 1.2. Aluminum honeycombs at wind tunnel entrance reduce the intensity of free-stream turbulence, and nominal turbulence intensity of test section is approximately 0.5%. Physical model and computational simplification processes are shown in Fig. 8.3.

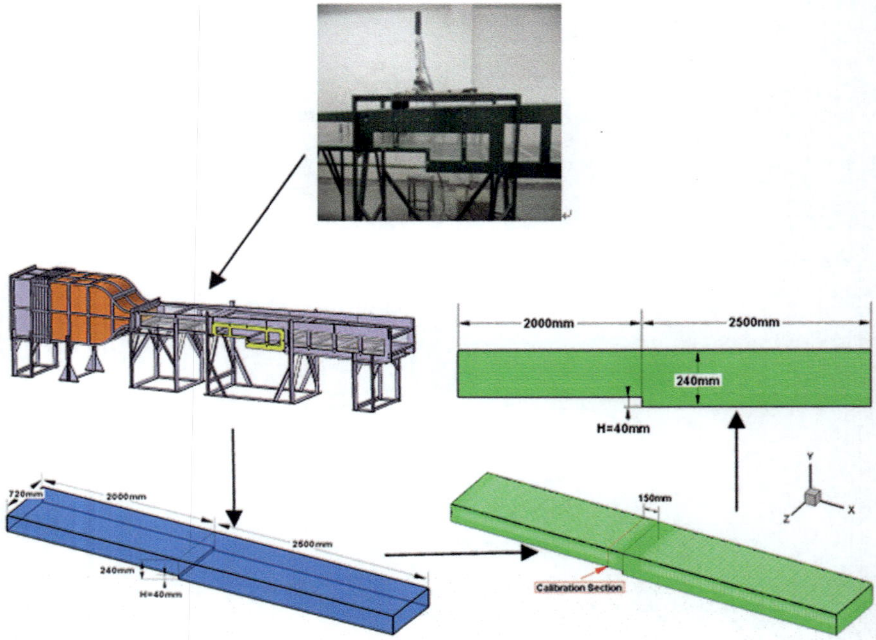

Fig. 8.3 Physical and computational model

In MARS project, our group is responsible for design optimization. Taking both computational cost and numerical precision into consideration, two-dimensional (2D) grid system is adopted. The 2D grid system is composed of 15 blocks with 18,221 grid points, as is shown in Fig. 8.4.

Temperature of the wind tunnel test section was T = 20 °C/68 °F (atmospheric temperature) and free stream velocity was 15.5 m/s. S-A turbulence model, third-order MUSCL reconstruction scheme, Riemann solver of the upwind ROE scheme, second order central difference scheme of viscous term and approximate factorization time marching are adopted in numerical simulation solver. Both numerical

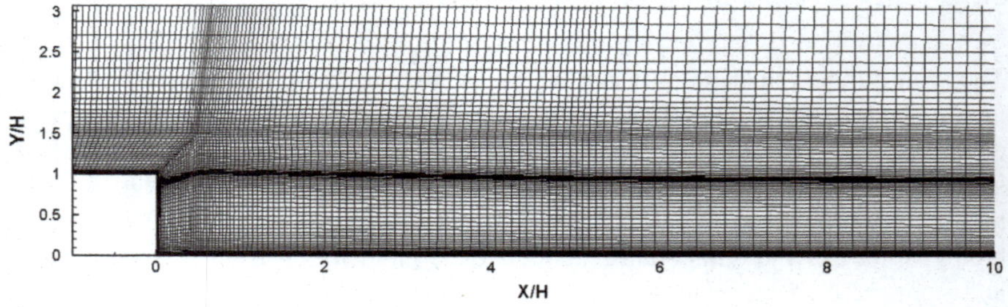

Fig. 8.4 Overview of 2D grid system

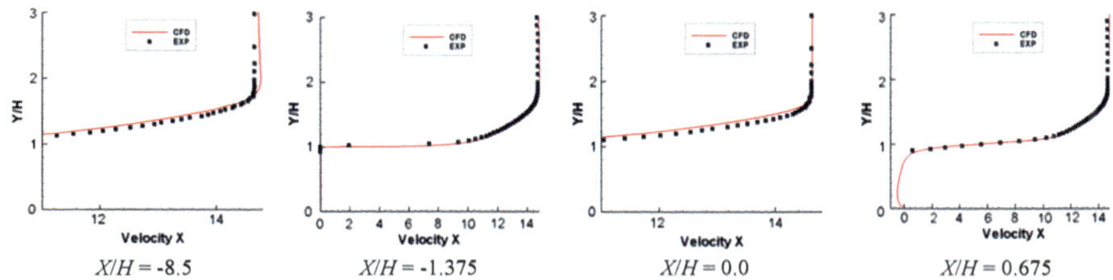

Fig. 8.5 Comparison of velocity profile

and experimental velocity profiles of four stations are shown in Fig. 8.5, and the distributions are in good agreement.

To analyze the effect of spanwise flow on flow field, we develop a three-dimensional (3D) grid system. The 3D grid system (18*H* spanwise) is composed of 15 blocks with 382,641 grid points and has the same cross-sectional mesh as 2D. Numerical simulations indicate that the velocity profiles are basically the same.

8.3.2 Design Optimization Model

Synthetic jet, which transports momentum with zero net mass flux and affects structure of the flow field with length scales higher than its characteristic length scales, is applied to control flow separation of backward-facing step. Computational model of synthetic jet is:

$$
\begin{aligned}
U &= U_{jet}\, f(\zeta)\sin(2\pi f t)\cos(\theta) \\
V &= U_{jet}\, f(\zeta)\sin(2\pi f t)\sin(\theta)
\end{aligned}
$$

where U is streamwise velocity and V is spanwise velocity, U_{jet} is the velocity magnitude of the jet slot, f is the vibration frequency of synthetic jet actuator, θ is the angle of the slot. Function $f(\zeta)$ represents the velocity distribution of the synthetic jet across the slot/orifice, we apply typical 'square wave' distribution $f(\zeta) = 1$. The optimization model is built:

$$
\begin{aligned}
obj:&\quad \min L(f, \theta, U_{jet}) \\
s.t.&\quad \begin{cases} 10\ \text{Hz} \le f \le 300\ \text{Hz} \\ 30^\circ \le \theta \le 150^\circ \\ 0 \le U_{jet} \le 10\ \text{m/s} \\ h = 5\ \text{mm} \end{cases}
\end{aligned}
$$

wherein, L represents time-averaged reattachment length, synthetic jet frequency f, velocity U_{jet} and angle θ are constrained.

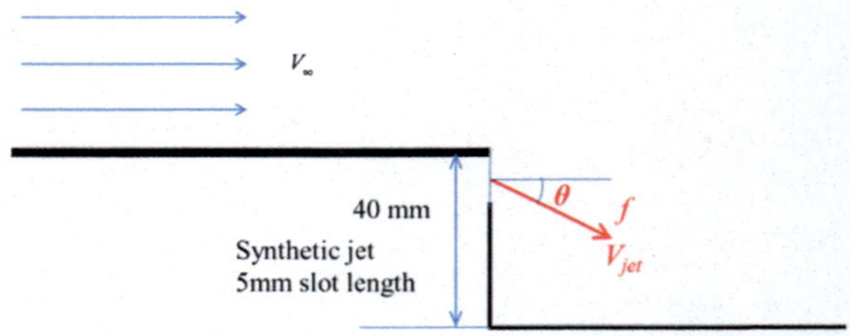

Fig. 8.6 Case A (synthetic jet after BFS)

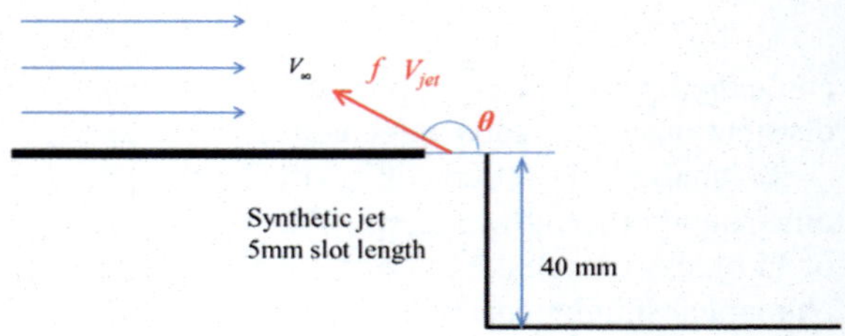

Fig. 8.7 Case B (synthetic jet on BFS)

8.3.3 Design Optimization of Synthetic Jet Installed After BFS

Empirically, flow control apparatus near the edge of the backward-facing step may have the greatest impact on the flow field. In order to explore the effects of various installation positions, we set up two schemes with different synthetic jet positions. Case A: synthetic jet after the backward-facing step, as is shown in Fig. 8.6. Case B: synthetic jet on the backward-facing step, as is shown in Fig. 8.7 DSM is used in the design optimization process. Design optimization based on surrogate model and hybrid particle swarm pattern search algorithm is adopted to search optimum parameters of synthetic jet actuator.

8.3.4 Reynolds Stress and Control Mechanism Analysis of Two Optimal Schemes

To reveal the flow control mechanism of synthetic jet device in different positions, flow field characteristics of the baseline and two optimum states are compared. The optimal synthetic jet parameters and controlling effects (time-averaged reattachment length) are shown in Table 8.1.

Figures 8.8 and 8.9 reveal the variation of reattachment position of case A and case B respectively. As can be seen in Fig. 8.8, the reattachment point of baseline flow is stable, while case A has only one reattachment point. The frequency is 172.7 Hz, which is the same as the synthetic jet. Figure 8.9 indicates a complicated process, with breakdown and shedding of the main vortex.

Table 8.1 Parameters and control effects of different schemes

Scheme	Velocity (m/s)	Angle (°)	Frequency (Hz)	Reattachment (X/H)
Case A	10.0	−31.8	172.7	5.48
Case B	5.6	−59.5	20.0	5.50

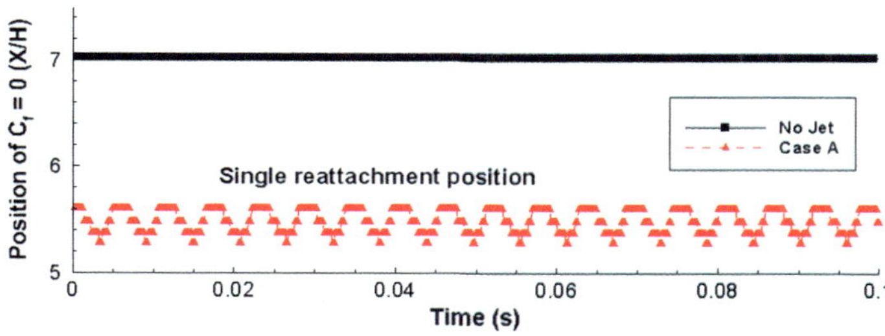

Fig. 8.8 Reattachment point variation of case A

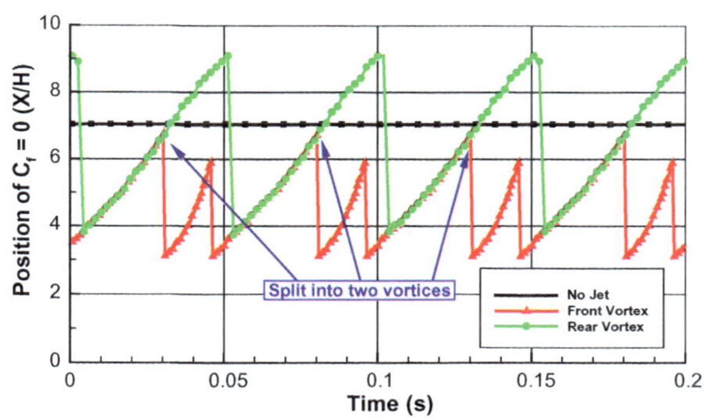

Fig. 8.9 Reattachment point variation of case B

Analysis of instantaneous flow field and streamlines of case B are shown in Fig. 8.10. The synthetic jet on the step produces a small separation vortex, and the small vortex moves downward. Then, enhanced by the small vortex, the secondary vortex breaks down and develops to main vortex.

Figure 8.11 shows the time-averaged skin friction of two control schemes, compared with baseline flow. As can be seen, though the time-averaged reattachment lengths are similar, the magnitude of case B is significantly larger than case A in the recirculation zone. Case B increases skin friction of zones both before and after the reattachment point, while case A only reduces skin friction of the recirculation zone. Since wall friction stress corresponds to the strength of swirling of vortex in the recirculation zone, synthetic of case B has greater influences on the flow field.

For streamline and pressure contour of baseline and two control schemes, in Case A the height of the recirculation zone decreases and in case B it increases. A region with strong negative pressure is formed in the vicinity downstream the backward-

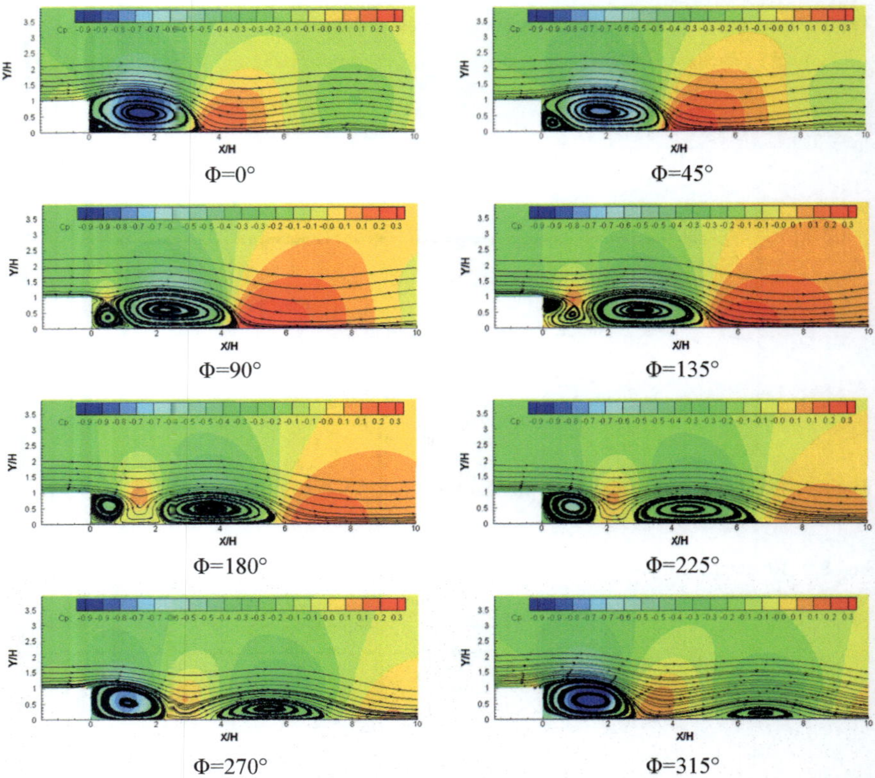

Fig. 8.10 Instantaneous flow field of case B

Fig. 8.11 Comparison of wall shear stress

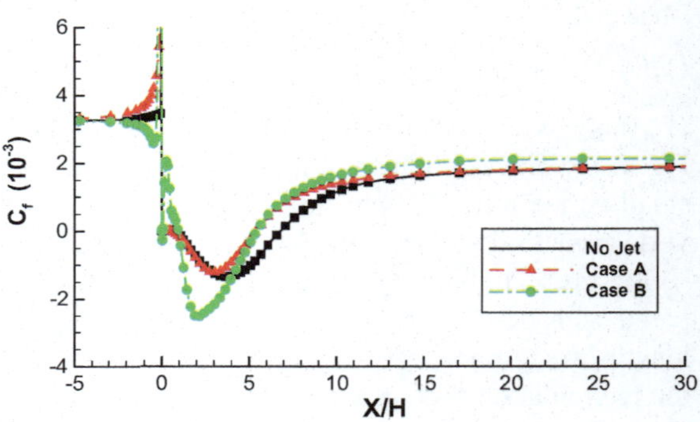

facing step of case B. For Reynolds shear stress contours of baseline and two control schemes, Case A has a minor change to the form of Reynolds stress, and shear layer structure can still be clearly observed at stations near the step. Case B dramatically changes the distribution of Reynolds stress, and shear layer structure is not obvious any more.

Though the time-averaged reattachment lengths of the time-averaged flow field of two schemes are almost the same, the distributions of Reynolds stress are dramatically different, revealing two inherently different flow control mechanisms. This indicates that the flow control scheme for one problem based on Reynolds stress may not be definite.

8.4 Design Optimization Based on Experimental Data

8.4.1 Experimental Apparatus and Baseline Flow Measurement

This experiment was conducted in the subsonic suction wind tunnel of Nanjing University of Aeronautics and Astronautics. Parameters of the wind tunnel are: 1998 mm from entrance to test section, 2035 mm from test section to the exit. Cross section of test section is 500 mm $(H) \times 300$ mm (W). The height of backward-facing step is 30 mm with expansion ratio of 1.06. Aluminum honeycombs at wind tunnel entrance reduce the intensity of free-stream turbulence, and nominal turbulence intensity of test section is approximately 0.3%.

The newly developed synthetic jet actuator was designed by Professor Ming of Nanjing University of Aeronautics and Astronautics. The reciprocator section is composed of controller, servo motor and cylinder, as shown in Fig. 8.12. Controller is responsible for controlling the servo motor via digital signal and then driving the reciprocating piston actuator in the cylinder. Piston actuator cycles up and down in

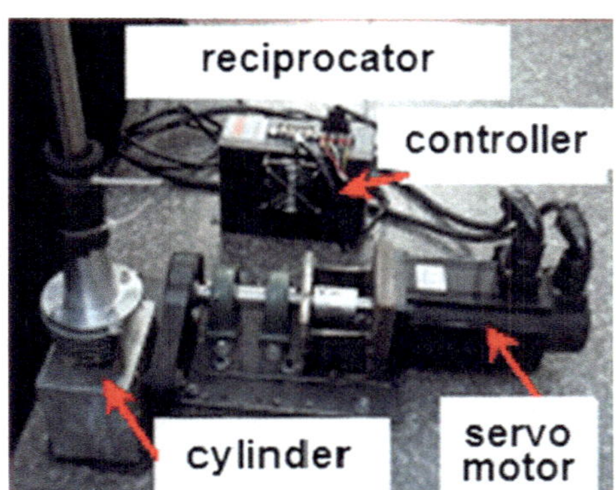

Fig. 8.12 Reciprocator section

the cylinder with the amplitude of 25 mm, causing velocity variation and pressure fluctuation at the bottom of the pipe. Configuration of the synthetic jet slot and connecting section is shown in Fig. 8.13. Jet flow is transmitted to the wind tunnel test section by a long inflected pipe connecting reciprocator and jet slot. Synthetic jet is ejected and sucked through the jet slot of 200 mm width and the operating frequency is between 0 and 35 Hz. The distance from centerline of the vertical jet slot to the step is 3.5 mm, as is indicated in Fig. 8.13. Considering the non-linear oscillation of the reciprocating piston actuator and the long pipe, expansion and convergence of pressure waves play a significant role in this new synthetic jet device.

Wall shear stress (skin friction), velocity profile and Reynolds stress were measured. Wall shear stress was collected by measuring the static pressure difference across a small fence mounted within the viscous sublayer, first introduced as classical sublayer (Stanton) fence by Konstantinov and Dragnysh [11]. Arrangements of the sublayer fences and silk threads are shown in Fig. 8.14. Silk threads, with middle section stuck on the wall, visually indicated length of recirculation zone and stability characteristic of reattachment point. Sublayer fences, as shown in Fig. 8.15, were connected to high-resolution pressure transmitter to obtain mean static pressure

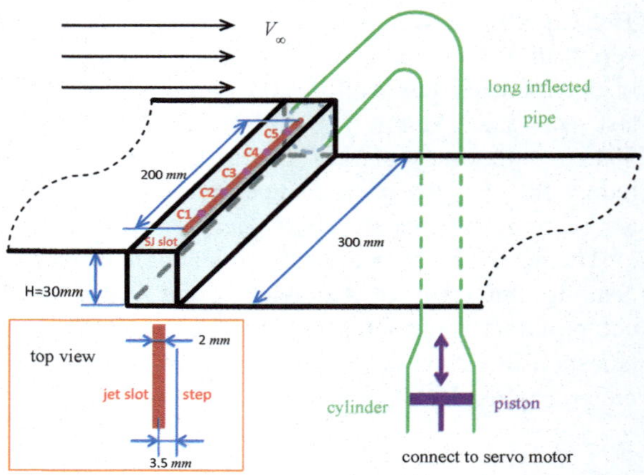

Fig. 8.13 Jet slot and connecting section

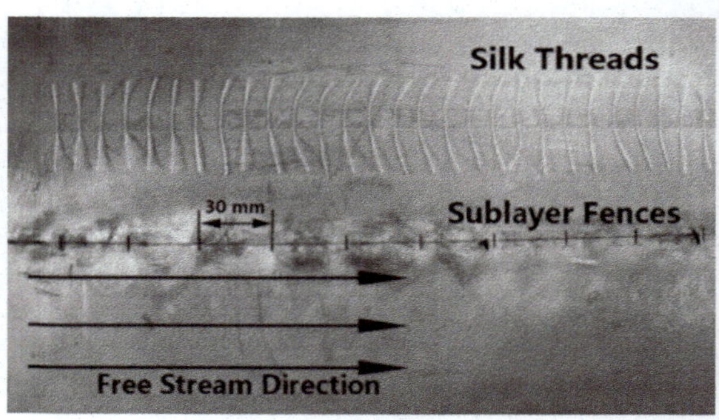

Fig. 8.14 Arrangements of sublayer fences and silk threads

Fig. 8.15 Schematic diagram of sublayer fences

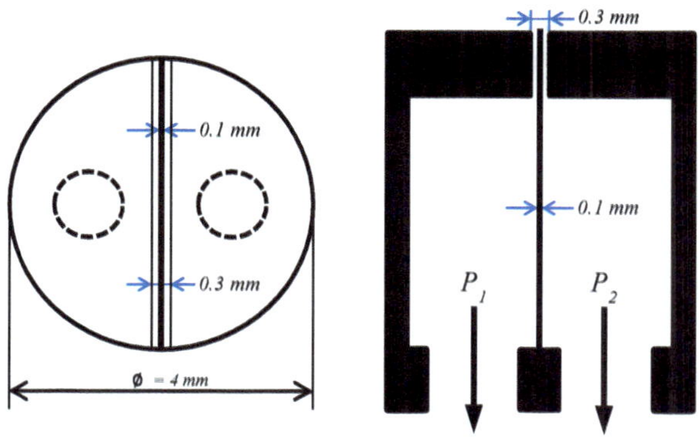

difference and corresponding wall shear stress. The accuracy of sublayer fences was first calibrated by channel flow with altering height and constant pressure drop to achieve conversion factors between pressure difference and corresponding wall shear stress. The conversion factor between wall shear stress and skin friction coefficient is 4.165×10^{-3}/pa. Time series of velocity were measured by constant temperature hot-wire anemometer *TSIIFA*-300 with sampling frequency of 10 kHz. Instantaneous velocities were then decomposed into time-averaged and fluctuating components and Reynolds stress was calculated by multiplying fluctuating ones.

Quasi linear relationship between frequency and velocity is substantially satisfied. From the mechanical perspective, higher actuating frequency leads to more intense velocity variation and pressure fluctuation of the piston actuator at the bottom of the pipe, and consequently higher velocity through the pipe and at the jet slot. Actuating frequency of 35 Hz corresponds to maximum velocity of approximately 31.5 m/s.

Since sublayer fences, silk threads and probes of hot-wire anemometer locate at different spanwise stations, two-dimensional characteristic of the flow field should be ensured. For every frequency, there is no significant difference among these monitoring points. Therefore, edge effect of synthetic jet slot is not obvious and the flow field of synthetic jet is two dimensional.

Figure 8.16 shows typical partitions of backward-facing step flow field and indicates recirculation zone, shear layer and potential flow region. Upstream perturbations are transferred to the reattachment zone through shear layer. The high-resolution pressure transmitter in Fig. 8.15 provides only time-averaged skin friction instead of instantaneous ones, since the sublayer fences cannot detect the flow intermittency with high frequency. The time-averaged skin friction makes sense and could be applied to the design optimization works. Skin friction in condition of free stream velocity of 19.8 m/s without synthetic jet control (baseline flow) is shown in Fig. 8.17. Temperature of the wind tunnel test section was $T = 21$ °C/70 °F (atmospheric temperature) and the Reynolds number based on free stream velocity and step height (*H*

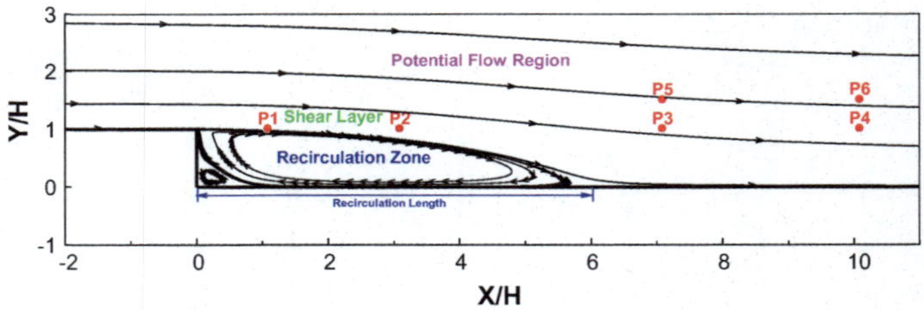

Fig. 8.16 Schematic diagram of typical partitions

Fig. 8.17 Time-averaged surface skin friction

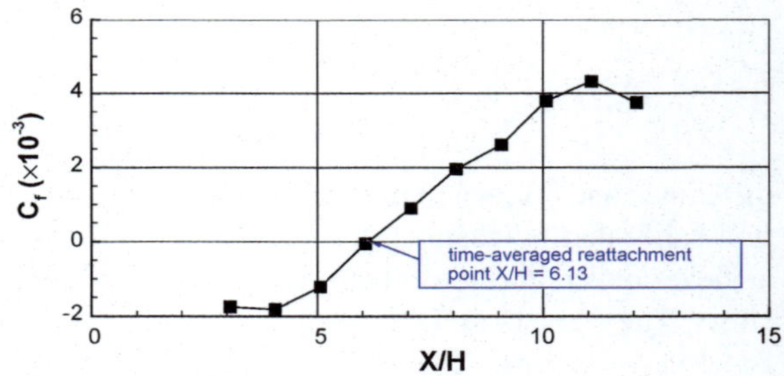

= 30 mm) was $Re_H = 40{,}665$. The time-averaged reattachment point of the primary vortex of baseline flow locates at $6.13H$.

Velocity profiles of baseline flow at ten stations are shown in Fig. 8.18. Shear layer structure, characterized by large vertical velocity gradient, is obvious near the

Fig. 8.18 Velocity profiles of baseline flow

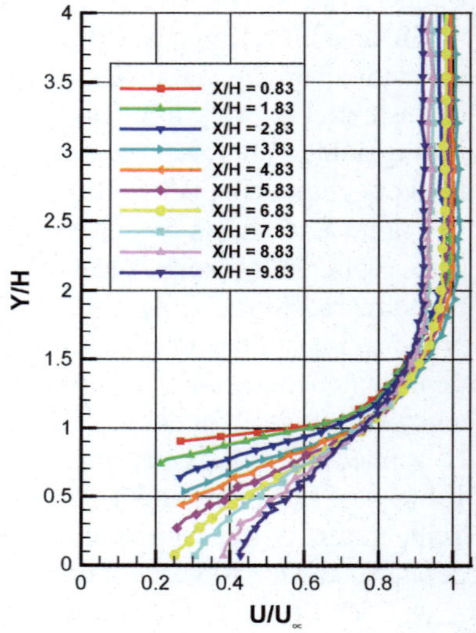

step ($X/H < 4$). Structure of shear layer is no longer apparent for downstream stations ($X/H > 4$) due to viscous dissipation and momentum transfer which mitigate vertical velocity gradient. For downstream stations, velocity deficit is clear in the potential flow region ($Y/H > 1.2$). Profiles of turbulence intensity of the baseline flow at these stations are shown in Fig. 8.19.

Profiles of Reynolds normal and shear stresses of the baseline flow at these stations are shown in Fig. 8.20. Both normal and shear stress share the same distribution pattern, with large magnitude for shear layer and virtually zero in potential flow region. Comparison between profiles reveals that when a station moves downstream, the peak value of Reynolds stress first increases and then declines, reaching maximum

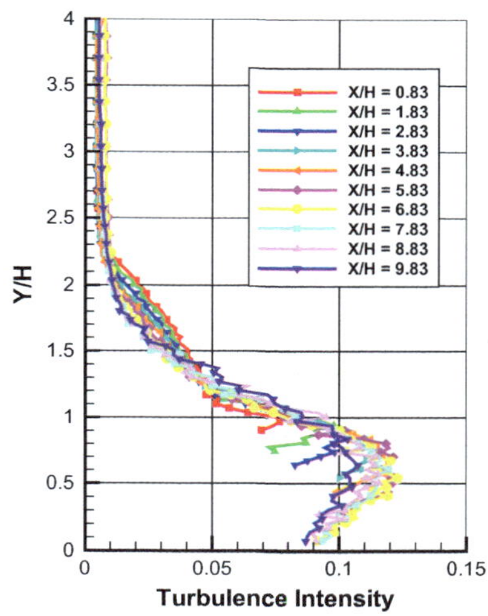

Fig. 8.19 Turbulence intensity of baseline flow

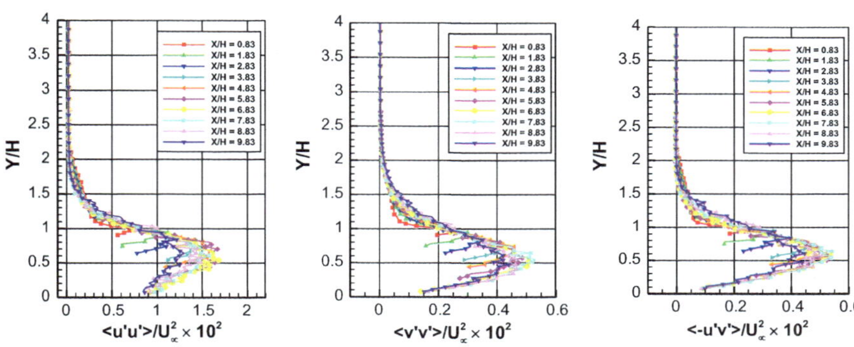

(*a*) Reynolds normal stressτ_{xx} (*b*) Reynolds normal stressτ_{yy} (*c*) Reynolds shear stress τ_{xy}

Fig. 8.20 Reynolds stress profiles of baseline flow

value at station of about 6H to 8H. Particular attention should be paid to shear stress τ_{xy} since it indicates the momentum transfer between potential flow region and recirculation zone through shear layer.

8.4.2 Optimization Model and In-Looped Update Surrogate Strategy

Optimization model contains design objective, design variables and constraints, as shown in Fig. 8.1. Mathematical description of a typical optimization model is:

$$obj: \quad \min f(x_1, x_2, x_3, \ldots)$$
$$s.t. \quad \begin{cases} a_1 \le g_1(x_1, x_2, x_3, \ldots) \le b_1 \\ a_2 \le g_2(x_1, x_2, x_3, \ldots) \le b_2 \\ h_1(x_1, x_2, x_3, \ldots) = c_1 \\ h_2(x_1, x_2, x_3, \ldots) = c_2 \end{cases}$$

wherein, f stands for design objective; x_1, x_2, x_3 are design variables; s.t. indicates constraints, including inequality and equality constraints.

Figure 8.21 shows skin friction of five frequencies (0 Hz—baseline/12.5 Hz/20 Hz/27.5 Hz/35 Hz) for slot angle of 90°. Frequency of 35 Hz corresponds to Strouhal number St = 0.053 based on free stream velocity and step height. Due to the structure of wind tunnel equipment, sublayer fences were not installed on stations (X/H < 3) that would reflect reattachment point for some experimental states. Considering unexpected excellent performance of this newly developed synthetic jet device, recirculation lengths of some high frequencies (27.5 Hz/35 Hz) were not always available. But as can be seen in Fig. 8.21, for one specific station downward of the reattachment point of baseline flow (X/H > 6.13),

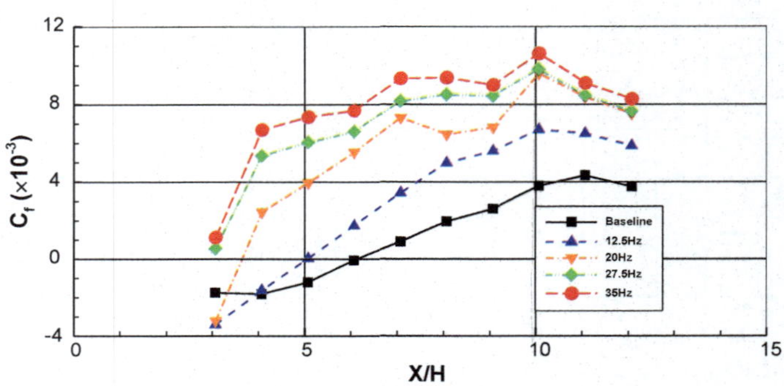

Fig. 8.21 Time-averaged surface skin friction at 90°

increase in skin friction is accompanied with reduction in length of recirculation zone. This discipline is also satisfied for angles of 120° and 150°. Therefore, for the certain circumstances of angular range between 90° and 150°, skin friction could be taken as the parameter reflecting flow control effect, substituting length of recirculation zone.

Maximizing surface skin friction at a station after the baseline reattachment point is selected as the objective of design optimization. Due to the deterministic mechanical structure of the reciprocator, frequency and velocity of the actuator cannot be decoupled and satisfy an equality constraint. Frequency and angle of the synthetic jet are two independent design variables. Thus, the mathematical expression of the optimization model is:

$$obj: \quad \max c_f(f, \theta, U_{jet}).$$
$$s.t. \quad \begin{cases} 0\text{ Hz} \le f \le 35\text{ Hz} \\ 90^\circ \le \theta \le 150^\circ \\ U_{jet} = U_{jet}(f) \end{cases}$$

wherein, f is the frequency of the synthetic jet; θ is the angle of the synthetic jet slot and U_{jet} is the maximum velocity of jet slot. The third equality constraint indicates the coupled relationship between frequency and velocity.

To construct the optimization strategy for parameter selection of the synthetic jet, construction of surrogate model using wind tunnel test data is needed. A new optimization strategy based on the in-looped update surrogate model with wind tunnel test data has been developed. Figure 8.22 shows the flow chart of this new iterative design optimization strategy. Firstly, wind tunnel experiment is designed to acquire data of the original sampling points; surrogate model is constructed with the original sampling points. Secondly, optimization search is performed with the surrogate model and the result is approved by wind tunnel experiment. Then, validated wind tunnel data are used to update the surrogate model within the iterative loop and the optimization search proceeds until optimum result is derived.

8.4.3 *Single-Objective Design Optimization*

Considering experimental machining precision and measurement error, the minimum variation of angle was set to 7.5° and optimum point was approximated to the same accuracy. To ensure identical mass flux for different angles, all synthetic jet slots were 2 mm in normal width. Definition of slot angle is shown in Fig. 8.23, with upstream pointing jet slot corresponds to $\theta > 90°$. To assure jet velocity directionality through the slot, a certain thickness of the synthetic jet slot should be guaranteed, leading to slight differences of the distance between slot and step for different angles. In this paper, the effect of minor differences of distance on the control effect is ignored.

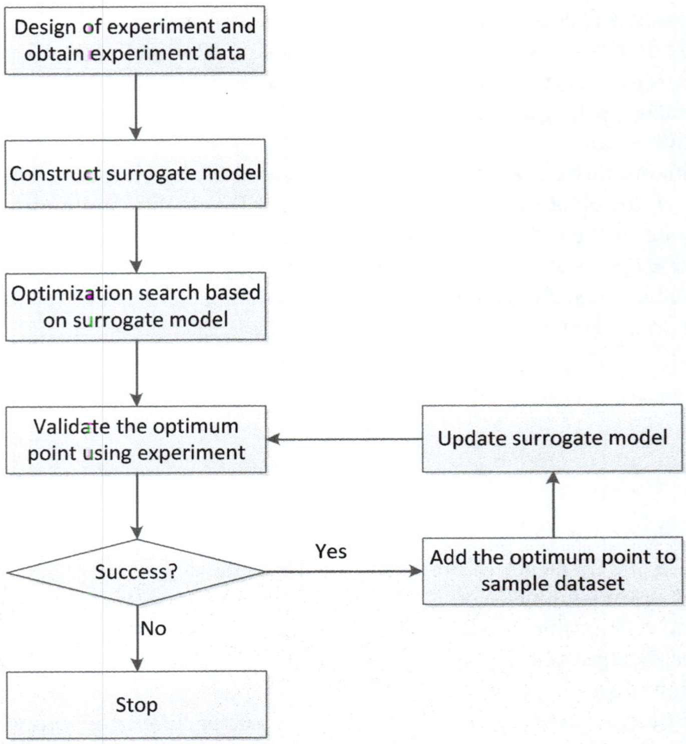

Fig. 8.22 In-looped optimization strategy

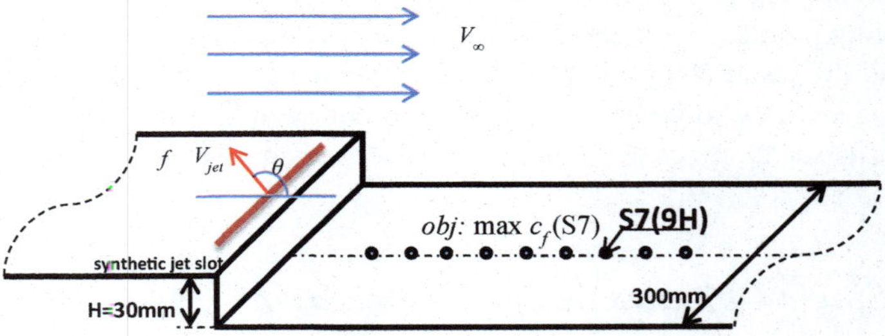

Fig. 8.23 Schematic diagram of the experimental model

As is proved above, skin friction at one specified station after baseline reattachment point could be taken as the parameter reflecting the effect of flow control, substituting the commonly used parameter—length of the recirculation zone. Skin friction at station $S7$ $(9H)$ was measured by sublayer fence and set up as the objective of design optimization.

Therefore, to control the backward-facing step flow using synthetic jet, the following optimization model was built:

$$\begin{aligned} obj: &\quad \max c_f(f, \theta, U_{jet}) \\ s.t. &\quad \begin{cases} 0\ \text{Hz} \le f \le 35\ \text{Hz} \\ 90^\circ \le \theta \le 150^\circ \\ U_{jet} = U_{jet}(f) \\ h = 2\ \text{mm} \end{cases} \end{aligned}$$

wherein, c_f represents skin friction at station $S7$; f is the frequency of the synthetic jet; θ is the angle of the synthetic jet slot and U_{jet} is the maximum velocity of jet slot. The third equality constraint indicates the coupled relationship between frequency and velocity and the fourth equality constraint ensures identical width of synthetic jet slot (identical mass flux). Frequency and angle of the synthetic jet were two independent design variables. The objective of design optimization is maximizing skin friction at the certain station $S7$ $(9H)$.

Considering pressure and temperature variations of laboratory environment, for each angle of synthetic jet slot, the corresponding baseline flow was subtracted to ensure data consistency. Relative values instead of absolute ones were used to exclude the influence of experimental environment (such as pressure change and temperature variation). Parameters of particle swarm pattern search algorithm were set to: population size of 20, maximum generation of 50 and convergence criteria of 10^{-4}.

Initial sampling angles were 90°, 120°, 150°, and measurements on five frequencies (0 Hz—baseline/12.5 Hz/20 Hz/27.5 Hz/35 Hz) were carried on at these three initial angles to obtain values of objective function. Based on these wind tunnel experimental data, surrogate model was constructed with original sampling points. Optimization search was performed with particle swarm pattern search algorithm based on surrogate model and the new sampling point was validated by experiment. Then the validated wind tunnel data were used to update surrogate model within the iterative loop and a new set of optimization search was carried on until no new sampling point appears (i.e., global optimum point derived). According to the strategy of in-looped design optimization shown in Fig. 8.22, sampling angles successively added at each turn were 135°, 105°, 112.5°, and 127.5°. After applying experimental data of all the seven angles to design optimization, it was found that the optimum design point remained at the angle of 127.5°, frequency of 35 Hz. The design optimization process terminated since no new sampling point turned up. Flow chart of design optimization is shown in Fig. 8.24. Skin friction (subtracting baseline flow) at all sampling points is shown in Fig. 8.25.

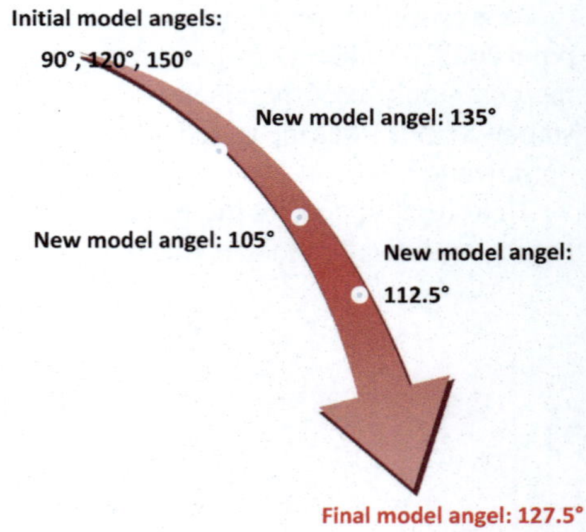

Fig. 8.24 Flow chart of design optimization process

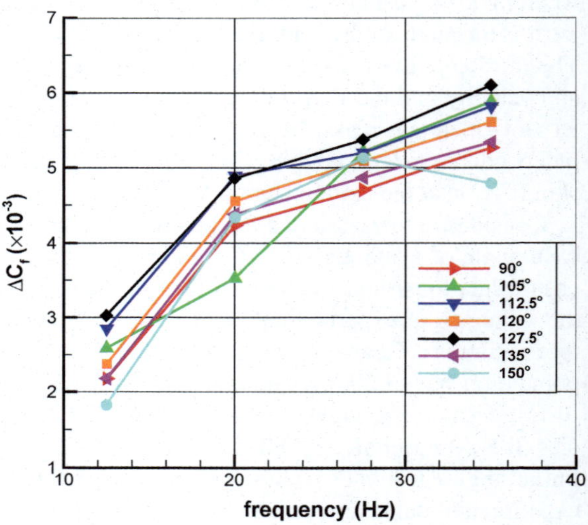

Fig. 8.25 Skin friction of all sampling points

Table 8.2 shows comparison of synthetic jet parameters and objective function values for both original and optimized states. As can be seen from the table, value of objective function increased from 5.601×10^{-3} to 6.097×10^{-3}, which corresponds to satisfactory control effect.

Table 8.2 Comparison of original and optimized state

	Frequency (Hz)	Angle (°)	Objective value (10^{-3})
Baseline flow	–	–	0
Original samples	35	120	5.601
Optimum point	35	127.5	6.097

8.4.4 Multi-objective Design Optimization Considering Energy Consumption

To get in line with practical engineering problems, design optimization process should consider the energy consumed in driving active flow control devices. Power consumed in driving reciprocator is taken into account to derive a multi-objective optimization scheme.

In this section, assumption that the energy (power) consumption varies linearly with the frequency of the actuator is made. The two objectives are combined into one and normalization of two design objectives is conducted at first.

$$obj_1 = (f_1 - obj_{1,\min})/(obj_{1,\max} - obj_{1,\min})$$
$$obj_2 = (obj_{2,\max} - f_2)/(obj_{2,\max} - obj_{2,\min})$$

Objective obj_1 is combined with obj_2 by linear weighting method to obtain the ultimate objective function. Weighting factors ω_1 and ω_2 represent weight of control effect and energy consumption respectively.

$$obj: \quad \max(\omega_1 \cdot obj_1 + \omega_2 \cdot obj_2)$$
$$s.t. \quad \begin{cases} 12.5\ \text{Hz} \le f \le 35\ \text{Hz} \\ 90^\circ \le \theta \le 150^\circ \\ U_{jet} = U_{jet}(f) \\ h = 2\ \text{mm} \\ \omega_1 + \omega_2 = 1 \end{cases}$$

Comparison between different optimum states provides guidance to the design process of synthetic jet devices that, for the multi-objective design optimization problem, the optimum state is not sensitive to frequency and thus the optimum angle at one frequency may also be applied to another frequency. For time-critical design tasks, optimum angle of existing optimization scheme could be chosen as an acceptable angle and only searching process of frequency is required.

8.4.5 Theoretical and Experimental Analysis on Reynolds Stress and Flow Control Mechanism

Analysis of the experimental data shows that, for 35 Hz synthetic jet frequency, jet slot angles of 105°, 112.5°, 120° and 127.5° lead to excellent control effects, whilst slot angles 90° and 150° lead to unsatisfactory outcome. A preliminary theoretical and experimental analysis of the mechanism of this phenomenon is conducted in the following section.

The operating (formation) formation frequency of synthetic jet is low and the periodic process can be divided into ejection phase and suction phase. Comparison of velocity profiles at four typical stations ($X/H = 1.07, 3.07, 7.07, 10.07$) of baseline and optimum states is shown in Fig. 8.26. It shows that the synthetic jet impacts velocity by 'suction' and velocity profiles in the potential flow region are notably decreased. As is indicated by the two dashed green lines, greater influence is observed for stations further downstream with clear deficits.

Profiles of Reynolds shear stress τ_{xy} are shown in Fig. 8.27, and please refer to [1, 12] for Reynolds normal stresses τ_{xx} and τ_{yy}. Magnitude of Reynolds stress significantly increases on the whole and the structure of the original shear layer is not yet obvious at all stations. Synthetic jet forms an enhanced shear layer and the influential region evolves upward with both wider areas and weakened intensity, leading to deficit of velocity in the potential flow region.

Power spectra of velocity magnitude at six monitoring points, as marked in Fig. 8.15 by $P1$ to $P6$, are calculated using Burg method and shown in Fig. 8.28. Flow field is dominated by evolution of low frequency vortices of 35 Hz (Strouhal number 0.053) induced by synthetic jet. By comparing $P5$ and $P3$, or $P6$ and $P4$, it can be known that synthetic jet introduces 'suction' effect (velocity profile) and more fluctuations (Reynolds stress) in high vertical region, thus restrains the downward

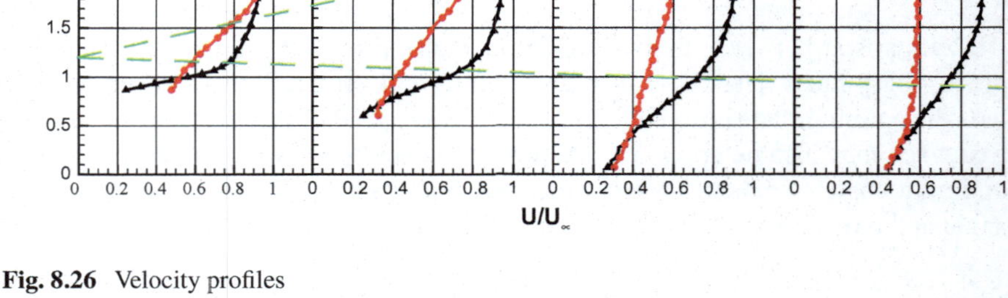

Fig. 8.26 Velocity profiles

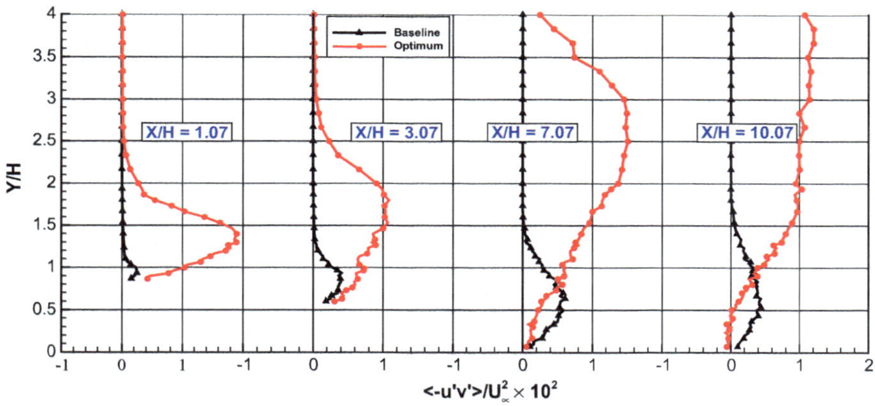

Fig. 8.27 Reynolds shear stress τ_{xy} profiles

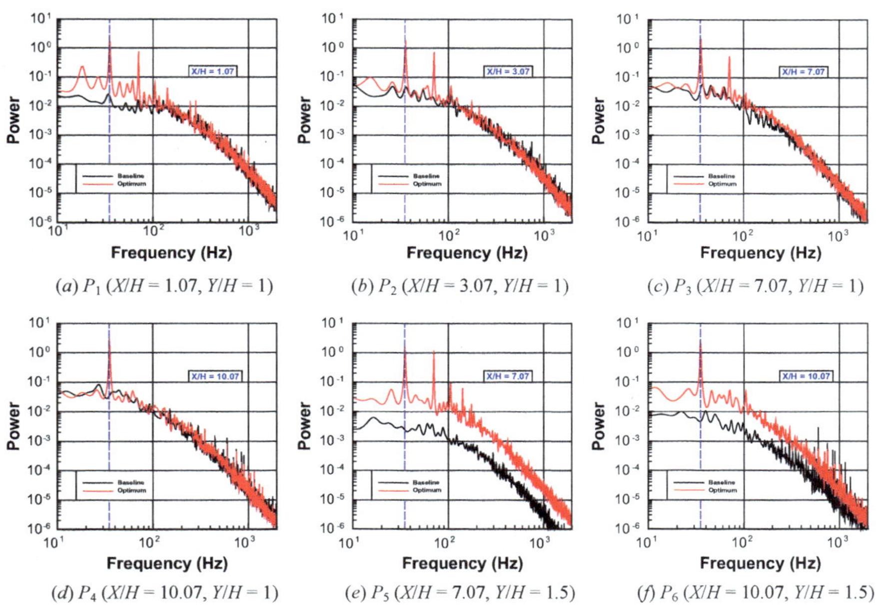

Fig. 8.28 Power spectra of velocity magnitude for baseline and optimum states

moving trend of power spectra (i.e., dissipation induced attenuation of fluctuation) in the potential flow region (*P*5 and *P*6).

8.5 Conclusion

(1) In the design optimization process, numerical simulation and wind tunnel experiment provide fast and reliable features in evaluating objective function. The optimization strategy based on the in-looped update surrogate model using wind tunnel test data is developed with hybrid particle swarm pattern search algorithm as the searching algorithm.
(2) Two optimal schemes with similar control effects reveals two inherently different flow control mechanisms. The flow control scheme for one problem based on Reynolds stress may not be definite. From this physical mechanism study, it is roughly proven that manipulation of Reynolds stress is effective for separation control and drag reduction.
(3) A newly developed synthetic jet device with non-linear oscillation of the reciprocating piston actuator into the pipe is introduced. Sublayer fences are used to measure wall shear stress and silk threads provide auxiliary visual monitoring of the transient flow field.
(4) Single-objective design optimization of synthetic jet parameters based on wind tunnel data leads to optimum result with excellent flow control effect.
(5) Influence of jet angle on flow field in both ejection and suction phases is preliminarily analyzed to reveal flow control mechanism. Both magnitude and distribution of Reynolds stress are significantly altered with enhanced shear layer structure in the potential flow region above.

Acknowledgements The current work is financially supported by MITT-EU within the research project MARS (Manipulation of Reynolds Stress for Separation Control and Drag Reduction) under contract No. FP7-AAT.2010-RTD-CHINA.

References

1. Xu F, Gao ZH, Ming X et al (2015) The optimization for the backward-facing step flow control with synthetic jet based on experiment. Exp Therm Fluid Sci 64:94–107
2. Troutt TR, Scheelke B, Norman TR (1984) Organized structures in an reattaching separated flow field. J Fluid Mech 143:413–427
3. Driver DM, Seegmiller HL, Marvin JG (1987) Time-dependent behavior of a reattaching shear layer. AIAA J 25(7):914–919
4. Heenan AF, Morrison JF (1998) Passive control of pressure fluctuations generated by separated flows. AIAA J 36(6):1014–1022
5. Armaly BF, Durst F, Pereira JCF et al (1983) Experimental and theoretical investigation of backward-facing step flow. J Fluid Mech 127:473–496

6. Smith B, Glezer A (1998) The formation and evolution of synthetic jet. Phys Fluids 10(9):2281–2297
7. Glezer A, Amitay M (2002) Synthetic jets. Annu Rev Fluid Mech 34:503–529
8. Ming X (1988) Characteristics and control of blunt body wakes. Ph.D. dissertation, Nanjing University of Aeronautics and Astronautics
9. Ming X (1992) New phenomenon of rectifying effect, Chinese. J Theor Appl Mech 24(1):52–60
10. Vaz AIF, Vicente LN (2007) A particle swarm pattern search method for bound constrained global optimization. J Glob Optim 1–31
11. Konstantinov NI, Dragnysh GL (1960) The measurement of friction stress on a surface. English translation DSIR RTS 1499
12. Xu F (2015) Research on numerical simulation of separated flow and synthetic jet flow control. Postgraduate thesis, Northwestern Polytechnical University

Chapter 9
Optimization of the Experimental Set-up for a Turbulent Separated Shear Flow Control by Plasma Actuator Using Genetic Algorithms

Nicolas Benard, Jordi Pons-Prats, Jacques Périaux, Jean-Paul Bonnet and Gabriel Bugeda

Abstract Since 1947, when Schubauer and Skramstad (J Res: 69–78, 1947 [1]) established the basis of the technology with its revolutionary work about steady state tools and mechanisms for the flow management, the progress of the flow control technology and the development of devices have progressed constantly. Anyway, the applicability of such devices is limited, and only few of them have arrived to the assembly workshop. The problem is that the range of actuation is still limited. Despite their operability limitations, flow control devices are of great interest for the aeronautical industry. The number of projects investigating this technology demonstrates the relevance of in the Fluid Dynamic field. The scientific interest focus not only on the industrial applications and the improvement of the technology, but also on the deep understanding of the physical phenomena associated to the flow separation, turbulence formation associated to the final drag reduction aim. A clear example of what has been mentioned is the EC MARS research project (MARS project, FP7 project number 266326, [2]). Its objectives are aimed to a better understanding of the Reynolds Stress and turbulent flow related to both drag reduction and flow control. The research was carried out through the analysis of several flow control devices and the optimization of the parameters for some of them was an important element of the research. When solving a traditional fluid dynamics optimisation problem numerical flow analysis are used instead of experimental ones due to their lower cost and shorter needed time for evaluation of candidate solutions. Nevertheless, in the particular case of the selected flow control plasma devices the experimental measurement of the performance of each candidate configuration has been much quicker than a numerical

N. Benard (✉) · J.-P. Bonnet
Institut PPRIME - UPR 3346 – CNRS - Université de Poitiers - ISAE/ENSMA - SP2MI
Téléport2, Bd Marie & Pierre Curie, BP 30179, 86962 Futuroscope Chasseneuil Cedex, France
e-mail: nicolas.benard@univ-poitiers.fr

J. Pons-Prats · J. Périaux · G. Bugeda
International Center for Numerical Methods in Engineering (CIMNE),
c/Esteve Terrades 5, 08860 Castelldefels, Spain

J. Périaux · G. Bugeda
Universitat Politècnica de Catalunya (UPC), c/Gran Capità S/N, 08034 Barcelona, Spain

N. Qin et al. (eds.), *Advances in Effective Flow Separation Control for Aircraft Drag Reduction*, Computational Methods in Applied Sciences 52,
https://doi.org/10.1007/978-3-030-29688-9_9

analysis. For this reason, the corresponding optimisation problem has been solved by coupling an evolutionary optimization algorithm with an experimental device. This paper discusses the design quality and efficiency gained by this innovative coupling.

Keywords Turbulent separated shear flow control · Surface plasma actuator · Experimental optimization · Genetic algorithm

9.1 Introduction

Despite their limitations, flow control devices remain of great interest to the aeronautical industry. The number of projects investigating this advanced technology confirms the relevance of the technology. The scientific outcomes are not focussed only on industrial applications and the continuous improvement of the technology, but also on a better understanding of the physical phenomena associated to the flow separation including the turbulence formation linked to the ultimate reduction of drag.

An example of the above approach can be found in the EC MARS project [2]. The objectives of MARS aimed to provide a better understanding of the Reynolds Stress and the turbulent flow related to drag reduction and flow control. The research was carried out through the analysis and comparison of several selected flow control devices.

On the framework of the research activities of European and Chinese partners involved in the MARS project, the optimization of the parameters of some selected devices was an important task.

When solving a fluid dynamics optimisation problem in aerodynamics full numerical analysis software are used in many cases instead of experimental analysis software due to their lower cost and shorter time execution to feed the optimizer. Nevertheless, in the particular case of a selected flow control plasma device the experimental measurement of the evaluation of different configurations has been much faster than with a numerical flow analysis code. This is because numerical simulation needs to be at least of the level of hybrid RANS-LES to resolve the unsteady turbulent scales. For this reason, it has been decided to solve the corresponding optimisation problem by coupling an evolutionary optimization algorithm to an experimental plasma device.

This paper evaluate from the results the design quality and efficiency of this coupling on a Backward Facing Step.

This paper describes in six (6) sections the experimental and, optimization setups through an efficient software and hardware interface and the analysis of optimized results obtained on a Backward Facing Step (BFS) test case. Sections 9.2 and 9.3 describe the experimental and optimization setup respectively. Numerical results obtained with the experimental optimization setup are discussed in Sect. 9.4 while Sect. 9.5 presents conclusions and future of the innovative research.

9.2 Experimental Setup

The selected test case is the optimization of a Backward Facing Step (BFS) with the Plasma device. Design variables and objective functions for the optimization are described in the following sections.

9.2.1 Description of the Wind Tunnel

The model is installed in a closed-loop wind-tunnel on Fig. 9.1 having a moderate turbulent intensity (1%). The dimension of the test section is $300 \times 300 \times 1000\ \mathrm{mm}^3$. All measurements are performed for a free-stream velocity of 15.6 m/s.

9.2.2 Description of the Step Model

The model includes a step having a height, h, of 30 mm and a span wise length of 300 mm. The model covers the full span wise and the expansion ratio equals to 1.1. This expansion ratio is high enough for reducing the influence of stream wise pressure gradient along the wall of the wind-tunnel on the mean flow reattachment position. The aspect ratio (channel width on step height) is larger than 10 as recommended by the literature to assure a two-dimensional flow in the centre of the wind tunnel. Measures are conducted for a fixed Reynolds number of 30,000 based on h, the height of the step.

9.2.3 Description of the Plasma Device

The step model is made of several pieces; in particular it has been designed in order to have a removable part (see Fig. 9.2). The removable part corresponds to the step geometry and is made of a 3-mm thick machined PMMA piece. So, the model can be

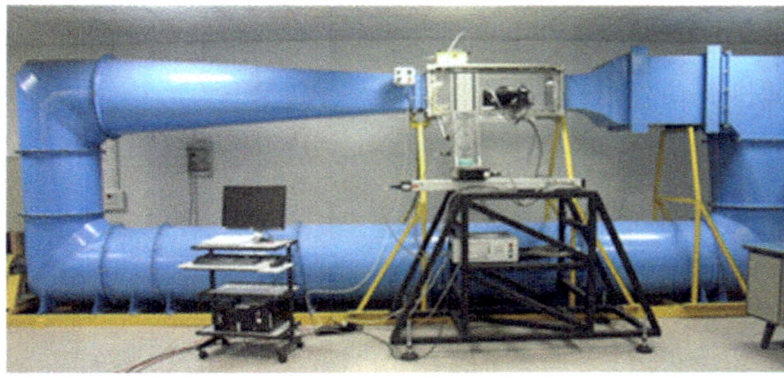

Fig. 9.1 Closed-loop wind tunnel dedicated to backward facing step measurements

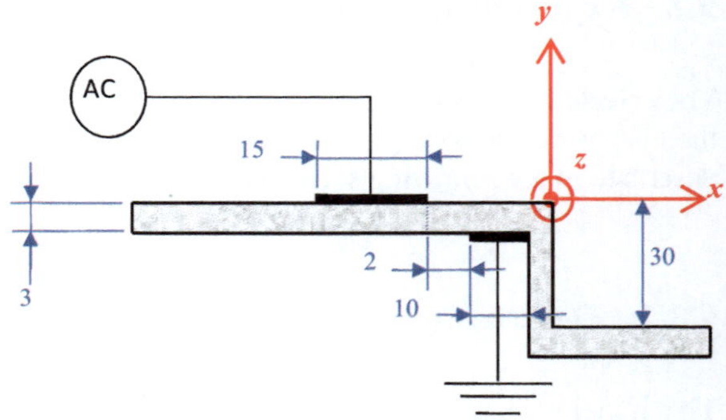

Fig. 9.2 Sketch of the DBD plasma actuator with details on electrode arrangement

used as dielectric barrier directly. This common electrode arrangement is introduced in Fig. 9.2. As shown, the active and the grounded electrode have a length of 15 and 10 mm respectively and the gap between them is fixed at 2 mm. By construction, the electrode protuberance into the flow is limited to the electrode thickness (i.e. 60 μm) and then the initial turbulent boundary layer characteristics are not modified by the actuator.

In all experiments, a TREK® (model 30/20A) high voltage power amplifier equipment is used as voltage source for the DBD. This amplifier generates a maximum output voltage of ±30 kV (tension gain 3000 V/V) and a maximum output current for AC peak of ±40 mA at 600 W maximum power (slew rate 500 V/μs).

9.2.4 *Description of the LabVIEW Software and Transducer Setup Interfaces to Operate the Wind Tunnel*

In the present investigation, an optimization study based on experimental results from the wind-tunnel is conducted. The bottom wall downstream of the step has been equipped with pressure taps (55 pressure tabs from x/h = 1 to x/h = 9 with a spatial resolution of 0.15 h) (Fig. 9.3). A set of 32 unsteady pressure sensors (12 bits) with bandwidth of 2 kHz for max pressure of 250 Pa has been designed and packed in a dedicated rack. Output voltage signals of the pressure sensors are recorded by PXI hardware (2.16 GHz dual core) with 32 channels acquisition card (PXI-6259). This system includes a signal generator card (PXI-5402) for supplying the actuator by the electrical parameters defined by the optimizer while the output responses of the 32 sensors are recorded. The LabVIEW code manages the acquisition of the pressure sensors, estimation of the objective functions, the electric command of the actuator and it makes the interface between the wind-tunnel and the optimizer. The interface between the optimizer (.exe file) and the wind-tunnel is done via DOS command under LabVIEW and the input/output interactions are managed through ASCII files.

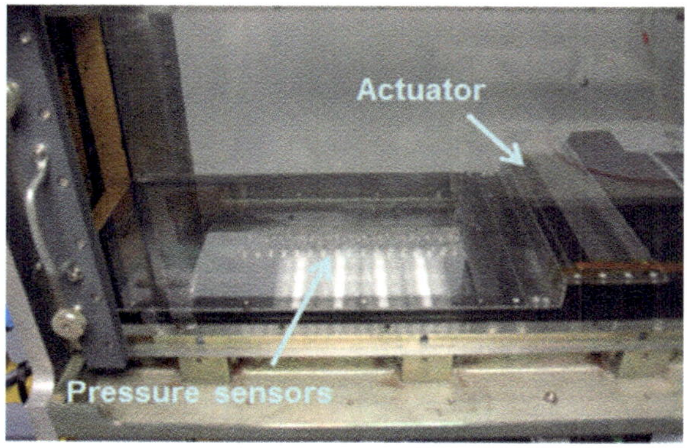

Fig. 9.3 Photograph of the step model equipped with a linear surface DBD actuator and 55 pressure taps

9.2.5 Description of the Design Variables/Input Parameters Selected

The control mechanism of plasma discharges is based on an electromechanical conversion meaning that all characteristics of the applied electrical signal have influence on the produced flow. As shown in literature [3], the voltage amplitude and driven frequency f_{ac} are responsible for the amplitude of the produced local flow, usually called electric wind. Higher are the applied voltage amplitude and frequency, higher is the induced maximum flow velocity. However, in all the cases the produced flow resembles a wall jet, tangent to the dielectric barrier. To maximize the induced flow velocity, frequency of kHz order is usually used. In order to amplify the interactions between the induced plasma flow and the natural flow having inner frequency content, the actuator is often operated in a modulated manner. Here, burst modulation is used meaning that the actuator is alternatively switched on and off at a frequency f_{bm} with a duty-cycle DC. Some illustrative results of the induced velocity profiles are introduced in Fig. 9.4 for 'steady' (Fig. 9.4a) and 'unsteady' forcing (unsteady forcing refers to burst modulation of the ac high-voltage signal, see Fig. 9.4b). Due to the fast response of any electromechanical converter in absence of any mechanical part, the driven frequency f_{ac} and the burst frequency f_{bm} induced periodic fluctuations in the produced flow, these fluctuations being at the supplied electrical frequencies.

In the framework of control optimization performed here, the design variables concern the voltage amplitude, the burst frequency and the duty-cycle considering that the driven frequency is fixed at $f_{ac} = 2\,\text{kHz}$. In these conditions, it is well known that the mean amplitude of the produced flow increase with the voltage amplitude. The motivation for changing the burst frequency is that the flow is very sensitive to periodic excitations at the right frequency (i.e., excitation frequency matching with the preferred mode of the considered flow). The last design variable is the duty-cycle. High duty-cycle value means that the produced flow approaches steady flow conditions (i.e., small fluctuation amplitude) while low duty-cycle will emphasize the amplitude of the fluctuations but in this case the mean produced flow velocity is small.

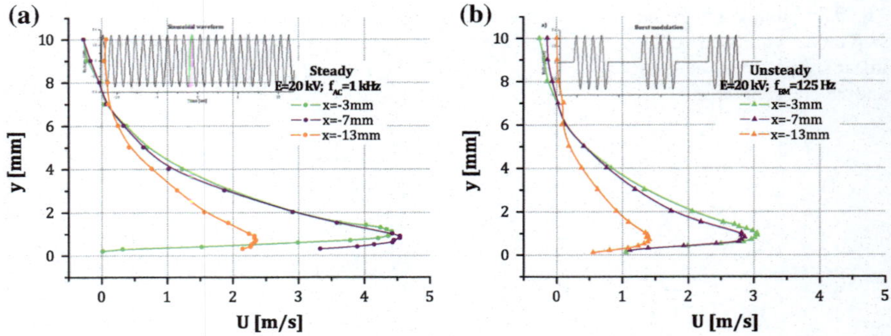

Fig. 9.4 Three stream wise velocity profiles induced by reference steady (**a**) and unsteady (**b**) actuation

Then, for flow control perspectives, a balance has to be found for producing a 'high' mean flow having 'high' fluctuation amplitude. All of these three parameters have strong influence on the performance of the control, this motivating for optimization studies using advanced algorithms.

9.2.6 Objective Functions Selected for Optimization

Two objective functions have been selected for the optimization. The first one concerns a mean quantity useful to demonstrate the potential of surface plasma to reduce reattachment downstream of the backward-facing step model. The location of the mean reattachment point is estimated by the position of the pressure fluctuation peak along the stream wise direction on the wall as recommended in literature [4] and illustrated in Fig. 9.5. The procedure simply consists of recording the 32 pressure signals on the bottom wall and computing each local pressure fluctuation amplitudes. Then a dedicated gradient-based algorithm seeks for the location of the maximal fluctuation, this point being related to the mean reattachment. Preliminary tests have shown that a correct estimation of the mean reattachment point requires 5–10 s to attain a converged value due to the unsteady character if the reattachment point. For this quantity, the estimated error is mainly caused by the spatial distribution of the sensors. Here, the reattachment is estimated within ± 0.15 h.

In order to extend the optimization to dynamic component of the flow a second objective function is considered. The spatial integral of the pressure fluctuations P' are performed all along the bottom wall (i.e., spatial integration of plot such as presented in Fig. 9.5). The pressure trace is assumed to be caused by the flow variations in the free shear layer developed from the step corner. Then, higher value of P', and thus higher value of the P' spatial integral, is supposed to be related to a more intense vortical flow activity. This is confirmed in Fig. 9.6 where the time-history of the pressure signals at different spatial position is plotted for the natural

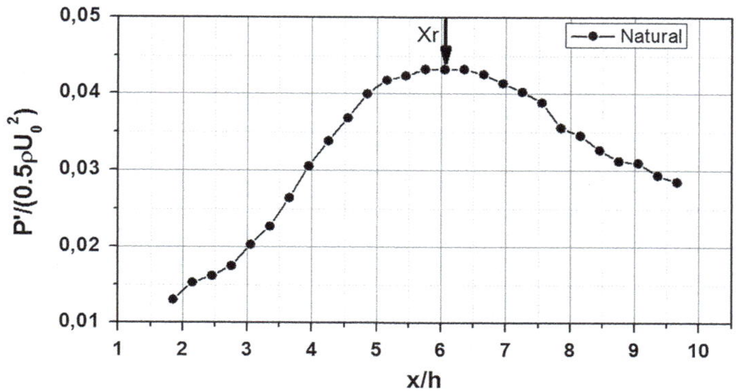

Fig. 9.5 Typical distribution of pressure fluctuations along the bottom wall downstream the step model

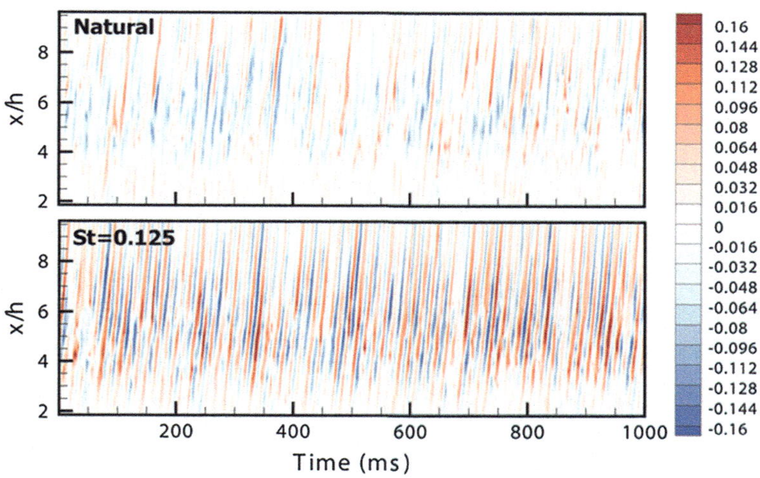

Fig. 9.6 Time-history of the wall pressure signature (normalized values) along the streamwise direction for natural and unsteady forcing by burst modulation at $f_{BM} = 65$ Hz ($St_h = 0.125$)

flow and flow controlled by a plasma discharge operated in burst modulation at 65 Hz. In both cases, the trace of the periodic fluctuations, i.e. the signature of the vortical flow structures, confirms that periodic flow is observed few step heights downstream the step corner. When forced at $St_h = 0.125$, large changes can be observed. The signature of the vortical flow structures is visible at a nearest position from the step corner (x/h ~ 3). The organization of these structures seems highly modified with a high coherence with downstream positions and a regularization in the periodicity of the vortex street. The spatial integral of the pressure fluctuation is largely increased in the case of actuation. Higher is the spatial integral, higher is the influence of the actuator on the vortical flow activity. The convergence rate of this integrated quantity is lower than the mean reattachment as only 5 s are needed to achieve a converged value while the estimation error is limited to 3%.

9.3 The Moga Evolutionary Optimizer

More and more often, engineering design problems require a simultaneous optimization of several objectives associated to a number of constraints. Unlike single objective optimization problems, the solution can be a set of non-dominated solution well known as the Pareto optimal Front Solutions are compared to other solutions using the concept of Pareto dominance. A multi-criteria optimization problem can be formulated as follows:

Maximise/Minimise

$$f_i(x), i = 1, \ldots, N_f, \tag{9.1}$$

Subject to constraints:

$$g_j(x) = 0 \text{ and } h_k(x) \geq 0, j = 1, \ldots, N_g, k = 1, \ldots, N_h, \tag{9.2}$$

where f_i g_j h_k are, respectively, the objective functions with a total number of N_f, the equality constraints with a total number of Ng and the inequality constraints with a total number of N_h. x is an N_x dimensional vector where its arguments are the decision variables.

For a minimisation problem using a Pareto game, a vector x_1 is said partially less than vector x_2 if:

$$\forall i \; f_i(x_1) \leq f_i(x_2) \text{ and } \exists j \; f_j(x_1) < f_j(x_2), \tag{9.3}$$

In this case the solution x_1 dominates the solution x_2.

In Evolutionary Computation, optimisation algorithms evaluate and classify the value of objective functions between several different candidates in order to detect best to worst performance. The selection of the designs to be evaluated is depending on the selected optimisation algorithm. In the case of Genetic Algorithms (GAs), they define and evaluate multiple populations of design points in order to capture a number of solutions belonging to the Pareto set. Pareto selection ranks the population and selects non-dominated individuals for building the Pareto Front. A Genetic Algorithm that has capabilities for multi-objective optimisation is termed Multi-Objective Genetic Algorithms (MOGAs). Theory and applications of MOGAs can be found in Ref. [5–7].

The single-objective case is formulated in the same way, but limiting the number of objective functions (N_f) to be equal to 1. The problem can include equality and inequality constraints, as defined in the multi-objective problem, and the concept of dominance is no longer used due to the fact that only one objective function is considered.

The selected optimizer used in this work is the Multi-Objective Genetic Algorithm (MOGA) module in Robust Multi-objective Optimization Platform (RMOP) developed at CIMNE. Details of RMOP can be found in Ref. [5–7]. The optimization platform enables the analysis of multi-objective problems using the hybridized techniques that combine Pareto-game and Nash-games.

Developing efficient optimization techniques is the most challenging task in the field of Evolutionary Algorithms (EAs) research due to the complexity of modern design problems. One of emerging techniques to improve an optimisation performance is the use of the Nash-equilibrium concept which will be acting as a pre-conditioner of global optimizer.

Lee et al. [8] introduced the concept of Hybridized Games (a Pareto and Nash coalition) coupled to a well-known MOEA. Non-dominating Sort Genetic Algorithm II (NSGA-II) [9] are used to solve Unmanned Aerial System (UAS) multi-objective Mission Path Planning System (MPPS) design problems and speed up with the Hybridized Games by 80% when compared to the original NSGA-II. In addition, Lee et al. [10] have also hybridized NSGA-II optimizer with Nash-Game strategies to study the dynamic speed up of Nash-Players via Hybridized Games in the capture of different shapes of Pareto fronts (convex, concave, discontinuous, …) with multi-objective mathematical functions A Hierarchical Asynchronous Parallel Multi-Objective Evolutionary Algorithm (HAPMOEA) [11] can be also hybridized to solve a real-world robust multidisciplinary design problem. Numerical results show that the Hybridized Games can improve up to 70% of the HAPMOEA performance while producing better Pareto optimal solutions. References [8] clearly describe merits of using Hybridized Games coupled to MOEA for engineering design applications which consider complex geometries with large number of design variables. The Hybridized Game concept has two major properties: the first is a decomposition of a multi-objective design problem split into several simpler ones handled by Nash-Players with their own design search space. The second one allowing Nash-Players to synchronise their elitist strategy with Global or Pareto Players and accelerate the optimization procedure by maintaining diversity by a dynamic strategy of the Nash-Players.

The optimizer used during the experimental optimization is a sub-version of the advanced genetic algorithm code [7, 11]. The simplified version of the optimization code is not using the parallel computing capabilities in order to avoid overloading the wind tunnel actuator with more than one individual at the same time.

9.4 The Plasma Wind Tunnel Optimization: Results

9.4.1 Inverse Reconstruction Problems with Plasma Wind Tunnel Analysis

The procedure of coupling an experimental Plasma active device with a numerical optimization code, as an analogic solver, has been carefully validated in order to ensure that the methodology is able to solve complex optimization problems. It has been decided to test the coupled experimental-numerical optimization procedure with the following three inverse problems:

(1) Problem 1: A single objective problem to recover a prescribed reattachment length
(2) Problem 2: A single objective problem to recover a prescribed P' integral value
(3) Problem 3: a multi-objective problem to recover both prescribed reattachment length and P' value.

All the three reconstruction problems (1), (2) and (3) use the same definition: Minimize

$$f_i(\boldsymbol{x}_1, \boldsymbol{x}_2, \boldsymbol{x}_3) = f_{pi}(\boldsymbol{x}_1, \boldsymbol{x}_2, \boldsymbol{x}_3) - f_{ei}(\boldsymbol{x}_1, \boldsymbol{x}_2, \boldsymbol{x}_3), i = 1, 2 \qquad (9.4)$$

The values $f_i(\boldsymbol{x}_1, \boldsymbol{x}_2, \boldsymbol{x}_3)$ will be the reattachment length or the P' integral value in Problem 1 and Problem 2, or both the reattachment length and the P' integral value for the Problem 3.

The values $f_{pi}(\boldsymbol{x}_1, \boldsymbol{x}_2, \boldsymbol{x}_3)$ and $f_{ei}(\boldsymbol{x}_1, \boldsymbol{x}_2, \boldsymbol{x}_3)$ are the prescribed value and the evaluated value for the objective functions respectively for each of the 3 problems defined, which will define the Reattachment length and/or P' integral value).

Subject to the definition of the vector search space three design parameters $(\boldsymbol{x}_1, \boldsymbol{x}_2, \boldsymbol{x}_3)$:

- $\boldsymbol{x}_1$: Voltage (V); V $\in \mathbb{N}$, $14 \leq \mathrm{V} \leq 20$
- $\boldsymbol{x}_2$: Burst Frequency (f_{bm}, in kHz); $f_{bm} \in \mathbb{N}$, $10 \leq f_{bm} \leq 200$
- $\boldsymbol{x}_3$: Duty Cycle (DC, in %); DC $\in \mathbb{N}$, $5 \leq \mathrm{DC} \leq 95$.

These inverse problems are solved as traditional reconstruction problems with the main characteristic that the final solution is known from a direct numerical simulation for given values of the triple $\boldsymbol{x}_1, \boldsymbol{x}_2, \boldsymbol{x}_3$. Thus, the recovery of the prescribed solution validates the operability of the optimization procedure.

Figures 9.7, 9.8 and 9.9 show the convergence history of the main objective functions for each of the three test case problems (1), (2) and (3). All three figures are plotting the convergence of the best individual along the iteration process of the optimizer. Figure 9.7 shows the deviation to zero of the difference between the real reattachment length and the evaluated one, Fig. 9.8 between the real P' and the evaluated one respectively. Figure 9.9 shows the main objective function of the multi-objective problem solved. It shows the convergence to zero of the difference

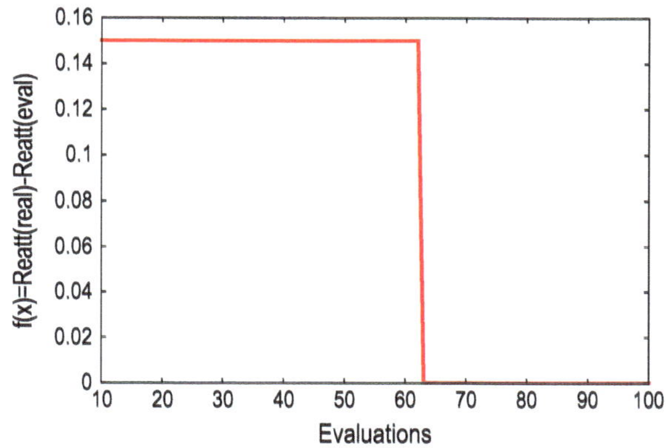

Fig. 9.7 Convergence history for problem: matching the reattachment length

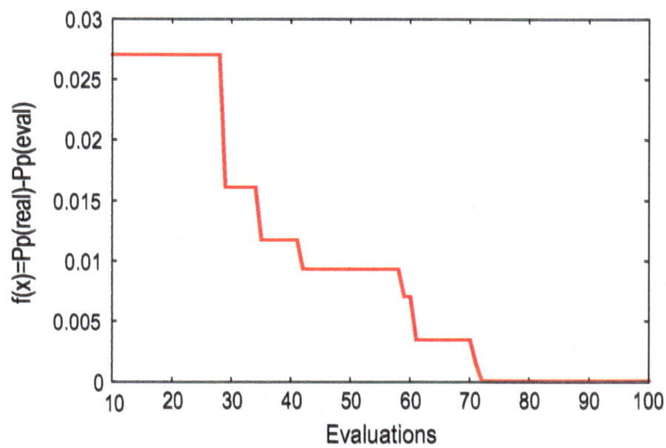

Fig. 9.8 Convergence history for problem 2, recover of P'

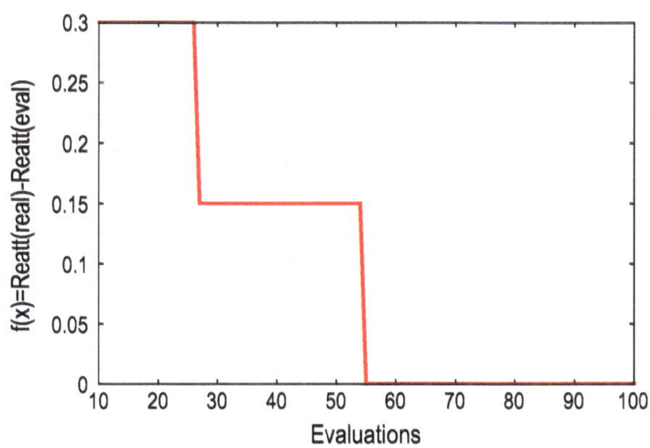

Fig. 9.9 Convergence history of the reattachment length for the problem 3; Reconstruction of the reattachment length and P'

between the real reattachment length and the evaluated one. In all three cases, the deviation value of the inverse problems is 0, and the prescribed design variables values of the direct simulation are fully matched by the converged solutions.

The GAs optimizer coupled with the Experimental Plasma active device captures the solution of Problem 1. The solution of the problem is capturing the evaluated value which compared to the prescribed value leads to a deviation of the objective function equal to zero. Regarding the design variables, the three parameters that recover the optimal individual:

- Voltage was 20 V and the obtained values is 20 V,
- Burst Frequency was 130 kHz, and the obtained one is 130 kHz,
- And finally the Duty Cycle was 60%, the same for the obtained one.

The solution of the Problem 2 also shows a good agreement with the prescribed values. The difference is 0.00004 in this case. Regarding the design variables, the three parameters that recover the optimal individual:

- Voltage was 20 V as well as the converged value,
- Burst Frequency was 70 kHz while the converged one is 72 kHz,
- And finally the Duty Cycle was defined between 60% and 70%, while the converged one is 52%.

The solution of Problem 3 also shows the best possible agreement with the prescribed values. The computed deviation with both the objective functions is zero. Regarding the design variables, the three parameters match quite well the prescribed variables:

- Voltage was 20 V and the converged values are 19 V,
- Burst Frequency was 130 kHz, and a similar converged value can be noticed,
- And the Duty Cycle was 60%, while the converged one is 55%.

The experimental-numerical methodology has been validated with a good agreement with the prescribed values defined by direct simulation. It should take into consideration that the precision of the wind tunnel is not able to distinguish among close values and the solution to obtain a prescribed value is not unique. It means that several combinations of V, f_{bm} and DC can lead to the same reattachment or P' integral values. It also should be taken into consideration the limit of operability of the Plasma Wind Tunnel. To avoid equipment damages, it has been requested to limit the time operability of the Plasma Wind tunnel to one hour during the optimization procedure. Considering that each wind tunnel shot evaluation necessitates about 5–10 s, only one hundred evaluations of the GAs population is allowed. The optimizer restarts the search to a pre-defined point to cope with the time limitation. As shown by convergence results obtained with best individuals of the GAs population, it was not necessary to use the restart option thanks to a fast convergence of all the three reconstruction problems.

9.4.2 Optimization Problems with Plasma Wind Tunnel Analysis

An optimization test case problem has been set up to complete the analysis of the efficiency of the experimentally coupled optimization procedure. It selects the P' integral value as the objective function. From the physical point of view of the plasma device set up analysis, the P' value is a sensitive parameter to control the flow, as described in Sect. 9.2.

The optimization problem is now defined as:

Maximize

$$f_o(\vec{\boldsymbol{x}}) = \max\left[P'(\vec{\boldsymbol{x}})\right], \tag{9.5}$$

where $\vec{\boldsymbol{x}}$ is the vector containing the three design parameters of the plasma wind tunnel device.

The optimization problem, defined by Eq. (9.5), is defined as a single-objective maximization problem, with the following definition of the lower-upper bounds search space of the three design parameters:

- Voltage (V); V $\in \mathbb{N},\ 14 \leq \text{V} \leq 20$
- Burst Frequency (f_{bm}, in kHz); fbm $\in \mathbb{N},\ 10 \leq f_{bm} \leq 200$
- Duty Cycle (DC, in %); DC $\in \mathbb{N},\ 5 \leq \text{DC} \leq 95$.

The optimized results using the Experimental- GAs optimizer are shown in Fig. 9.10, where the convergence of the P' value to the optimum is shown. As it happened in the previous reconstruction problems, a time limitation had to be set up for the safe operability of the Plasma Wind Tunnel. Once again, a fast convergence of the P' values to the optimum made unnecessary to use the restart option. The plot shows the comparison between the maximized value and the baseline design. The optimized value of P' is 2.247, while the baseline design defines a P' equal to 2.211. It means that the improvement is of 1.6%. The converged best individual of

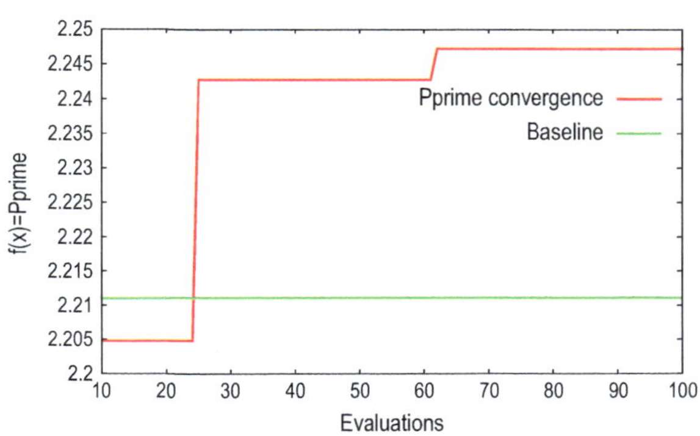

Fig. 9.10 Convergence history of the P' in the GAs maximization with Plasma Wind Tunnel analysis

the population is defined by a Voltage equal to 20 V, a burst frequency of 81 Hz, and a duty cycle of 63%.

9.5 Conclusion and Future

The Plasma Wind tunnel analysis coupled to an evolutionary optimization procedure is an innovative research. It uses the wind tunnel as an analogic solver, mastering the difficulties of coupling Plasma experimental facilities with an evolutionary optimizer. This approach proposes an efficient alternative to the expensive CFD analysis cost. In the selected BFS test case, the coupling between the Plasma Wind tunnel and the optimizer enables a very fast and quasi-real-time estimator of active candidate flow conditions only constrained by limitations in the Plasma wind tunnel equipment which were not necessary in these simple inverse and optimization problems.

The obtained results are relevant to demonstrate the reliability and fast convergence of the approach dealing with inverse and optimization problems. The definition of a multi-objective inverse optimization does not present a major problem and it also converges quickly and accurately.

From the results achieved further research activities on this dual experimental-Numerical approach are planned namely the management of the precision of wind tunnel parameters and the coding of design variables, the choice of other design variables and/or objectives functions.

In this context other more complex multi objective optimization test cases are envisaged with advanced hybridized EAs and Game strategies.

Acknowledgements This research has been partially funded by the European Commission (EC), though the Framework Programme 7 (FP7) and the Ministry of Industry and Information Technology of the People Republic of China (MIIT), Project # 266326 entitled: "Manipulation of Reynolds Stress for Separation Control and Drag Reduction" (MARS).

References

1. Schubauer GB, Skramstad HK (1947) Laminar boundary-layer oscillations and transition on flat plate. United States Bureau of Standards. J Res, 69–78
2. MARS project, FP7 project number 266326. www.cimne.com/mars
3. Benard N, Moreau E (2012) EHD force and electric wind produced by plasma actuators used for airflow control. AIAA paper 3136
4. Eaton JK, Johnston JP (1981) A review of research on subsonic turbulent flow reattachment. AIAA J 19
5. Lee DS, Gonzalez LF, Periaux J, Bugeda G (2011) Double shock control bump design optimisation using hybridised evolutionary algorithms. Special Issue J Aerosp Eng 225(10):1175–1192. https://doi.org/10.1177/0954410011406210

6. Lee DS, Morillo C, Bugeda G, Oller S, Onate E (2011) Multilayered composite structure design optimisation using distributed/parallel multi-objective evolutionary algorithms. J Compos Struct. https://doi.org/10.1016/j.compstruct.2011.10.009
7. Lee DS, Periaux J, Onate E, Gonzalez LF (2011) Advanced computational intelligence system for inverse aeronautical design optimisation. In: International Conference on advanced software engineering (ICASE-11), Proceedings of the 9th IEEE international symposium on parallel and distributed processing with applications workshops, ISPAW 2011—ICASE 2011, SGH 2011, GSDP 2011, Busan, Korea, 26–28 May, 2011, pp 299–304. ISBN 978-1-4577-0524-3. https://doi.org/10.1109/ispaw.2011.46
8. Lee DS, Periaux J, Gonzalez LF. UAS mission path planning system (MPPS) using hybrid-game coupled to multi-objective optimiser. J Dyn Syst Meas Control. DS-09-1135
9. Deb K, Agrawal S, Pratap A, Meyarivan T (2002) A fast and elitist multi-objective genetic algorithm: NSGA-II. IEEE Trans Evol Comput 6(2):182–197
10. Lee DS, Gonzalez LF, Periaux J, Srinivas K. Hybrid-game strategies coupled to evolutionary algorithms for robust multidisciplinary design optimization in aerospace engineering. IEEE Trans Evol Comput. TEVC-00213-2009
11. Periaux J, Lee DS, Gonzalez LF, Srinivas K (2009) Fast reconstruction of aerodynamic shapes using evolutionary algorithms and virtual nash strategies in a CFD design environment. Special Issue J Comput Appl Math 232(1):61–71. ISSN 0377-0427

Chapter 10
Transient Process of Separation Control on a NACA0015 Airfoil

W. L. Siauw and Jean-Paul Bonnet

Abstract Experiments concerning the control of flow separation were conducted on a NACA0015 airfoil model with a chord length of 0.35 m at a Reynolds number of 1 million. We analyzed the impact of the transient attachment and separation process on the mean and turbulent fields. To control the flow, angled fluidic vortex generators through 44 orifices were positioned in a single array at a position of 30% from the leading edge of the airfoil. PIV measurement was applied to the vicinity of the wake with the airfoil set at an incidence of 11° with a separation length of 1/3 chord from the trailing edge. Optimization was not performed, although the Cd reduced from 0.0275 to 0.02 and the Cl indicated an improvement of ~17%. Phase-averaged energy distributions were performed in the near wake. When the actuators were impulsively deployed, the data showed the trace of the passage of a large eddy associated to the beginning of the attachment process over the airfoil. It had a strong transient effect on the turbulent energy. When the FVGs were deactivated, a gradual increase in the turbulent energy was observed. Detailed analyses of the mean and fluctuating fields were provided.

Keywords Unsteady separation control · NACA 0015 airfoil · Fludic Vortex Generator · Transient phenomena · Unsteady wake · Turbulent fluctuations · Turbulent shear stress · PIV · Hot Wire

10.1 Introduction

It has been of great technological importance to improve the efficiency and performance of an aircraft in terms of lift enhancement and drag reduction. Depending on the type of aircraft (military/civil), it would have impact on the maneuverability, fuel

W. L. Siauw
DSO National Laboratories, 12 Science Park Drive, Singapore 118225, Singapore

J.-P. Bonnet (✉)
Institut PPRIME (CNRS UPR3346, Université de Poitiers, ISAE-ENSMA), TSA 51124 Cedex 9, Poitiers 86072, France
e-mail: jean-paul.bonnet@univ-poitiers.fr

N. Qin et al. (eds.), *Advances in Effective Flow Separation Control for Aircraft Drag Reduction*, Computational Methods in Applied Sciences 52,
https://doi.org/10.1007/978-3-030-29688-9_10

consumption efficiency, the emission of gas and production of noise. In this paper, we focused on the use of fluidic vortex generator (FVG) to suppress flow separation. In particular, the transient in wake which corresponded to flow attachment and separation over the airfoil surface was studied in detail. The impact on the turbulent fields was of primary importance for physical understanding, flow control strategies and validation of CFD computations.

Fluidic vortex generator had been shown to produce longitudinal vortices [6] that negotiated along a surface. Such a system was desirable as it does not introduce parasitic drag when not in use as compared to passive mechanical vortex generators. The key to efficient control utilizing such a system depended on the sustainability of vortical coherence, strength of the vortices, proximity towards a wall surface and the relative proximity between vortices. All these depended on condition of its origin at the orifices: angle of the jets, jet peak velocities and momentum. These technical facts were well documented in Abe et al. [1], Johnston and Nishi [7], Milanovic and Zaman [10], Pauwley and Eaton [11]. From these studies, we deduced and adopted fixed characteristics for the actuators. Concerning the study of the transient process of attachment/separation, Amitay and Glezer [2], Darabi and Wygnanski [3, 4], Mathis et al. [9] and Deb.en et al. [5] respectively conducted studies on a modified 4-digit NACA airfoil (synthetic jets deployed by piezo material), generic flap (synthetic jets deployed by loud speakers), a beveled splitter plate (continuous jet controlled by high speed valve) and a sharp edge ramp (by Active Vortex Generators). From these authors, the ratio of the time taken for flow separation to attachment was found to vary between ~1.2 and 5 (at almost optimal $C\mu$ of the control jets). In all cases, time taken for separation to recover was always greater than the case for attachment. An evacuation of the separated region was observed in their experiment. A reduced order model had been successfully developed in Siauw et al. [13]. It was shown form POD analysis that a first order system adequately represented the dynamics of the controlled process of flow attachment and separation.

The baseline flow characteristics corresponded to a fully turbulent boundary layer on a NACA0015 airfoil with a 2 dimensional flow separation occurring at 1/3 of chord from the trailing edge at 11° of angle of attack. First, the evolution of the near wake turbulent flow corresponding to the phase of impulsive FVGs deployment was presented. It described a transient between separated and attached flow over the airfoil. Second, the article provided an analysis of the flow transient when the FVGs were suddenly deactivated. In this case, the flow evolved between a fully attached state back to the baseline configuration of flow separated state.

10.2 Test Facility, Model and Instrumentation

The wind tunnel used for the study was a closed loop type, with a 2.4 m by 2.6 m test section, with a turbulence intensity of 0.5% at $U_\infty = 40$ m/s. A NACA0015 model of chord length c = 0.35 m with a full span of 2.4 m was installed across the test section. 80 microns carborandum grits had been applied at a location corresponding

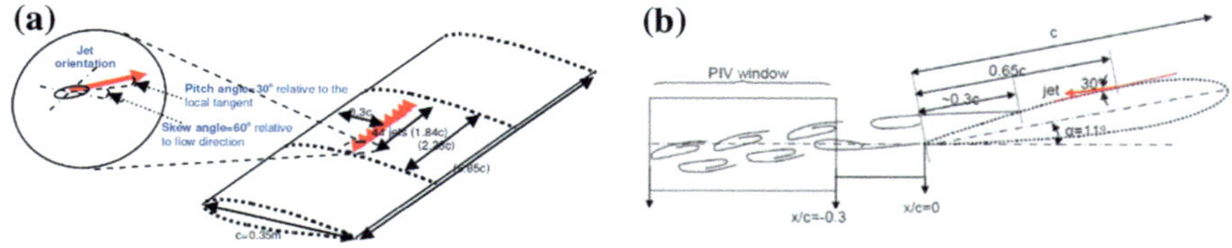

Fig. 10.1 **a** Actuator arrangement and **b** location of the PIV window relative to the NACA0015 airfoil. Note that the x-axis was oriented from the trailing edge to the leading edge

to 0.4% of chord from the leading edge in order to achieve fully developed turbulent boundary layer at separation. Test condition corresponded to a Reynolds number of approximately 1 million. 44 orifices were installed in the vicinity at the central one third portion of the airfoil. Each orifice had a diameter of 1 mm (of the order of the boundary layer thickness at test condition) so that efficient exchange of momentum can take place between free the stream flow and the boundary layer. The total span covered by the array of actuators was 1/3 of chord. They were pitched 30° and yawed at 60°, see Fig. 10.1. The peak velocity of the jet was set at ~200 m/s. PIV measurements had been performed in the near wake of the airfoil as shown in Fig. 10.1. The resolution of the camera was 1375 by 1039 pixels. Velocity vectors were calculated on an interrogation window of 16 by 16 pixels with a 50% overlap ratio.

In order to characterize the transients during deployment and deactivation, the FVGs deployment system must respond faster than the characteristic times of attachment and separation. This was achieved by installing ASCO solenoid valves (CM25-5W) inside the airfoil model. These valves possessed an average response time (time to open/close the valve) of 3 ms ($T^+ = 1$, see the definition in Sect. 10.3). The valves were operated by controlling the on/off state of a solid state relay that was triggered via a square wave.

Conditional sampling technique was employed during FVG deployment or removal. 300 independent snapshots were obtained for 40 different time delays relative to the start of the transient. The phase-locked ensemble averaged velocity was

$$\langle U_i(X_i, \Delta\tau)\rangle = \left[\sum_{m=1}^{N=300} \frac{U_i(X_i, \Delta\tau + mT)}{N}\right] \tag{10.1}$$

where $T = 1$ s was the period of the square wave, which was chosen to be sufficiently long for the flow to reach an asymptotic steady state. The fluctuations statistics were obtained from the 300 individual snapshots.

Figure 10.2a depicts the evolution of the wake before and after jet deployment. It was obtained by hot wire measurement at $x = -1c$. It should be noted that there is a momentary spike at $T^+ \sim 8$. This corresponded to the signature of a passing structure and was not observed in the separation process which commenced at $T^+ \sim 190$. The wake survey technique (calculation of the momentum deficit) was applied to estimate the Cd. As shown in Fig. 10.2b, the Cd reduced from 0.0275 to 0.02. The time taken

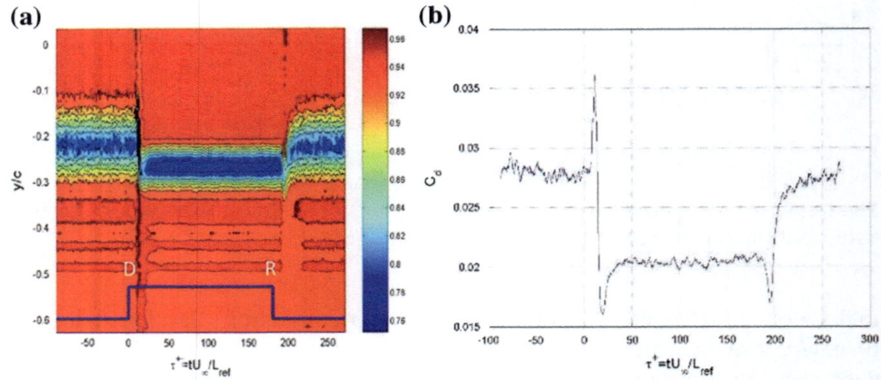

Fig. 10.2 **a** Velocity contour depicting the changes in its size during jet deployment (D) and deactivation (R); and **b** the corresponding estimated drag coefficient evolution

for attachment (T^+ ~ 10) was half that of separation (T^+ ~ 20). Details are given in Siauw et al. [13]. The Cl calculated from the integration of mean surface pressure measurement indicated an improvement of ~17% at incidence of 11° Siauw [12].

The transient evacuation of a strong vortex had been observed in the unsteady control for a pitching airfoil by Woo and Glezer [14]. Even without pitching, these authors showed a clockwise vortex generated during the attachment process. This effect clearly explained the transient increase of drag and lift were influenced by vortex emission.

10.3 Analysis of Flow Behavior During the Activation of FVG

This section presented the results concerning analysis by averaging of scalar fields at a particular time delays. Changes in mean and turbulence levels during the transient process were described. The chronological events were made with reference to a non-dimensional time defined as:

$$T^+ = \frac{t \cdot U_\infty}{L_{sep}} \tag{10.2}$$

with t representing the physical time and L_{sep} (= 0.3c) was the length of separation over the airfoil at incidence of 11°. $T^+ = 0$ was the reference time when the electrical signal was sent to the valves that controlled the jet. There was small time lag for the jet to initiate at the exit of the pitched and skewed orifices on the surface of the airfoil. This time lag was related to the opening/closing of the valve before any changes could be felt at the orifice exit.

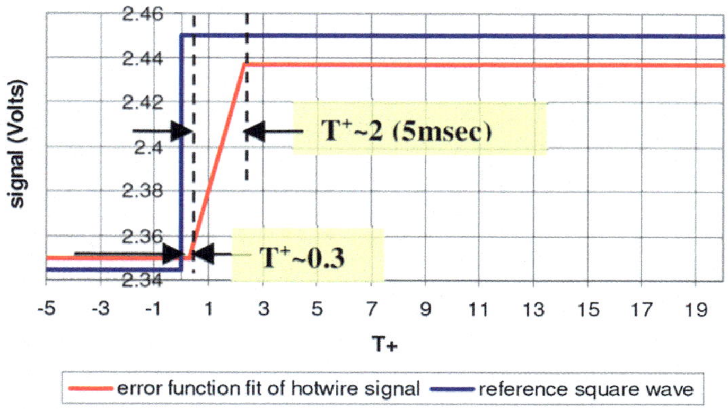

Fig. 10.3 Graph showing the relative time lag between trigger signal to the electro valves and jet deployment at the orifice

After that, another time delay was expected for the controlled jet to reach asymptotic state; this time delay is 5 ms ($T^+ \sim 2$). This is schematically represented in Fig. 10.3.

10.3.1 Analysis of Changes in Mean Velocity Field

With reference to Fig. 10.4, it was observed from the contour of U/U_∞ velocity that the wake started to undulate between $T^+ = 6.96$ and 8.82 without much increase in size. The wake started to increase in size from $T^+ = 8.82$ to $T^+ = 10.2$; this increase was observed to be rapid and progressed from upstream to downstream position. From $T^+ = 10.7$ to 13.5, the wake reduced in size progressively from upstream to downstream. After which, the wake tended asymptotically to the final reduced size corresponding to fully attached flow. During the entire process, the velocity in the region of the wake axis had been redistributed such that the initially higher velocity deficit at $x/c = -0.3$ was reduced (i.e. higher velocity) and conversely the initially lower velocity deficit at $x/c = -0.9$ was increased (i.e. lower velocity). The overall result was a reduction in drag as shown on Fig. 10.2b. The enlargement of the wake could be better understood by the wake profile at $x/c = -0.85$ plotted for the period of $T^+ = 0.464$–11.6 as shown in Fig. 10.5a. The initial undulation could be observed in the shift of the minimum velocity towards the positive y direction (i.e. upwards with reference to the PIV window). Figure 10.5b showed the sequence of the reduction of the wake size still at $x/c = -0.85$. There was little change in the minimum velocity U_D ($U_D/U_\infty \sim 0.835$) at $T^+ = 0.464$ and $T^+ = 16.3$ at this position. Using the wake function, the half wake width δ_{wake} was estimated as 0.053c and 0.038c for the steady uncontrolled and controlled state as shown in Fig. 10.6a

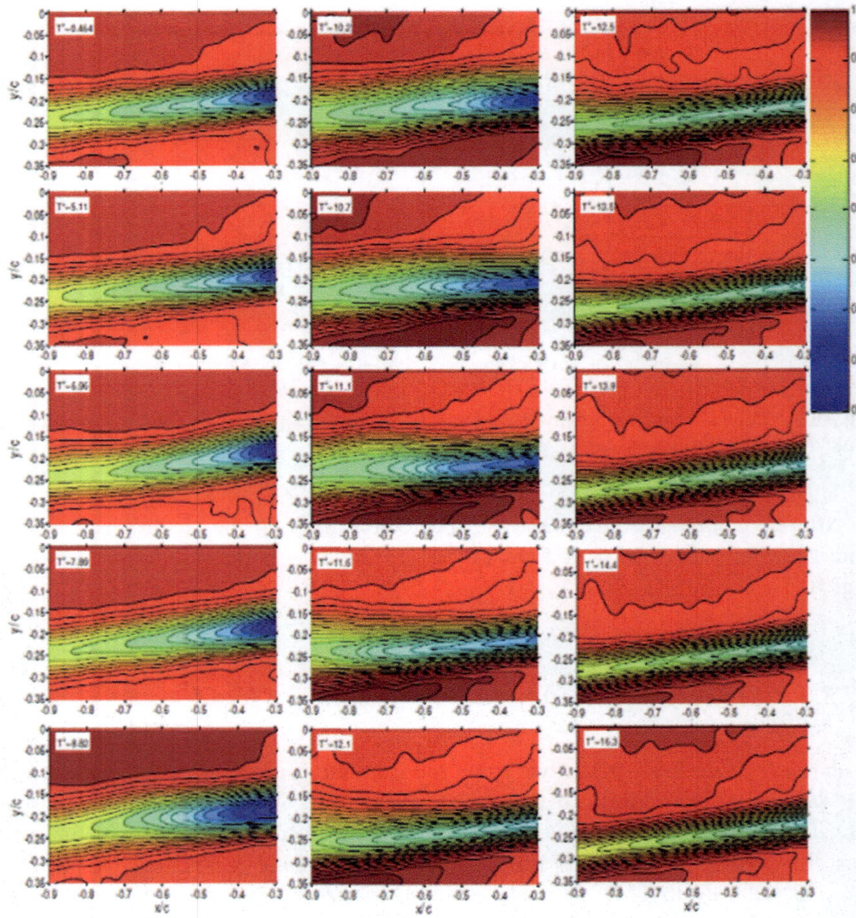

Fig. 10.4 Evolution of the mean velocity $\langle U\rangle/U_\infty$ during jet deployment (flow from right to left)

and b. The ratio of the velocity deficit to the free stream velocity (U_D/U_∞) was 0.847 and 0.857 respectively.

The estimated peak of Cd, calculated at $x/c = -0.85$, for the transient process of jet deployment observed on Fig. 10.2b was consistent with the observation of the transient increase in wake size (increase in Cd) followed by a decrease in size (decrease in Cd). It was important to note that the value of Cd was not very accurate during $T^+ = 7.89$ and 13.5 because curve fitting using the wake function becomes more difficult due to an unbalanced wake. Nevertheless, the transient physics of jet deployment was revealed.

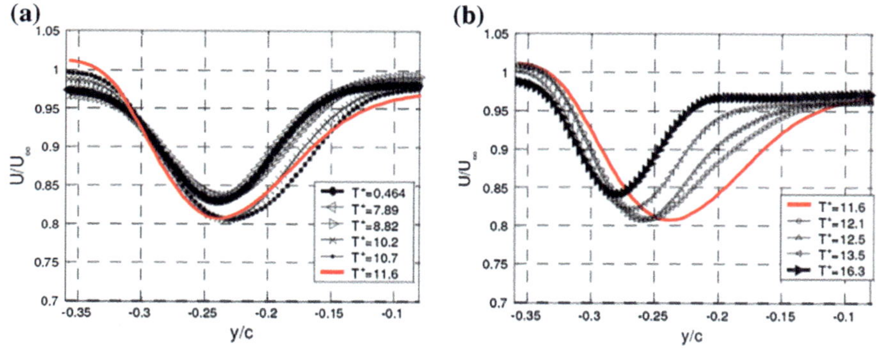

Fig. 10.5 Evolution of $\langle U\rangle/U_\infty$ at $x/c = -0.85$ in the wake of the NACA0015 airfoil during jet deployment

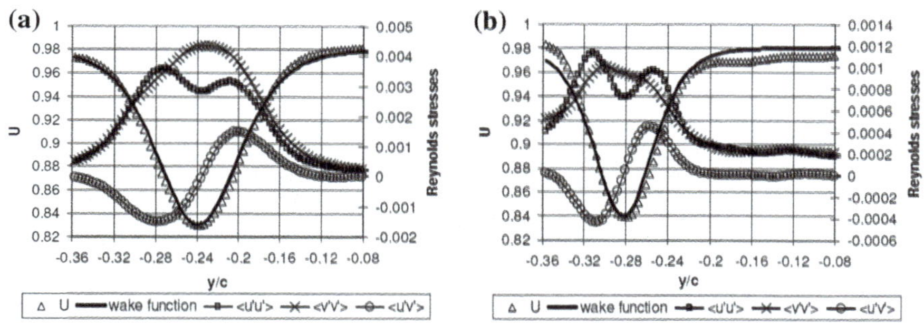

Fig. 10.6 Mean streamwise velocity and Reynolds stresses non-dimensionalized by U_∞ and $(U_\infty)^2$ respectively at T^+ of: **a** 0.464 (wake corresponding to uncontrolled flow separated steady state); **b** 16.3 (wake corresponding to controlled flow attached steady state)

10.3.2 Analysis of Changes in Turbulence Fields

10.3.2.1 Evolution of $\langle u'u'\rangle/U_\infty^2$ Reynolds Stress

With reference to Fig. 10.7, a parcel of flow with high turbulent intensity started to enter the PIV window at $T^+ = 6.96$. As schematically described on the figure, it appeared as two connected parcels (top and bottom) of high intensity regions at $T^+ = 8.82$. At $T^+ = 10.2$, the bottom parcel began to split from the top parcel and eventually left the PIV window at $T^+ = 12.1$. For $T^+ > 12.1$, the remaining top parcel was progressively evacuated from the PIV window and disappeared completely at $T^+ = 13.9$.

This can be viewed as the passage of a series of convecting large eddies on the top of the wake followed by another on the bottom during the transient process. The eddies on the top portion of the wake should correspond to the evacuated separated

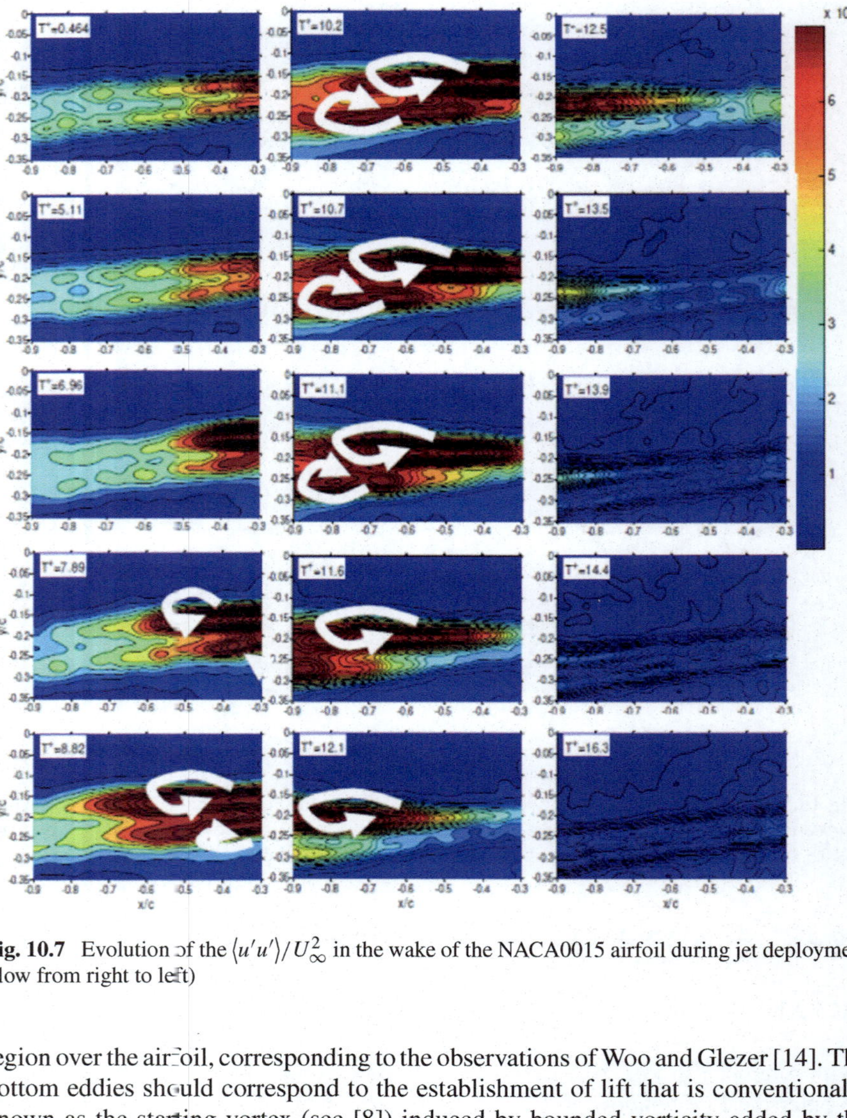

Fig. 10.7 Evolution of the $\langle u'u' \rangle / U_\infty^2$ in the wake of the NACA0015 airfoil during jet deployment (flow from right to left)

region over the airfoil, corresponding to the observations of Woo and Glezer [14]. The bottom eddies should correspond to the establishment of lift that is conventionally known as the starting vortex (see [8]) induced by bounded vorticity added by the FVG on the airfoil.

This process could be more clearly represented by the $\langle u'u' \rangle$ Reynolds stress profile at x/c = −0.85 in Fig. 10.8a and b. As expected in wakes, double peak was observed at all T^+; the intensity of the bottom peak dominated the initial stages (0.464 < T^+ < 10.7), see Fig. 10.8a. The peak associated to the upper part of the wake assumes a high intensity at $T^+ = 11.6$ and reaches a maximum value at $T^+ = 12.1$. After which, the overall intensity started to decrease drastically and the $\langle u'u' \rangle$ distribution tend asymptotically towards an almost symmetrical double peak at $T^+ = 16.3$, see

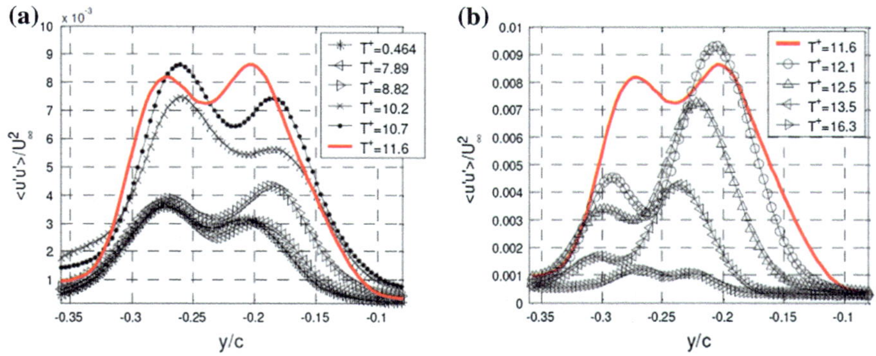

Fig. 10.8 Variation of $\langle u'u'\rangle/U_\infty^2$ as a function of $T^+ = -0.85$

Fig. 10.8b. These observations reflected the complex process of entrainment of fluid from the free stream into the wake during the jet deployed transient. The peak value of $\langle u'u'\rangle$ was higher at the bottom of the wake for $T^+ = 0.464$–7.89; the contrast was observed in the later stages where entrainment at the top of the wake dominated.

10.3.2.2 Evolution of $\langle u'v'\rangle/U_\infty^2$ Reynolds Stress

With reference to Fig. 10.9, the shear stress $\langle u'v'\rangle$ contours showed a positive and negative distribution on the top and bottom part of the wake respectively. The shear stress was more intense in the upstream position at $T^+ = 0.464$, before the effects of the jet deployment was felt. From T^+ between 6.69 and 8.82, the shear stresses increased in intensity in the downwards direction. At $T^+ = 10.7$, the high shear stress region began to decrease from the upstream direction. The negative shear stress region (wake bottom) had been evacuated from the PIV window more quickly than the positive shear stress regions (wake top); this occured for a T^+ between 11.1 and 135. The same observation could be seen from Fig. 10.10a, b.

Considering the shear stress distribution at $T^+ = 0.464$ at the reference profile at $x/c = -0.85$, the shear stress increased by more than 50% during the transient and decreased by about 50%, at $T^+ = 16.3$, when the flow had been attached over the airfoil.

10.4 Analysis of Flow Structures During Deactivation of FVG

This section focused on the changes in flow structure in the wake during the deactivation of the control jets. The flow evolved from a full attached situation to a separated

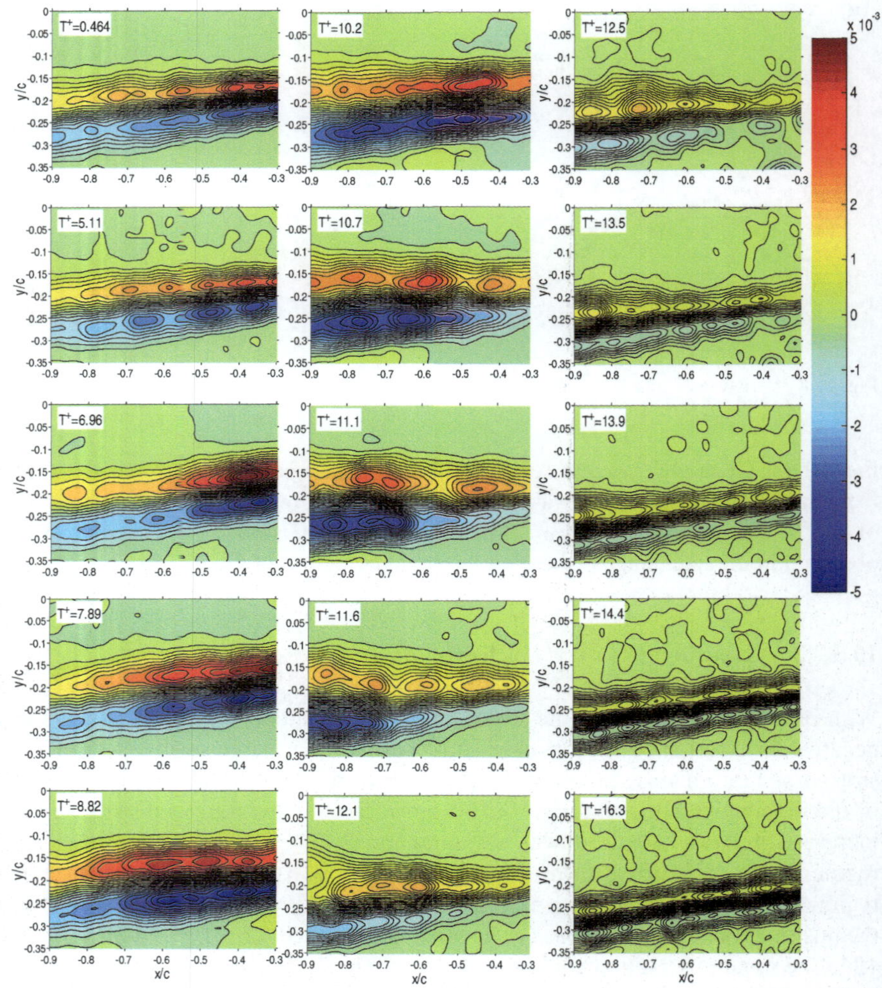

Fig. 10.9 Evolution of $\langle u'v\rangle'/U_\infty^2$ in the wake of the NACA0015 airfoil during jet deployment

base flow. In the PIV experiment 38 time delays are used to define the process of jet deactivation. Similar to the case of jet deployment, 300 snapshots are used to define statistics at each time delay.

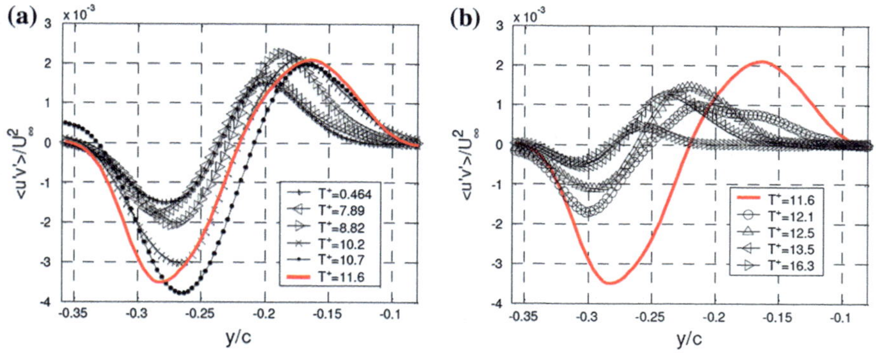

Fig. 10.10 Variation of $\langle u'v' \rangle / U_\infty^2$ as a function of T^+ at x/c = −0.85

10.4.1 Analysis of Changes in Mean Velocity Field

With reference to Fig. 10.11, the contour of the mean U velocity field showed a gradual enlargement of the wake and an upwards shift of the wake axis which was comparable to the jet deployment process. However, in the process of the transient, no undulation (wave like behavior) in the wake was observed. Thus, the passage of large eddies with spatial scales larger than that of the uncontrolled vortex streets in the downstream direction was probably not present. This statement would be substantiated in the analysis of the mean turbulence field. Based on the contour levels at the wake axis, the velocity deficit decreased at the downstream position at x/c = −0.85. The inverse was observed at the upstream position at x/c = −0.3. The movement of the wake axis proceeded by a slight downwards movement (see Fig. 10.12a) followed by a gradual upwards movement towards asymptotic position (see Fig. 10.12b).

The wake width determined by the wake function was observed to be of the scale of the spacing of the maximum and minimum peaks in the shear stress distribution as shown in Fig. 10.13. This was consistent with the estimated drag coefficients using the curve fitted wake profiles presented in Fig. 10.2b. Its evolution was such that there was a slight decrease followed by a gradual increase. The decreased in Cd can be attributed to a decrease in the size of the wake between $T^+ = 11.8$ and 13.6 as shown in Fig. 10.12a. This was dramatically different from the jet deployment process (see Fig. 10.6) where there was a rapid change (increase) in Cd before it settled down in the asymptotic state with a lower Cd compared to the uncontrolled state.

There was a difference between the "flow attached" wake profiles of the jet deployment (see Fig. 10.5b) and deactivation process (see Fig. 10.12a). The velocity deficit is higher for the jet removal process as compared to the jet deployment process. A more balance wake fringe was also observed for the jet removal process and this makes the curve fitting of a wake function easier. This could be attributed to the fact that there was inherent hysteresis in the two processes.

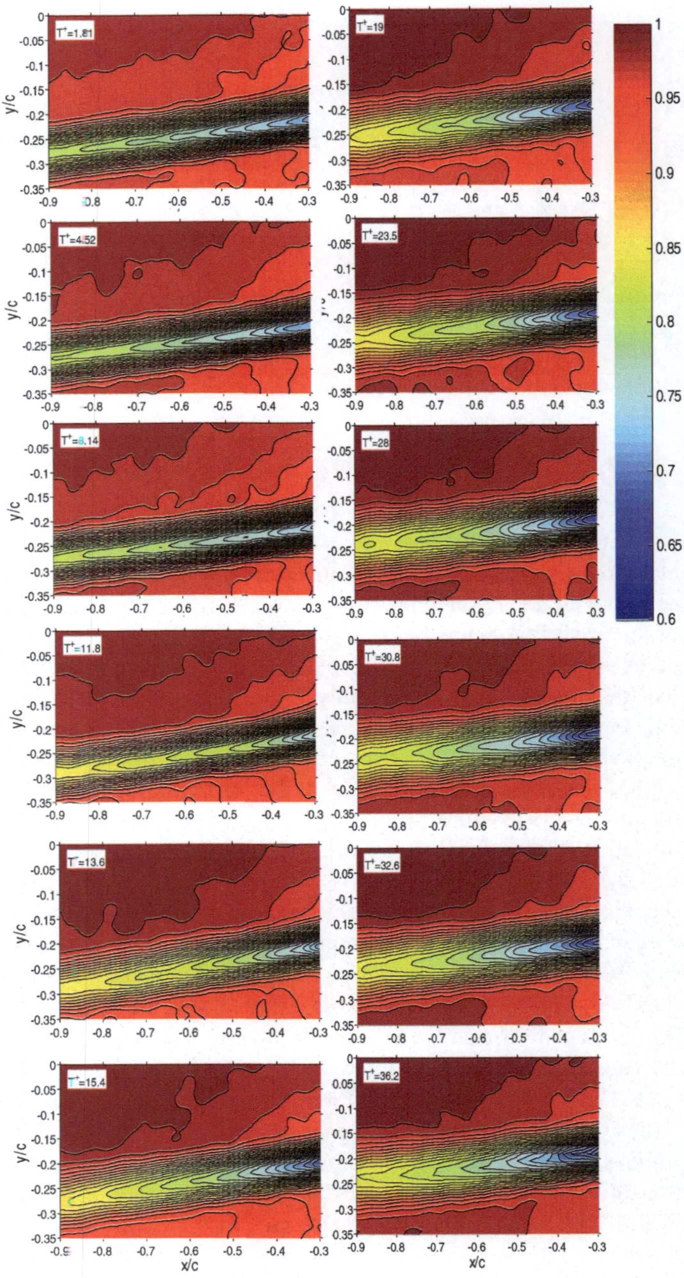

Fig. 10.11 Evolution of $\langle U \rangle / U_\infty$ in the wake of the NACA0015 airfoil during jet removal

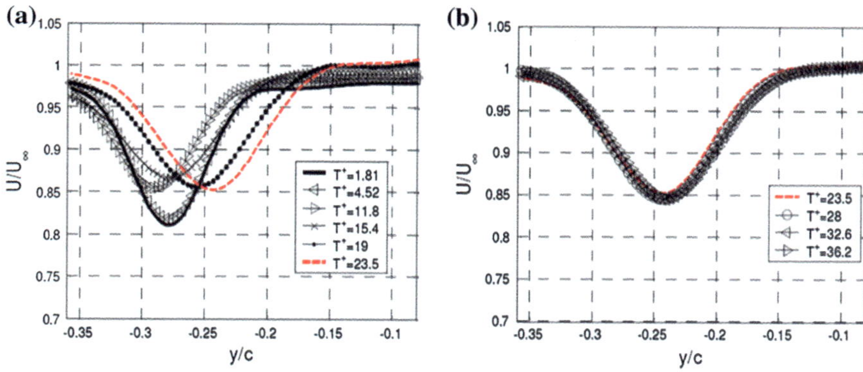

Fig. 10.12 Evolution of ⟨U⟩/U$_\infty$ as a function of y/c at x/c = −0.85 in the wake of the NACA0015 airfoil during jet removal

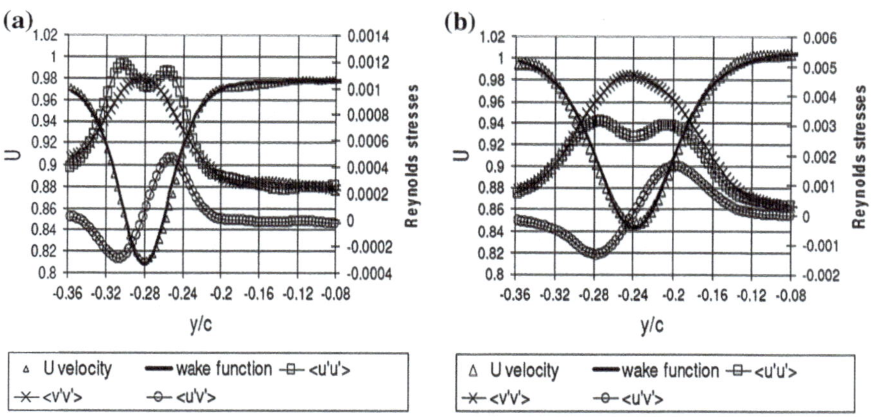

Fig. 10.13 Mean streamwise velocity and Reynolds stresses non-dimensionalized by U$_\infty$ and (U$_\infty$)^2 respectively at T$^+$ of: **a** 1.81 (wake corresponding to flow attached state over the airfoil); **b** 36.2 (wake corresponding to flow separated state over the airfoil)

10.4.2 Analysis of Changes in Turbulence Fields

10.4.2.1 Evolution of $\langle u'u' \rangle / U_\infty^2$ Reynolds Stress

With reference to Fig. 10.14, the $\langle u'u' \rangle$ Reynolds stress started to increases from the upstream position at x/c = −0.3 at about T$^+$ = 4.52 when the wake feels the influence of jet removal. This was accompanied by an increase in wake size (observed from extent of the spread of the contour levels) and the upward movement of the wake axis. This is a big difference as compared to the jet deployment case where the change

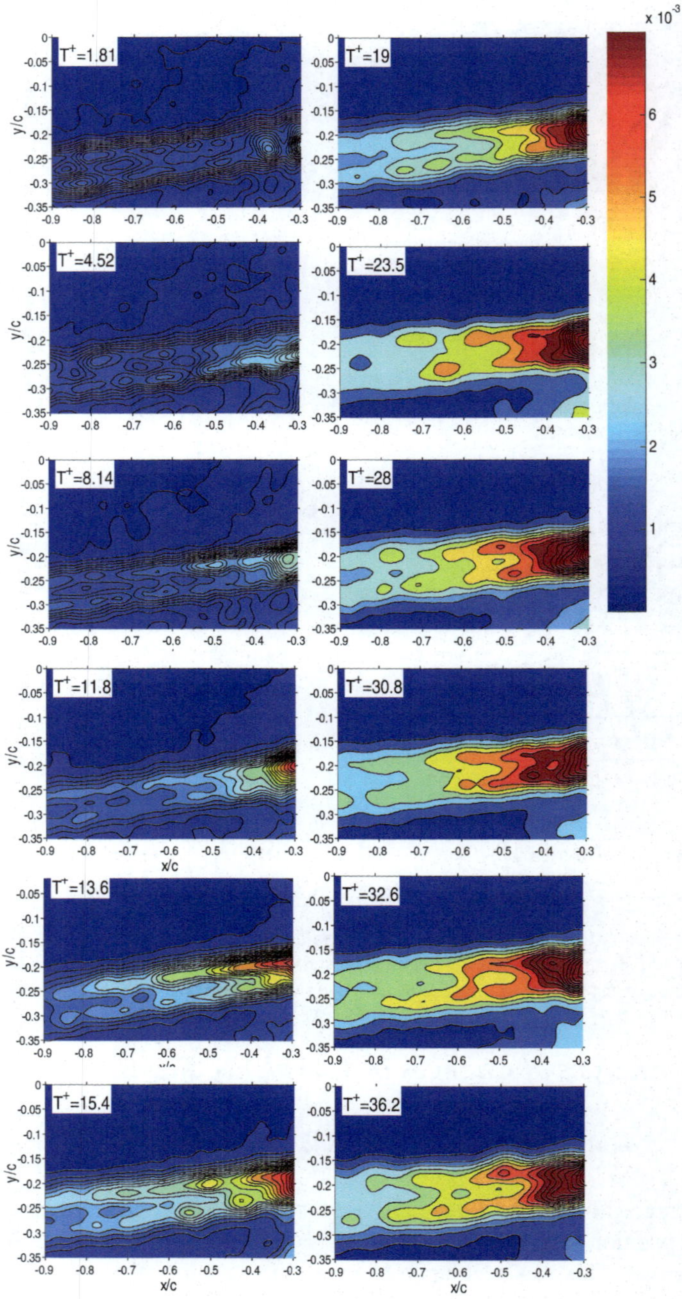

Fig. 10.14 Evolution of $\langle u'u' \rangle / U_{\infty}^{2}$ in the wake of the NACA0015 airfoil during jet removal

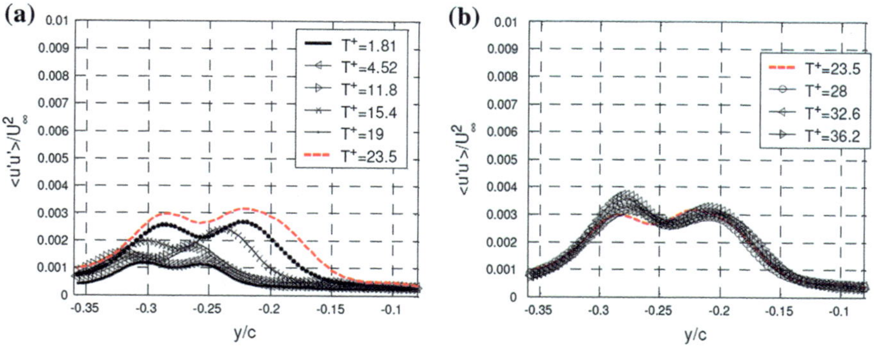

Fig. 10.15 Variation of $\langle u'u' \rangle / U_\infty^2$ as a function of T^+ at x/c = −0.85

(decreasing size) in wake size started distinctly from the upstream to downstream position (see Fig. 10.7).

The increase in the intensity of $\langle u'u' \rangle$ occurred from the upstream towards the downstream position. This was explained by the progressive increased in the size of the separation on the airfoil when the control jet has been removed. Figure 10.15 shows the evolution of the $\langle u'u' \rangle$ Reynolds stress profile as a function of y/c at x/c = −0.85. The profiles were typically double peaked and was symmetric at $T^+ = 1.81$. It progressively became asymmetric with one of the double peak, associated to the upper part of the wake, developing a higher magnitude.

The observation was reversed after $T^+ = 23.5$ when the peak related to the lower wake started to become higher. This continued to persist until the asymptotic state. There was clearly no convection of a parcel of high intensity region as seen in jet deployment process described in Fig. 10.7.

10.4.2.2 Evolution of $\langle u'v' \rangle / U_\infty^2$ Reynold Stress

The evolution of the shear stress distribution during deactivation was depicted in Fig. 10.16. As usual in wakes, the contour levels in the wake was characterized by a positive (upper part of the wake) and negative (lower part of the wake) bounded by a region of zero shear. The axis of the wake was determined by the line of zero shear at the interface of the positive and negative regions. Unlike the case of jet deployment, undulation in the line of zero shear was observed during the transient (see Fig. 10.9).

The magnitude of the negative peak (lower part of the wake) was smaller as compared to the positive peak during the initial transient ($T^+ = 1.81$–19). This signified that the entrainment was stronger at the upper part of the wake when the jet was deactivated. After $T^+ = 23.5$, the magnitude of the two peaks tended towards equality (Fig. 10.17b).

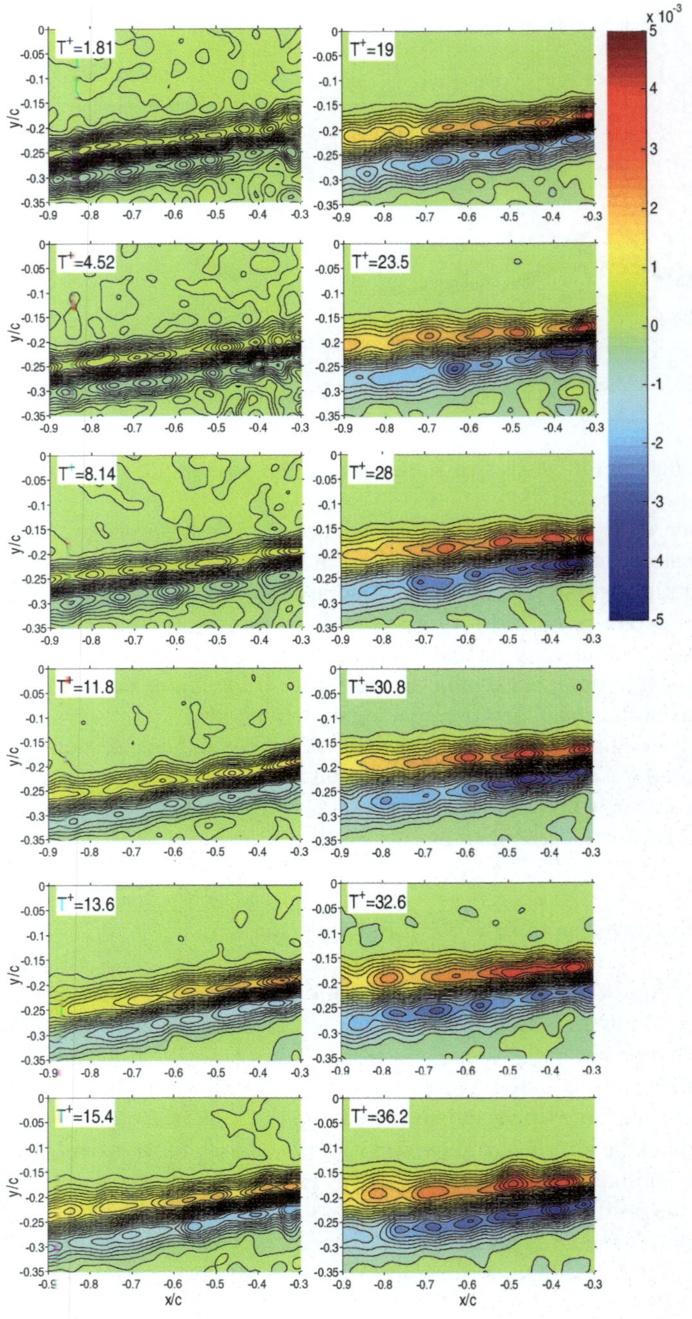

Fig. 10.16 Evolution of $\langle u'v' \rangle / U_\infty^2$ in the wake of the NACA0015 during jet deactivation

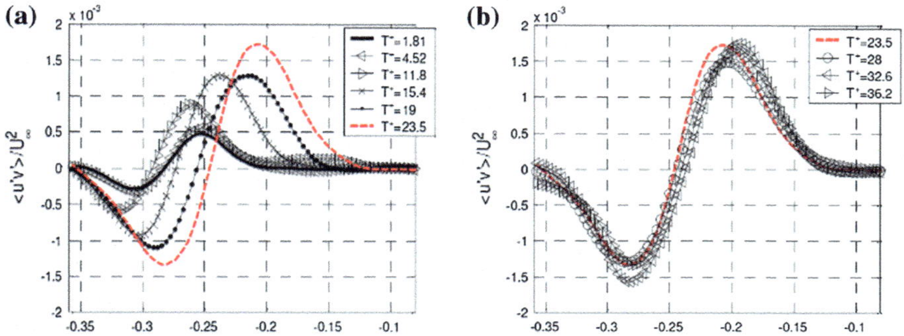

Fig. 10.17 Variation of $\langle u'v' \rangle / U_\infty^2$ as a function of T$^+$ at x/c $= -0.85$: **a** T$^+$ from 1.18 to 23.5; and **b** T$^+$ from 23.5 to 36.2

10.5 Conclusion

The transient dynamics of attachment and separation on a NACA0015 at high Reynolds number using impulsive fluidic vortex generators (FVG) deployment or deactivating was studied. The airfoil was orientated at an AoA of 11°. This condition corresponded to a stable and 2D separation occurring at 0.3 chord upstream of the trailing edge.

The flow was analyzed via conditional PIV measurements in the near wake of the airfoil. The conditional statistics were provided, such as ensemble averages of longitudinal velocity and longitudinal and shear components of the turbulent stresses. Details of the two process were described.

The attachment process corresponding to the activation of the FVG showed a strong transient effect associated with the passage of a starting vortex. On the other hand, when the FVGs were deactivated, the flow recovered its baseline (separated) situation, and a more progressive separation process was observed.

The data allowed for a better physical interpretation of the transient process. In particular, the increase of drag due to the transient passage of a starting vortex. The details of the turbulent fields provided could be used for validation of computations.

Acknowledgements The authors gratefully acknowledge Dr. J. Tensi for the wind tunnel tests and the financial support for FP7/2010-2013, MARS programme.

References

1. Abe H, Segawa T, Matsunuma T, Yoshida H (2000) Management of a longitudinal vortex for separation control. In: Symposium of the National Maritime Research Institute of Japan
2. Amitay M, Glezer A (2002) Controlled transients of flow reattachment over stalled airfoil. Int J Heat Fluid Flow 25

3. Darabi A, Wygnanski I (2004a) Active management of naturally separated flow over a solid surface. Part 1. The forced reattachment process. J Fluid Mech 510
4. Darabi A, Wygnanski I (2004b) Active management of naturally separated flow over a solid surface. Part 2. The separation process. J Fluid Mech 510
5. Debien S, Aubrun S, Mazelier N, Kourta A (2015) Active control separation process over a sharped edge ramp. In: TSFP9, Melbourne, Australia
6. Johnston JP, Compton DC (1992) Streamwise vortex production by pitched and skewed jets in a turbulent boundary layer. AIAA J 30(3)
7. Johnston JP, Nishi M (1990) Vortex generator jets—means for flow separation control. AIAA J 28(6)
8. Kuethe AM, Chow CY (1986) Foundation of aerodynamics: bases of aerodynamic design, 4th edn. Wiley
9. Mathis R, Lebedev A, Collin E, Delville J, Bonnet JP (2009) Experimental study of the transient forced turbulent separation and reattachment on a beveled trailing. Exp Fluids 46
10. Milanovic IM, Zaman KBMQ (2004) Fluidic dynamics of highly pitched and yawed jets in crossflow. AIAA J 42(5)
11. Pauley WP, Eaton JK (1987) Experimental study of the development of longitudinal vortex pairs embedded in a turbulent boundary layer. AIAA J 26(7)
12. Siauw WL (2008) Transient process of separation and attachment over a NACA0015 airfoil by fluidic vortex generators. PhD thesis, University of Poitiers
13. Siauw WL, Bonnet JP, Tensi J, Noack BR, Cordier L, Cattafesta L (2010) Characterization of typical response in the wake of a NACA0015 under action of fluidic vortex generators. Int J Heat Fluid Flows 31
14. Woo GTK, Glezer A (2013) Controlled transitory of stall on a pitching airfoil using pulsed actuation. Exp Fluids 54:1507. https://doi.org/10.1007/s00348-013-1507-5

Chapter 11
Numerical Investigation on Flow Control with Moving Surface Over a NACA0015 Airfoil

Li Li

Abstract Moving surface boundary-layer control (MSBC) appears a promising flow control technique to enhance the airfoils aerodynamics prominently with little energy input. In this paper numerical investigation on flow control with moving surface over a NACA0015 airfoil is conducted with a Reynolds Averaged Navier-stokes (RANS) based approach, in which the moving surface is implemented with mounting two rotating cylinders at leading edge and leeward side of the airfoil separately. In order to do that, a quasi-steady method has been proposed for modelling the rotating cylinder, and is used to study the effects of MSBC technique for the airfoil aerodynamic performance. The numerical results show that the airfoil aerodynamic performance would get worse with the installation of local moving surface devices; nevertheless, its performance can be greatly compensated with rotation of the cylinders. It is found that the boundary layer of airfoil can be suppressed with an effective rotating strategy such that the flow separation on the airfoil is greatly delayed and the aerodynamic performance is improved.

Keywords Moving surface · Active boundary-layer control · NACA0015 · Quasi-steady method · Computational fluid dynamics (CFD)

11.1 Introduction

Moving surface related boundary layer control (MSBC) techniques were paid widely attentions as effective flow control strategies in most recent years, which had brought out remarkable results for lift-enhancement of airfoils and drag-reduction of blunt bodies [1, 2] with advantages as simplicity of device, low energy consumption, and so on.

In this paper a quasi-steady method for modeling the moving surface is proposed, and is combined with a Reynolds Averaged Navier-Stokes (RANS) based approach to study the effects of MSBC for the airfoil aerodynamic characteristics. The basic

L. Li (✉)
AVIC Aeronautics Computing Technique Research Institute, Xi'an 710068, Shaanxi, China
e-mail: westlili@163.com

N. Qin et al. (eds.), *Advances in Effective Flow Separation Control for Aircraft Drag Reduction*, Computational Methods in Applied Sciences 52,
https://doi.org/10.1007/978-3-030-29688-9_11

idea is to introduce an equivalent boundary condition for the moving surface based on the theoretical analysis for the work principles of MSBC. With this method, the aerodynamic characteristics of a NACA0015 airfoil near stall with and without MSBC are numerically studied. The emphasis is to investigate the effect of different MSBC strategies (e.g. the setup position of the rotating cylinders, the rotating velocity, etc.) on the aerodynamic characteristics near stall. Compared with previous studies, as part work in MARS project [3], one of highlights the reader can find in this paper is that an analysis on effect of Reynolds stress in mechanism of flow control with moving surface is included, which lays a solid basis to reveal the mechanism of MSBC in depth.

11.2 Numerical Methods

11.2.1 An Equivalent Boundary Condition Model for MSBC

It is well know that in general cases non-slip condition has to be satisfied for a viscous wall boundary, which means that velocity of flow particles on such boundary will be enforced to be zero. In 1995, Bechert and his co-authors proposed to ease the non-slip condition, where the viscous wall was taken as a belt driven by shear stress of fluid. In this way, they found that velocity difference between mean flow and that near the wall was reduced, which resulted in reduction of skin friction [4]. The primary work principle of MSBC is similar to idea from Bechert. Nevertheless, in this case it is not necessary to take the viscous wall as a belt driven by free shear stress but that the wall can move at a constant speed clockwise or anti-clockwise by actively adding external forces. There are two types of typical moving surfaces, which can be illustrated in Fig. 11.1. One type is the whole driven moving surface, in which the actuator won't do changes in the aero shape of airfoil. The other type is the partially driven moving surface, in which local wall surface is motivated with installation of rotating cylinders. In the latter case the cylinders may bring changes in shape of airfoils, and therefore a passive control may be implemented if we keep these cylinders no movement. This paper mainly contributes to discuss the latter type of moving surface.

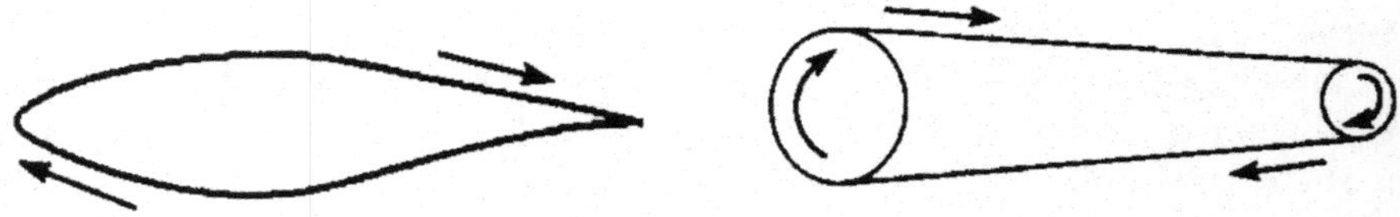

(a) whole-driven moving surface (b) partially-driven moving surface

Fig. 11.1 Typical classifications of moving surface

It can be notice that as a type of particular actuator moving surface will enforce fluid on wall surface to move along with the wall and results in strong shear effect. If the gap effect between the actuator of moving surface and the airfoil wall surface is ignored, the velocity of fluid on moving surface can be enforced to equal the rotating velocity of moving surface in term of the principle for relative motion, and then an equivalent boundary condition model for moving surface can be naturally established. When using the rotating cylinder as the actuator, it is easy to deduce that

$$\vec{\mathbf{u}}_{fluid} = \vec{\mathbf{u}}_{solid} = \vec{\boldsymbol{\omega}} \times (\vec{\mathbf{r}} - \vec{\mathbf{r}}_c) \tag{11.1}$$

where $\vec{\boldsymbol{\omega}}$ is the angular speed for rotation of the actuating cylinder; $\vec{\mathbf{r}}_c$ is the center for rotation; and $\vec{\mathbf{r}}$ is the position of arbitrary fluid particles on the moving surface. The above condition in essence is a type of slip boundary condition for a wall type surface, which can be called as the MSBC wall boundary condition. Following the idea, on the moving wall surface, the tangential velocity of fluid equals the moving velocity of the wall, while the normal velocity of fluid satisfies the non-penetration condition.

From the formulation from Eq. (11.1) for moving surface, one can obtain an equivalent solution with no direct simulation for movement of the moving surface such that a quasi-steady numerical modeling method for moving surface can be effectively established.

11.2.2 Flow Equation and Solution Strategies

The steady Reynolds averaged Euler/Navier-Stokes (RANS) equation in conserved form is taken as the fluid control equation, which is solved with a cell-centered finite volume method based on structured grid. Under the framework of curvilinear coordinates, the control equation in general form can be written as

$$\frac{\partial \mathrm{Q}}{\partial t} + \frac{\partial \mathrm{F}}{\partial \xi} + \frac{\partial \mathrm{G}}{\partial \eta} + \frac{\partial \mathrm{H}}{\partial \zeta} = \frac{1}{\mathrm{Re}}\left(\frac{\partial \mathrm{F}_v}{\partial \xi} + \frac{\partial \mathrm{G}_v}{\partial \eta} + \frac{\partial \mathrm{H}_v}{\partial \zeta}\right) \tag{11.2}$$

where $Q = (\rho, \rho u, \rho v, \rho w, \rho E)^{\mathrm{T}}$ is the conserved solution variable, F, G and H are the in-viscid flux vectors, F_v, G_v and H_v are the viscous flux vectors in three dimensions.

Equation (11.2) then can be solved numerically. In this paper a structured mesh based finite volume method is used, in which a Roe's flux difference splitting (FDS) scheme [5] based 3rd order MUSCL [6] extrapolation for the in-viscid flux item, while a standard central difference scheme is used for the viscous flux item, and an implicit approximation factorization (AF) algorithm for time integration. In calculations, full turbulence simulations are performed, and the Menter's κ-ω SST model [7] is used as the default turbulence model.

11.2.3 Geometry, Mesh and Computational Condition

A NACA0015 airfoil defined in [8] is taken as the baseline geometry, which has a chord length 0.35 m. The MSFC are implemented with installation of rotating cylinders at leading edge or leeward side of the airfoil. The front cylinder at leading edge is located at $C_1(0.05c, 0)$ with a radius $R_1 = 0.05c$, while the back cylinder at leeward side is located at $C_2(0.52c, 0.036c)$ with a radius $R_2 = 0.036c$. As noticed, since there are two different rotating cylinders, one can use them alone or together for boundary layer flow control. The cases include: single cylinder control at leading edge, single cylinder control at leeward side, and a combined control with cylinders at the leading edge and the leeward side both.

For the baseline case with a clean airfoil, the mesh is generated with a C-type topology, where the far field is with 50 times chord length, and a total number of volume elements about 90 thousands. The first level height from the wall is kept an order of magnitude around 10^{-6}c. For the control cases, all meshes are generated with the same strategy and based on the baseline case to ensure a similar size and a similar distribution for all meshes. As an example, in Fig. 11.2, illustration of the computational meshes used for the baseline case and the single leading-edge MSFC control case are shown.

The flow control effect of cylinder is investigated under the flow condition at free-stream inflow velocity $U_\infty = 40$ m/s, angle of attack $\alpha = 11^0$, and the Reynolds number $\mathrm{Re}/c = 1 \times 10^6$, which is a state near stall according to the experiment [8].

11.3 Numerical Results for Passive Flow Control Cases

In Table 11.1 the variation of total aerodynamic forces due to different strategies in passive flow control case is shown. It is found that both the front cylinder and the back cylinder will bring loss of the aerodynamic performance of the airfoil near stall. In particularly the cylinder mounted at the leading edge results in an obvious loss of aerodynamic performance. The lift coefficient gets smaller, and the drag coefficient gets larger. The main reason is that the cylinder mounted at the leading edge has brought a change in aerodynamic shape. From pictures in Fig. 11.3, one

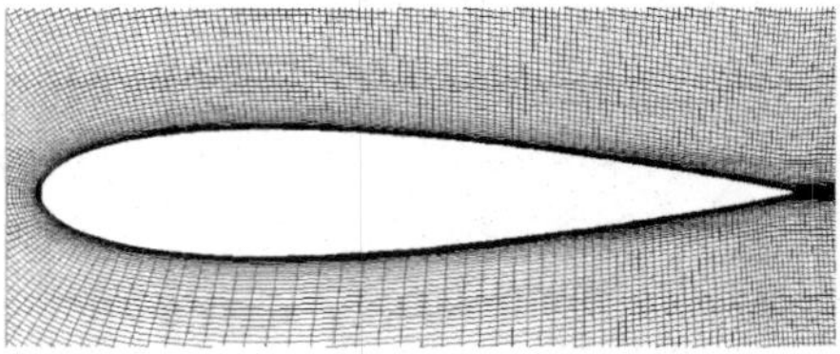

(a) clean NACA0015 airfoil

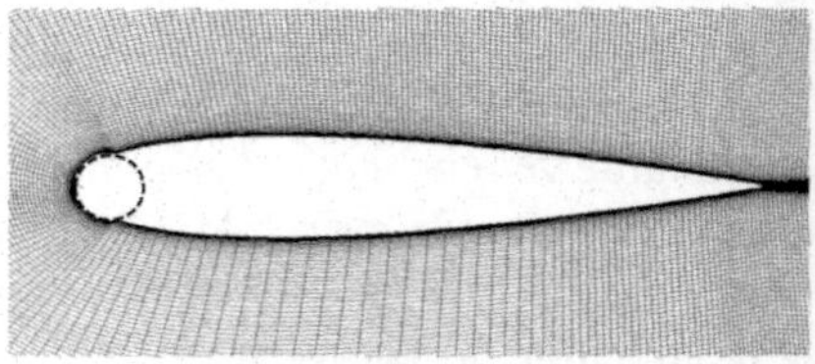

(b) NACA0015 airfoil with moving surface

Fig. 11.2 Illustration of computational meshes

Table 11.1 Variation of total aerodynamic forces in passive flow control case

Model/strategy for control	Cl	Cd
Baseline	1.025	0.0225
With single leading-edge cylinder	0.823	0.0432
With single leeward-side cylinder	1.001	0.0283
With a combined cylinders control	0.778	0.0480

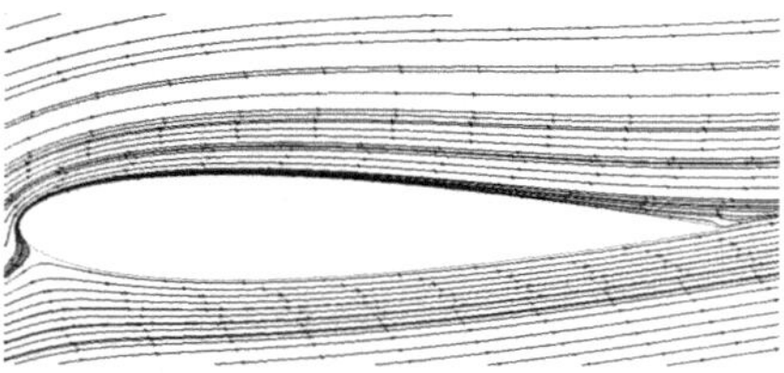

(a) baseline airfoil

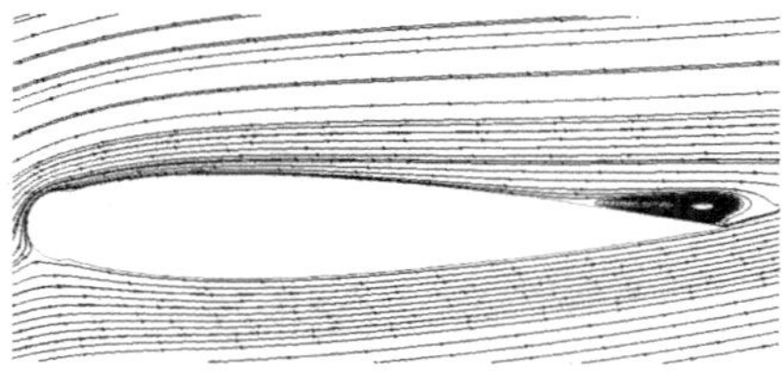

(b) single leading-edge cylinder control case

(c) single leeward-side cylinder control case

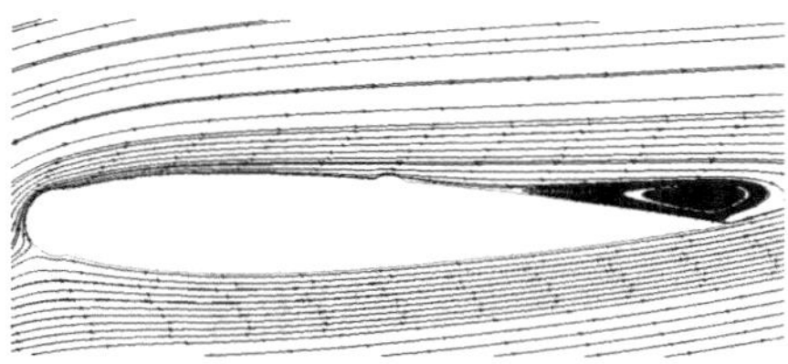

(d) a combined cylinder control case

Fig. 11.3 Comparison of streamlines with cylinder as the passive flow control device

can find that an obvious separation occurs with the cylinder mounted at leading edge, and the separation will be further amplified with installation of the cylinder at the leeward side. But for the baseline case and the single leeward side cylinder control case, there is no obvious separation. These observations are consistent with the MSFC experiment by Chinese partner in project [9]. In Fig. 11.4, comparisons of turbulence quantities (including turbulent energy $K \propto \langle u'u' \rangle$ and shear stress $\langle u'v' \rangle$) between different cases in wake at $x/c = 1.98$ is shown in order to observe the effect of Reynolds stress to flow control. From the figure, one can see that the Reynolds stresses in the wake will vary with the separation coincidently, i.e., the bigger the Reynolds stress in the wake the larger the separation. From this observation, it seems that actions to reduce the Reynolds stress in wake have to be taken in order to eliminate the separation.

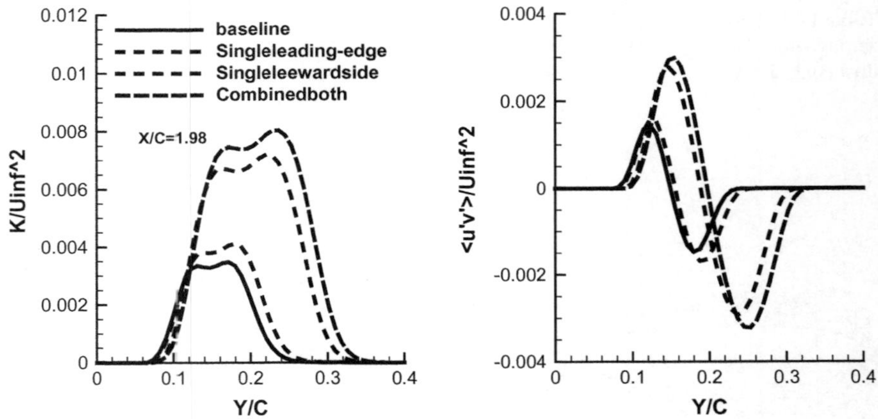

Fig. 11.4 Comparison of turbulence quantities with cylinder as the passive flow control devices

11.4 Numerical Results for Active Flow Control Cases

11.4.1 Effect of Rotating Strategy

When using the rotating cylinder as an active flow control device, one can rotate the cylinder in a clockwise direction, or in an anticlockwise direction. These different control strategies will result in different results. In order to study the effect of the rotating strategy, we have enforced the cylinder installed at the leading edge rotating in different directions with a constant speed at $V_1 = 40$ m/s. As shown in Fig. 11.5, when the cylinder rotating anticlockwise, separation due to shape change will get larger and result in a heavy loss of aerodynamic performance. However, when the cylinder is enforced to rotate in clockwise with the same constant speed, the separation due to shape change will get disappeared so that the aerodynamic performance can be effectively compensated as illustrated in Fig. 11.6. In Fig. 11.6, the control off case is corresponding to the passive control case, while the control on case to the active control with clockwise rotating at a constant speed. Again, in Fig. 11.7, the

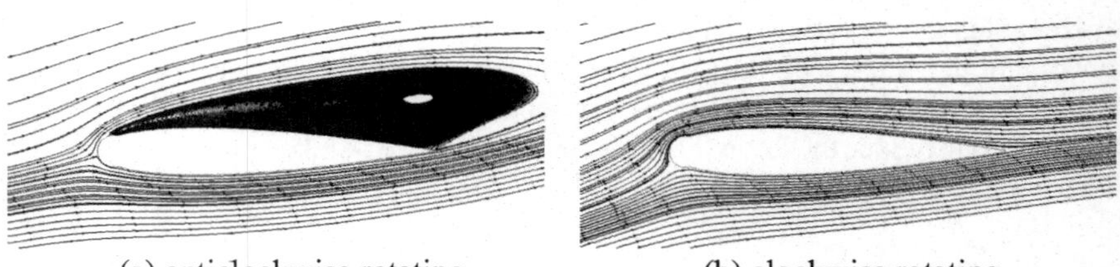

(a) anticlockwise rotating (b) clockwise rotating

Fig. 11.5 Effect of rotating strategies for flow past airfoil

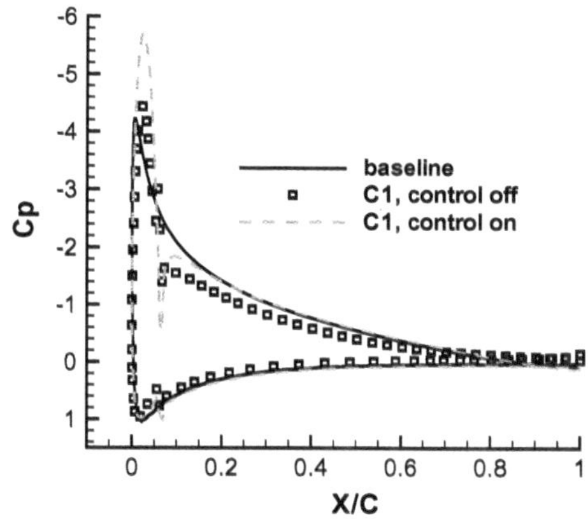

Fig. 11.6 Comparison of pressure coefficient distribution with different control strategies

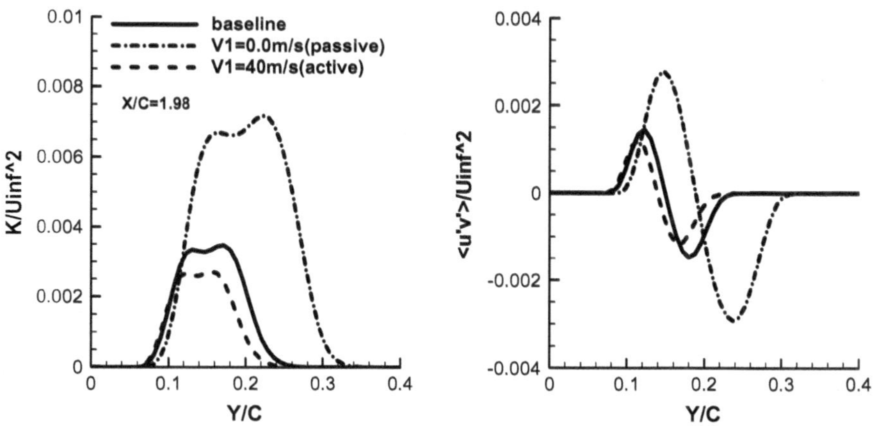

Fig. 11.7 Comparison of turbulent quantities (K, and ⟨u′v′⟩) between with and without control

turbulence qualities (i.e., K, and ⟨u′v′⟩) in wake at $x/c = 1.98$ are compared between the active control case with clockwise rotating between the passive control case as well as the baseline case. It is concluded that with external energy input by rotating, Reynolds stresses in wake have been reduced effectively, which is helpful to eliminate the separation. The reason may ascribe that external momentums with rotating have speed up the wake to enforce the turbulence smoother.

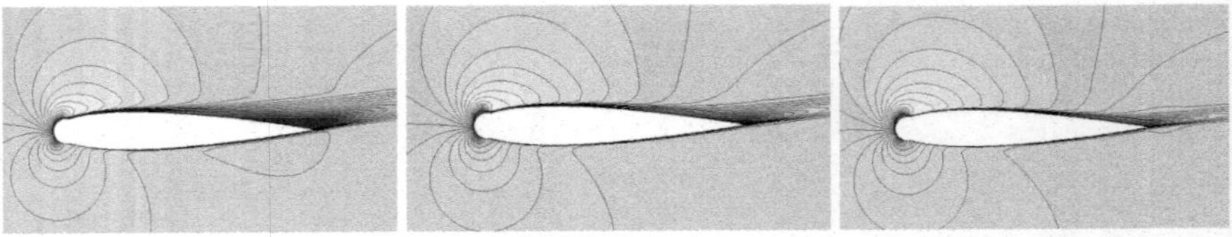

(a) V_1=0 (b) V_1=20m/s (c) V_1=40m/s

Fig. 11.8 Effect of rotating speed for flow past airfoil

11.4.2 *Effect of Moving Speed*

In order to investigate effect of moving speed with MSFC, four cases have been considered for the single leading-edge cylinder control case rotating in clockwise at different rotating speed $V_1/U_\infty = 0.5, 1, 2$. In Fig. 11.8, Mach number contour of the flow past airfoil are shown with different rotating speed. It can be found that with increase of the rotating speed, a stronger energy will be introduced into the flow field such that the separation at the trailing edge will be greatly suppressed as expected. In our case, when the rotating speed equals 20 m/s (i.e. $V_1/U_\infty = 0.5$), the separation due to shape charge cannot be removed completely. From this observation it can be concluded that the moving speed is one of the determining factors for MSFC in a practise use.

11.4.3 *Results for the Combined Control Case*

In Fig. 11.9 effect of rotating speed in the combined control case is investigated, in which the cylinders rotate in clockwise at a constant speed $V_1 = V_2 = 40$ m/s and $V_1 = V_2 = 80$ m/s, separately. Compared with the passive flow control case, it is found that with MSFC the separation at leading edge is disappeared completely, and the separation at trailing edge, moreover, gets smaller. It is also shown that the faster the cylinders rotate, the smaller the separation bubble becomes. Nevertheless, when compared with results for the single leading edge cylinder control case, it is

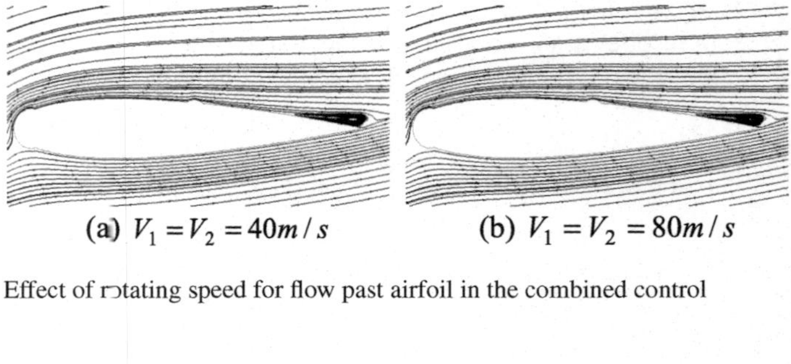

(a) $V_1 = V_2 = 40 m/s$ (b) $V_1 = V_2 = 80 m/s$

Fig. 11.9 Effect of rotating speed for flow past airfoil in the combined control concluded that more external energy will have to be added for the combined control case. In fact, as pointed out in simulation for the passive cases, the cylinder installed at the leeward side has amplified the separation due to shape change with cylinder at the leading edge.

11.5 Further Discussions on Control Mechanism

As shown in above (in is suppressed !), when the MSFC is used as a passive flow control devices, it may bring changes in aerodynamic shape such that a separation at trailing edge will be observed near the stall. The difference between with and without control is that the Reynolds stress in wake gets larger in a passive control case. However, if the MSFC is used as an active flow control device, it can be effectively used to eliminate the separation. The difference between with and without control is that Reynolds stress in wake gets smaller in an active control case. Therefore, one can conclude that MSFC will result in Reynolds stress in wake smaller by adding external momentums. In Fig. 11.10, comparisons of skin friction on the upper wall surface between the baseline case and the passive control case as well as the active control case are shown. It can be found that, with the passive control, the skin friction is smaller, while, with the active control, the skin friction is larger. If we look at the definition of skin friction, this observation means that MSFC has

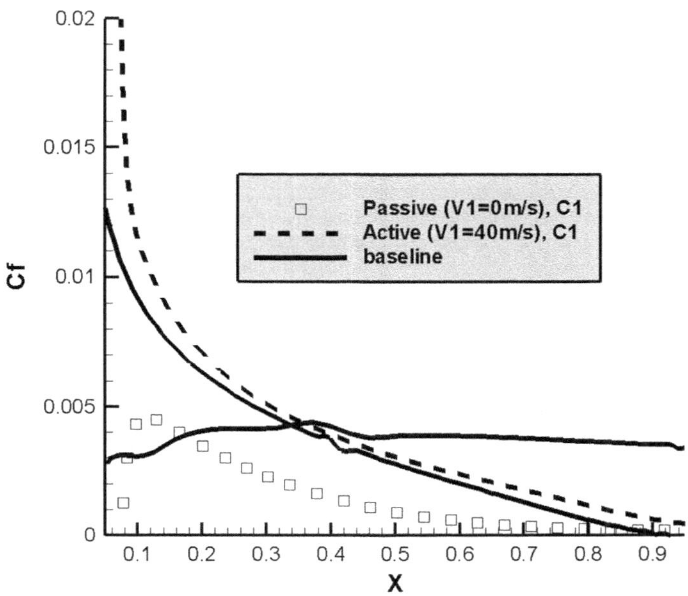

Fig. 11.10 Comparisons of skin friction on wall surface between with and without control

already increased the Reynolds stress on wall surface in the active control case. It is concluded that Reynolds stresses have played different roles during the procedure for flow control. The numerical results show that MSFC will increase the Reynolds stress in boundary layer but decrease the Reynolds stress in space. So if one wants to control the separation, one should take means to increase the Reynolds stress in the boundary layer but decrease the Reynolds stress in space.

11.6 Conclusions

A quasi-steady method based on an equivalent boundary condition is proposed for modelling moving surface, and is used for numerical study of effect on flow control with moving surface over a NACA0015 airfoil near stall. The moving surface is physically implemented with installation of two rotating cylinders at the leading edge and the leeward side of the airfoil. The relationship between Reynolds stress and the flow control mechanism of MSFC is discussed. Based on numerical results, some conclusions can be done as followed: first, MSFC can be used as a passive flow control device. In this case, there is some loss in aerodynamic performance for the airfoil due to the change in shape; secondly, MSFC can be also used as an active flow control device. In this case, aerodynamic performance of the airfoil can be compensated through clockwise moving to introduce external momentums into the flow field. The moving strategy and the moving speed are two determining factors in a practical use. When talking about relationship between Reynolds stress and the control mechanism of MSFC, it is found that there are different roles for Reynolds tress in boundary layer and in space, where the active MSFC results in a decrease of Reynolds stress in wake but creates an increase of Reynolds stress near the wall so that the separation in the wake is eliminated, and the aerodynamic performance is improved.

References

1. Modi VJ, Deshpande VS (2000) A Joukowski airfoil with momentum injection. In: AIAA atmospheric flight mechanics conference, Dnever, CO, 14–17 Aug 2000
2. Zhiyong LU (2000) Separation control for a strake-wing by rotating cone placed near the leading edge. In: 38th AIAA aerospace sciences meeting and exhibit, Reno, NV, 10–13 Jan 2000
3. MARS project website. http://www.cimne.com/mars/
4. Bechert DW, Brusek M, Hage W (1995) Drag reduction with the slip wall. AIAA J 34(5):1072–1074
5. Roe RL (1981) Approximate Riemann solvers, parameter vectors and difference schemes. J Comput Phys 43(2):357–372
6. Leer BV (1976) Towards the ultimate conservative difference scheme—V: a second order sequel to godunov's method. J Comput Phys 32(1):101–136
7. Menter R (1994) Two-equation eddy-viscosity turbulence models for engineering applications. AIAA J 32(8):1598–1605

8. Siauw WL, Bonnet JP, Tensi J (2009) Physics of separated flow over a NACA 0015 airfoil and detection of flow separation. In: 47th AIAA aerospace sciences meeting including the new horizons forum and aerospace exposition, Orlando, Florida, 5–8 Jan 2009
9. Liu GZ, Sun L (2013) PIV test research of NACA0015 airfoil and its moving surface flow control. In: Proceeding for MARS M36 meeting from Chinese partners, 31 Oct–1 Nov 2013 (Edited by Chinese Academy Establishment)

Chapter 12
Advanced Measurement Techniques Used in the Joined Measurement Campaign

Reinhard Geisler, D. Schanz and J. Agocs

Abstract A joint measurement campaign has been performed by the MARS project partners NUAA, CNRS (Poitiers) and DLR. The objective was to apply advanced measurement techniques to the actuator types which turned out to be most promising in the previous investigations. The selected devices were synthetic jet actuators from NUAA and plasma actuators from CNRS. Both actuator types were investigated in a backward facing step (BFS) geometry. Additional measurements took place using the synthetic jet actuator in a NACA-0015 profile from NUAA. For the measurements, a tomographic PIV and a long-distance microscopic stereo PIV measurement system—both capable to provide time-resolved data—have been set up by DLR in the one meter wind tunnel (1MG) at DLR Göttingen. The successful completion of this intercontinental experimental cooperation provides results which are a valuable data base for comparison with and validation of numerical methods.

Keywords PIV · Time-resolved PIV · Tomographic PIV

12.1 Introduction

The DLR department of experimental methods contributed to the project MARS with advanced measurement techniques applied in a four-week joint measurement campaign and the corresponding data evaluation. The campaign took place in July/August 2012 in the one meter wind tunnel Göttingen (1MG). DLR provided the setup and operation of long-distance microscopic stereo PIV [1] and tomographic PIV [2], both time resolved, for the application to models and actuators of the project partners. Due to their complexity and long evaluation time, these measurement techniques have been limited to the most promising actuator types. The tests can provide a direct comparison of actuators from different project partners under the very same experimental conditions using high-performance measurement techniques.

R. Geisler (✉) · D. Schanz · J. Agocs
Institute of Aerodynamics and Flow Technology, German Aerospace Center (DLR), Göttingen, Germany
e-mail: Reinhard.Geisler@dlr.de

N. Qin et al. (eds.), *Advances in Effective Flow Separation Control for Aircraft Drag Reduction*, Computational Methods in Applied Sciences 52,
https://doi.org/10.1007/978-3-030-29688-9_12

The original scope of these tests has been extended and the focus was shifted from the well-established stereo PIV (2-dimensional, 3-component) to the more advanced, but also more complex and evaluation time consuming tomographic PIV. This is beneficial since tomo-PIV can instantaneously capture 3-dimensional, 3-component velocity vector fields of a flow enabling access to the complete 3-dimensional velocity gradient tensor as well as the analysis of spatial structures inside the flow.

Also, additional high resolution mono PIV (2-dimensional, 2-component) measurements have been performed to characterise the incoming flow as well as the actuators.

The models used for the test are the backward facing step (BFS) geometry defined by CNRS and the NACA-0015 air foil profile from NUAA. Active flow control was implemented by the plasma actuator from CNRS and the synthetic jet actuator from NUAA.

12.2 Experimental Setup

12.2.1 Models

In the first two weeks of the four week measurement campaign a backward facing step (BFS) geometry defined by CNRS was installed together with an adapted tomo-PIV setup (Fig. 12.2). The BFS has a step height of 30 mm, a span of 1300 mm and a total stream wise length of 700 mm including an initial ramp of 300 length and was mounted on top of a 1300 mm $\times$ 1500 mm base glass plate with an elliptical leading edge. The flow was tripped to establish a well-defined turbulent boundary layer (TBL).

Following the BFS experiments, the last two weeks of the joint campaign were devoted to the NACA-0015 measurements. NUAA provided the model with a chord length of 350 mm and a span protruding the test section. Circular end plates were installed to ensure two dimensional flow conditions (Fig. 12.4). 5 mm behind the leading edge, the flow was tripped to force a reproducible TBL development. The model was set to a geometric angle of attack of 11°.

12.2.2 Actuators

Two actuators types have been tested with the BFS model. The plasma actuator provided by CNRS consists of two 10 mm stripe electrodes glued in close connection to the surface near the step position. A high voltage pulsed with a frequency of 83 Hz and a duty cycle of 50% is applied to the electrodes (Fig. 12.1, left). The synthetic jet (SJ) actuator provided by NUAA is a slit of 2 mm height located in the step and facing in flow direction (Fig. 12.1, centre). Its position is either at the top of the step

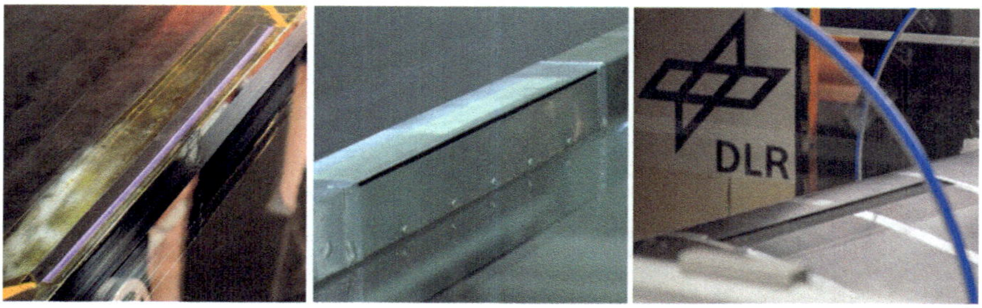

Fig. 12.1 Plasma actuator (left), synthetic jet actuator in upper position in BFS (centre) and synthetic jet actuator in NACA-0015 (right)

or at its bottom close to the base plate. The slit is integrated into a square profile cavity connected by tubes to a motor-driven piston. The piston is driven at several selected frequencies, thus giving a sinusoidal variation of the air volume which leads to a periodic blow from the slit and suction back into it.

The NACA-0015 model is equipped with an integrated synthetic jet actuator similar to the one installed in the BFS model. A step of 3 mm is embedded into the suction side of the profile (Fig. 12.1, right). While the upper millimetre of the step is solid, the lower two millimetres form a slit leading to a cavity inside the model. Tubes connect this cavity to the driving oscillating piston.

Details about the actuators can also be found in the chapters of the corresponding project partners CNRS and NUAA.

12.2.3 Metrology

To characterise the incoming flow in the BFS experiments and gain a first estimate of the actuator influence, a high-resolution mono PIV setup is used. A dual-cavity BigSky CFR400 Laser emits a sequence of double laser pulses with an energy of 2×200 mJ at a temporal separation of $dt = 150$ µs and a repetition rate of 3.3 Hz. The laser beam is used to form a stream wise light sheet of 60 mm height perpendicular to the base plate (Fig. 12.2). DEHS (Di-Ethyl-Hexyl-Sebacat) droplets of about 1 µm diameter are dispensed into the flow as tracer particles. The illuminated particles are recorded at a resolution of 4008×1000 pixels by a PCO.4000 double shutter camera equipped with a Nikkor 1.4/85 lens operated at F#2.8. Data evaluation is done using a multi-grid cross-correlation algorithm with image deformation. The final correlation window size is 16×16 pixels (0.8 mm $\times$ 0.8 mm) and the vector spacing 5 pixels (0.25 mm).

For the tomographic BFS measurements, a measurement volume of about 50 mm $\times$ 50 mm $\times$ 12 mm centred in the shear layer 50 mm behind the step is selected. A dual-cavity LEE LDP-200MQG high repetition rate laser is used to illuminate a volume light sheet perpendicular to the flow at a repetition rate of 1 kHz with a

Fig. 12.2 BFS with laser light sheet and PCO.4000 camera for mono PIV

double-pulse separation of $dt = 60$ μs and an average power of 2 × 13 W. Two parallel mirrors bounce the light back and forth throughout the volume to enhance the light sheet intensity (Fig. 12.3). An arrangement of 4 high-speed video cameras (2 × Photron APX-RS with Micro Nikkor 105/2.8 and 2 × Photron SA-1 with Zeiss Macro-Planar T100/2.8, all operated at F#11) with a resolution of 1024 × 1024 pixels record the DEHS-particles illuminated by the light sheet at a frame rate of 2 kHz (frame-straddling mode). The recorded images are processed by a SMART (simultaneous multiplicative algebraic reconstruction) Algorithm from DLR with OTF (optical transfer function [3]) compensation to reconstruct the volume intensity. Three-dimensional multi-grid cross-correlation with volume deformation (LaVision DaVis) down to a correlation volume of 48 × 48 × 48 voxels (3 mm × 3 mm × 3 mm) and a vector spacing of 12 voxels (0.75 mm) is applied to the reconstructed volumes to determine the velocity vectors.

The field of view for the NACA-0015 profile stereo-PIV measurement is located around 58% of the chord length. To gain a sound insight into the TBL flow, an image

Fig. 12.3 BFS with volume light sheet for tomographic PIV

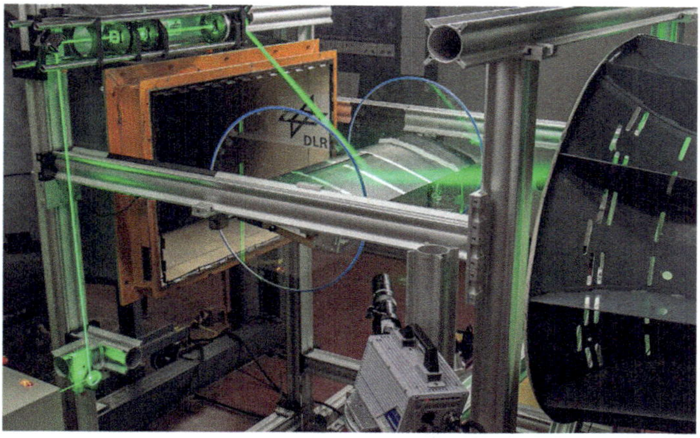

Fig. 12.4 NACA-0015 with laser light sheet and Photron SA1 camera with long-distance microscope Infinity K2 for stereo-PIV

scale of 1:1 has to be realised. Since the cameras have to be located outside the flow, special long-distance microscope lenses (Infinity K2) are used (Fig. 12.4). With the Photron SA-1 cameras set to a reduced resolution of 1024 × 496 pixels, a field of view of 25 mm × 10 mm is achieved at a recording rate of up to 12,000 frames per second. A laser light sheet is generated using double pulses ($dt = 10\ \mu s$) from the LEE LDP-200MQG Laser at a repetition rate of 3 kHz or 6 kHz with a single pulse energy of 10 mJ or 6 mJ, respectively. Thus, the temporal resolution of the measured velocity vector fields is 3 and 6 kHz, where the former is used for statistics, whereas the latter gives a better temporal resolution for the investigation of the development of flow structures. Again, data evaluation is done using multi-grid cross-correlation with image deformation at a final correlation window size of 24 × 24 pixels (0.47 mm × 0.47 mm) and a vector spacing of 6 pixels (0.12 mm).

12.2.4 Flow Conditions

Organised and coordinated by DLR, all experiments of the joint measurement campaign have been performed in the one meter wind tunnel (1MG) at DLR Göttingen (Fig. 12.5). This facility is an open jet test section, closed-circuit (Göttingen type) wind tunnel with a rectangular nozzle of 1 m × 0.7 m and a free flow turbulence level below 0.5%.

In the BFS experiments the free flow velocity is set to about 10 m/s. The flow is tripped at the leading edge of the base plate as well as in front of the slope of the BFS model (Fig. 12.6, left). As a result, an incoming TBL flow of 15 mm thickness is established on top of the model. The velocity profile derived from the high resolution mono PIV measurement is depicted in Fig. 12.6 (right).

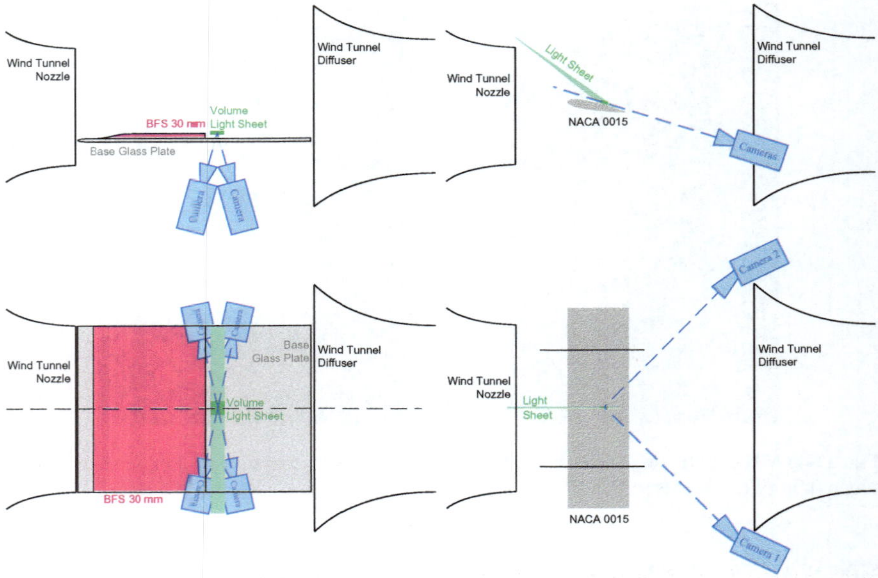

Fig. 12.5 Sketch of wind tunnel setup for BFS (left) and NACA-0015 (right) in side view (top) and top view (bottom)

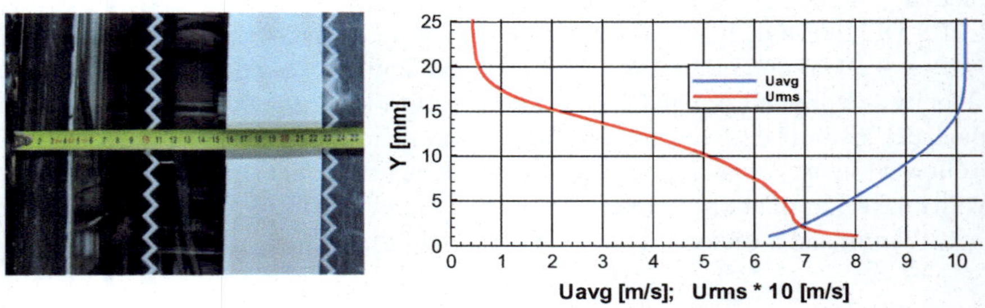

Fig. 12.6 Incoming flow of BFS tripped at leading edge ($x = 0$ mm), before ($x = 50$ mm) and on the ramp ($x = 230$ mm); measured incoming TBL flow characteristics (right)

The NACA-0015 measurements took place in the open jet test section at a free flow velocity of 40 m/s. The flow is tripped near the leading edge of the profile and develops a TBL thickness in the measurement area of about 10 mm.

12.3 Results

12.3.1 BFS Configurations

From the mono PIV measurement, the incoming turbulent boundary layer properties (Fig. 12.6, right) and the position of the reattachment point (Fig. 12.7) are determined. Starting with the step flow without actuation, the reattachment point is increasingly forced closer towards the step with the plasma actuator, the synthetic jet in its lower position at 9 and 14 Hz and the synthetic jet in the upper position at 9 and 14 Hz (Fig. 12.7a–f, respectively). The turbulent boundary layer thickness behind the step

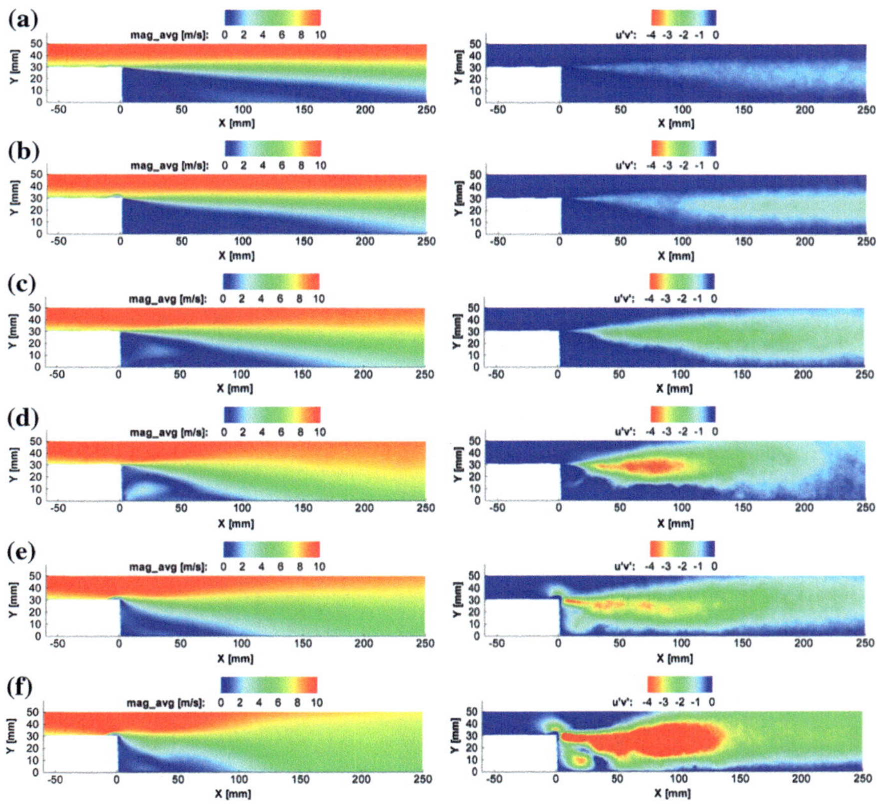

Fig. 12.7 Average velocity magnitude (left) and Reynolds stress u′v′ (right) of the BFS flow. **a** Without actuator, reattachment point: x = 215 mm; **b** with plasma actuator, reattachment point: x = 164 mm; **c** with synthetic jet (lower position, 9 Hz), reattachment point: x = 118 mm; **d** with synthetic jet (lower position, 14 Hz), reattachment point: x = 88 mm; **e** with synthetic jet (upper position, 9 Hz), reattachment point: x = 68 mm; **f** with synthetic jet (upper position, 14 Hz), reattachment point: x = 50 mm

grows correspondingly and increased negative Reynolds stresses $u'v'$ (Q2- and Q4-events) can be observed visualizing the production of turbulence (Fig. 12.7, right).

From the tomographic PIV results, the 3-dimensional distribution of Reynolds stresses can be calculated (Fig. 12.8). For the plain step flow as well as the plasma actuator no difference to the corresponding 2D-recordings is apparent. In contrast, the synthetic jet actuator shows an asymmetry in z-direction. This is not unexpected because the cavity behind the actuator slit is supplied single-sided (negative z-direction) from the piston while its opposite side is closed. Thus, a pressure gradient along the slit induces the asymmetry. However, it has to be kept in mind that due to the time-resolved nature of the recordings, convergence of the Reynolds stresses is still limited—even for the used number of more than 3000 velocity vector fields.

In the time-resolved snapshots the turbulent structures generated by the actuators can be observed and tracked in time. Without actuation, no predominant period is visible (sample snapshot Fig. 12.9a). Only small fluctuations in the structures are visible within one cycle of the plasma actuator (Fig. 12.9b–e). In contrast, the strong impact of the synthetic jet dominates the flow in the shear layer both for the lower (Fig. 12.10, left) and for the upper (Fig. 12.10, right) position of the actuator slit.

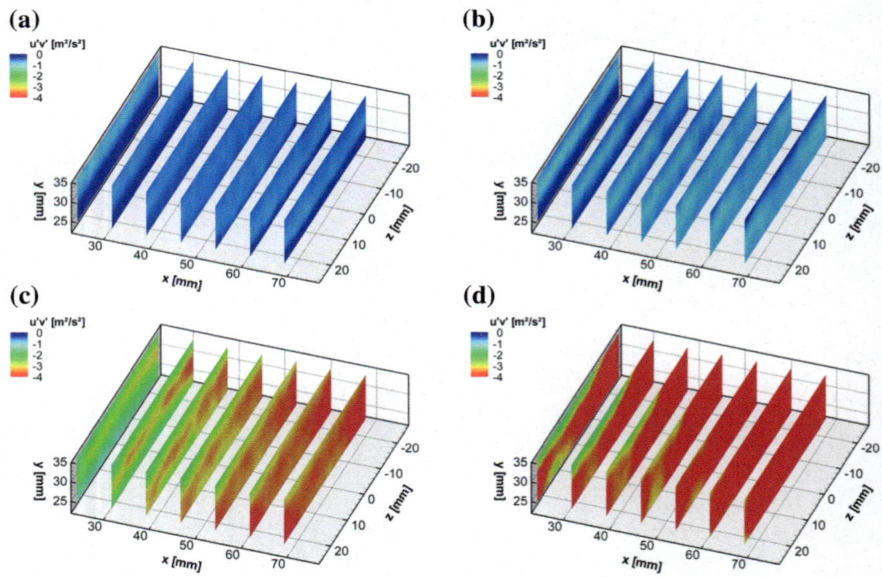

Fig. 12.8 Flow in the BFS shear layer depicted as selected planes colour-coded with the Reynolds stress $u'v'$; without actuator (**a**), with plasma actuator (**b**) and with the synthetic jet actuator at 14 Hz in lower (**c**) and upper (**d**) position

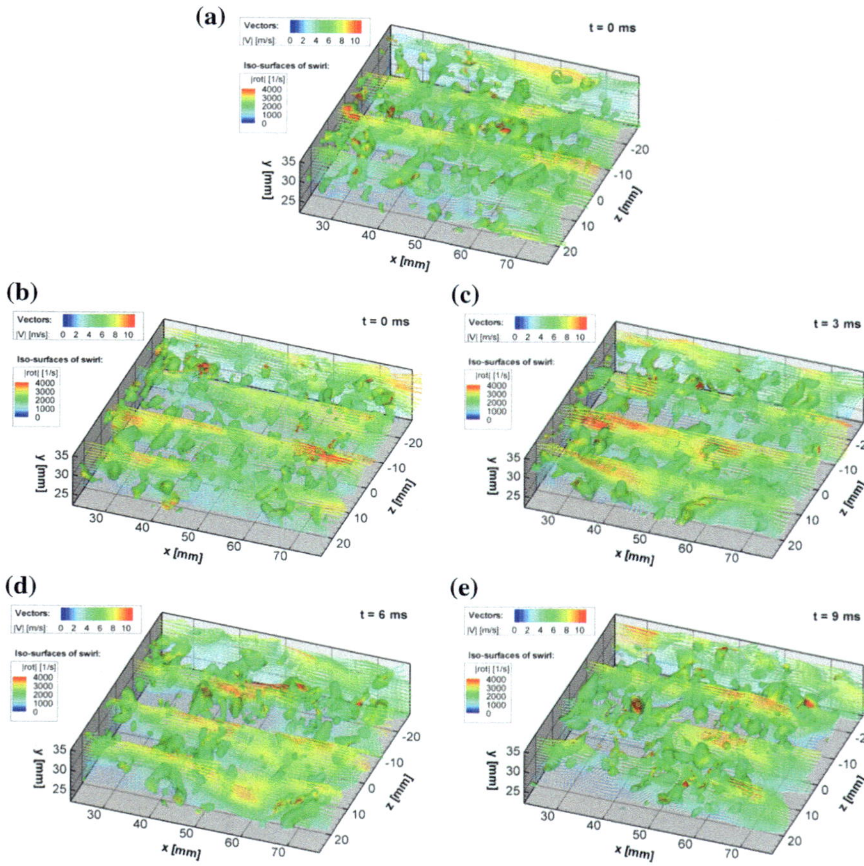

Fig. 12.9 Flow in the BFS shear layer depicted as four selected velocity vector planes colour-coded with the velocity magnitude and iso-surfaces of the swirl colour-coded with the vorticity; without actuator (**a**), one cycle of the plasma actuator with relative phases 0°, 90°, 180° and 270° (**b**–**e**)

12.3.2 NACA-0015 Configurations

The synthetic jet configuration of the NACA-0015 profile can be regarded as a combination of backward facing step and synthetic jet actuator. Hence, the actuator influence is compared to the plain NACA-0015 profile as well as to the profile with a passive BFS of the actuator geometry. The BFS alone already increases RMS and Reynolds stress $u'v'$ significantly and boundary layer thickness slightly. An even more pronounced effect can be observed with the active synthetic jet (Figs. 12.11 and 12.12). Although the field of view is located in a considerable distance to the actuator (~40% chord), the periodicity of the actuation is clearly visible (Fig. 12.13).

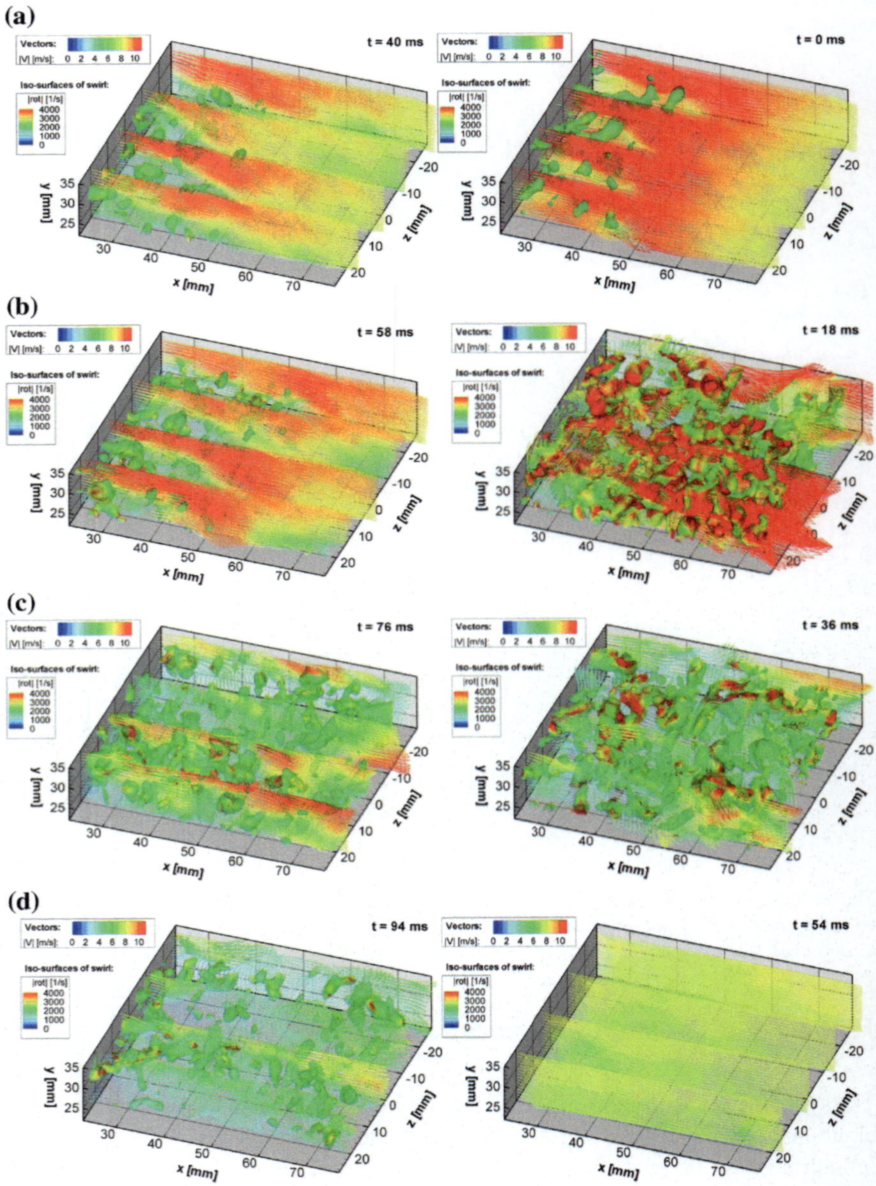

Fig. 12.10 Flow in the BFS shear layer depicted as four selected velocity vector planes colour-coded with the velocity magnitude and iso-surfaces of the swirl colour-coded with the vorticity; one cycle of the synthetic jet actuator at 14 Hz in lower (left) and upper (right) position with relative phases 0°, 90°, 180° and 270° (**a**–**d**, respectively)

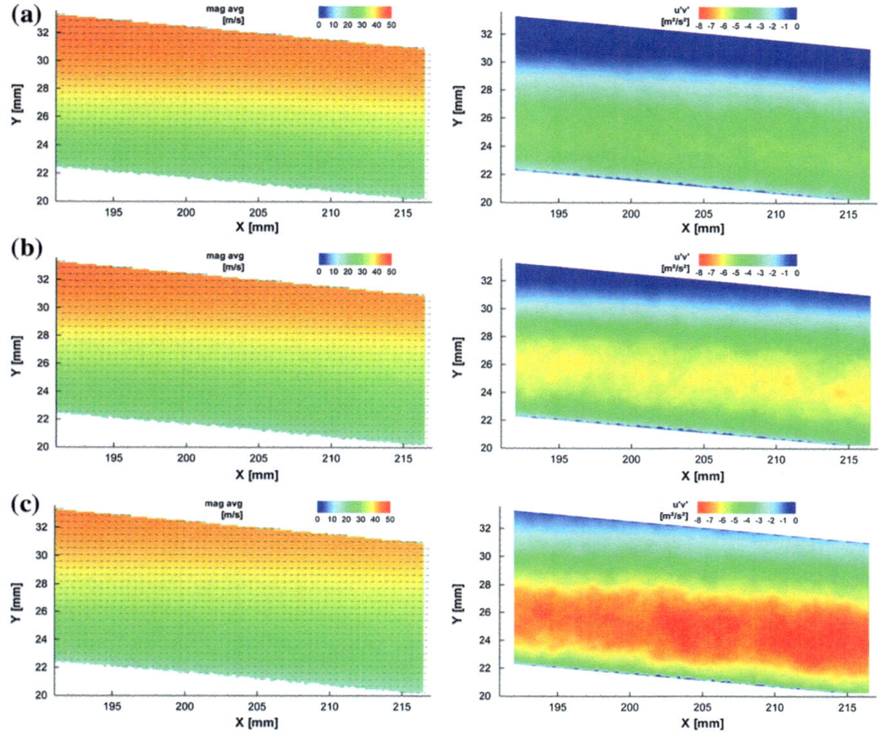

Fig. 12.11 Average velocity magnitude (left) and Reynolds stress u′v′ (right) of the TBL on NACA-0015: plain NACA without actuator (**a**), NACA with passive step (**b**), and NACA with synthetic jet at 45 Hz (**c**)

Furthermore, as in the BFS case, an asymmetry in z-direction occurs due to the asymmetric pressure supply system.

12.4 Conclusions

A joint four week measurement campaign coordinated by DLR has successfully been performed by the project partners NUAA, CNRS (Poitiers) and DLR. Careful preparation and intensive communication between the partners prior to the campaign ensured a smooth integration of models and measurement technique and finally a prosperous completion of this intercontinental campaign within the limited wind tunnel time.

The results of the applied advanced measurement techniques show a striking impact on the flow especially of the synthetic jet actuator. The gained time-resolved

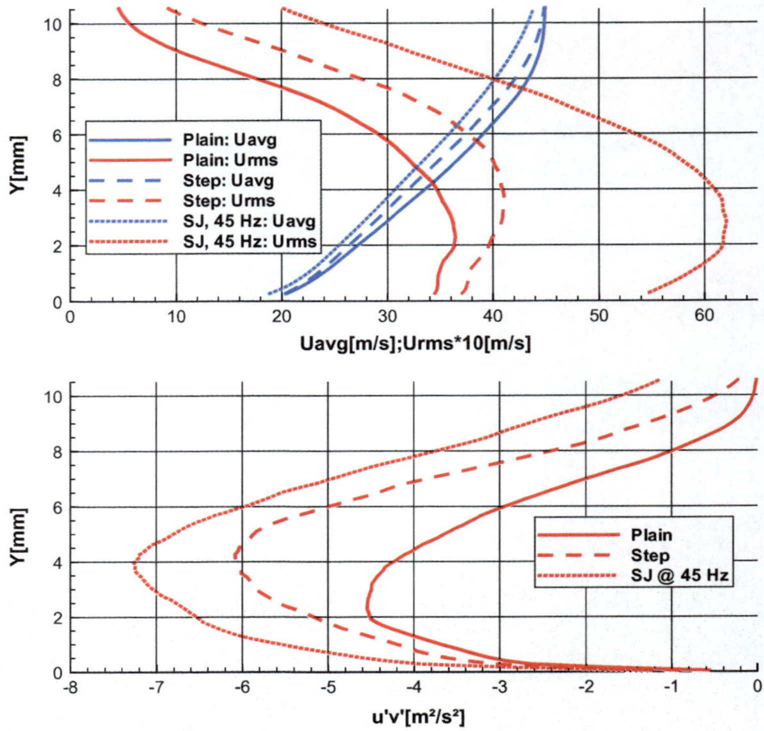

Fig. 12.12 Velocity u (top) and Reynolds stress u′v′ (bottom) of the TBL on NACA-0015 at 56% chord; plain profile (solid), profile with passive step (dashed) and with synthetic jet at 45 Hz (dotted)

velocity vector field data give a valuable data base for comparisons to computational fluid mechanics (CFD) which is made available to all project partners.

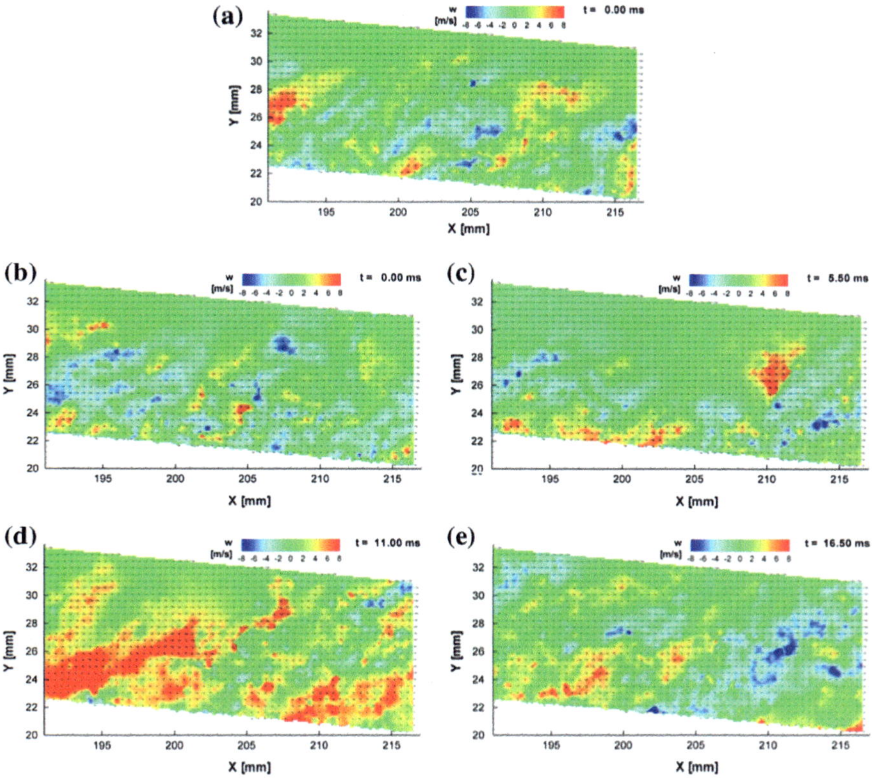

Fig. 12.13 Velocity w of the TBL on NACA-0015; plain profile (**a**), one cycle of the synthetic jet at 45 Hz with relative phases 0°, 90°, 180° and 270° (**b**–**e**, respectively)

References

1. Willert CE, Gharib M (1991) Digital particle image velocimetry. Exp Fluids 10:181–193
2. Elsinga GE, Scarano F, Wieneke B, van Oudheusden BW (2006) Tomographic particle image velocimetry. Exp Fluids 41(6):933–947
3. Schanz D, Gesemann S, Schröder A, Wieneke B, Novara M (2013) Non-uniform optical transfer function in particle imaging: calibration and application to tomographic reconstruction. Meas Sci Technol 24:024009

Chapter 13
Application of Active Flow Control on a High Lift Wing-Body Combination: An Experimental Investigation

Guozheng Liu, Jun Dong and Nan Sun

Abstract Loads and shear stress measurements, as well as PIV flow field visualization, of a high lift wing-body half model using jets/synthetic jets flow control methods are conducted in a low speed wind tunnel. Results show that the maximum lift coefficient of the taking-off/landing configuration can be increased by 40.8%, and the stall angle of attack can be increased by 4.34° with jets flow control devices on wing; while the maximum lift coefficient can be increased by 19.0% with jets flow control devices on flap, and 0.87% with synthetic jets on flap. Flow control with the energy cost goes the similar way, which means that the flap synthetic jets flow control costs least. The shear stress is increased with the separation effectively controlled, accordingly the lift coefficient is increased.

Keywords High lift wing-body configuration · Flow control · Jets · Synthetic jets

13.1 Introduction

The purpose of aircraft design is to obtain best aerodynamic performances which meet the design requirements by choosing the optimal combination of aerodynamic measures. Among all the requirements, drag reduction and lift promotion are the most basic and important ones. For example, lift promotion can increase the aircraft payload and shorten the taxi distance during taking-off and landing procedure. As Wimpress estimated, 5% lift promotion can increase payloads by 20% at taking-off state, while 65% at landing state.

Active flow control technology means any mechanism or method which can make boundary layer flow behavior different to natural development, including laminar transition control, boundary layer separation control, vortex control and etc. with the control purpose of increasing lift, reducing drag, as well as suppress noise and etc. Therefore, it is one of the effective methods to increase lift.

G. Liu · J. Dong (✉) · N. Sun
AVIC Aerodynamics Research Institute (AVIC ARI), P.O.B. 701, 110034 Shenyang, China
e-mail: dongcaria@hotmail.com

N. Qin et al. (eds.), *Advances in Effective Flow Separation Control for Aircraft Drag Reduction*, Computational Methods in Applied Sciences 52,
https://doi.org/10.1007/978-3-030-29688-9_13

Loads and shear stress measurements, as well as PIV flow field visualization, of a high lift wing-body half model using jets/synthetic jets flow control methods are conducted in AVIC ARI FL-8 low speed wind tunnel to obtain the separation control and lift promotion performance/mechanisms of the two control methods on a high lift wing-body configuration.

13.2 Test Devices and Technology

13.2.1 Wind Tunnel

The test is conducted in FL-8 low speed single circuit wind tunnel of AVIC ARI, with a closed test section of 3.5 m (width) × 2.5 m (height) × 5.5 m (length), while the effective cross section area is $S = 7.685\ \text{m}^2$. The maximum wind speed of the empty wind tunnel is 73 m/s, and the average turbulence intensity of the wind tunnel is $\varepsilon = 0.19\%$.

The test wind speed range is $V_\infty = 20\text{~}60$ m/s, while the angle of attack $\alpha = -4°\text{~}20°$, sideslip angle $\beta = 0°$.

13.2.2 High Lift Wing-Body Configuration Half Model

The wind tunnel test model [1] is a half model of a high lift transport civil aircraft wing-body configuration (Fig. 13.1), with the model length of 3.251 m and mean aerodynamic chord of 0.3658 m. The wing span of the half model is 1735 mm, with a leading edge sweep angle of 10.977°. The flap deflection angle is $\delta = 20°$. The

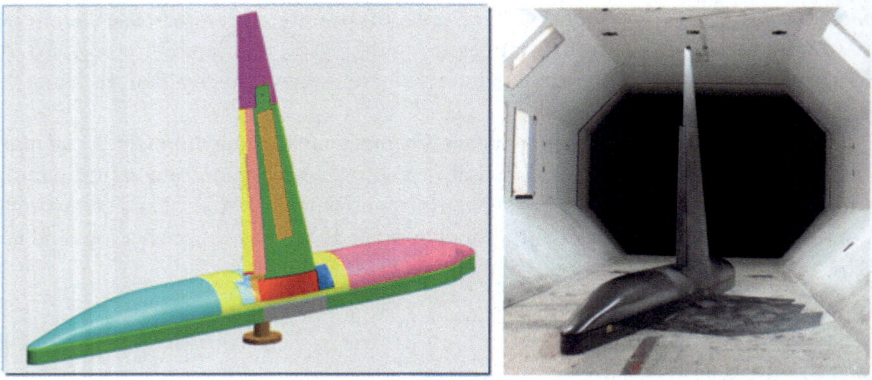

Fig. 13.1 High lift wing-body configuration half model

Reynolds number is $Re = 1.51 \times 10^6$ based on the mean aerodynamic chord length, under the condition that wind speed $V_\infty = 60$ m/s and standard atmosphere at sea level.

13.2.3 Flow Control System

Flow control methods like wing jets, flap jets and flap synthetic jets are applied on the high lift model.

The jets control system [2, 3] consists of high pressured air source, digital regulating valve, flow meter and pipes. The high pressured air is regulated by the digital valve, and the airflow measured by flow meter flows into wing/flap inner cavities by pipes. The flux of the wing jets are $q = 0.36, 0.40, 0.50$ kg/m^3, and the flux of the flap jets are $q = 0.31, 0.36, 0.44$ kg/m^3.

The synthetic jets control system consists of motor and controller, piston, cylinder stator and pipes. The motor controlled by the controller drives the piston reciprocating moving to form periodic synthetic jets at the upper exit of the cylinder stator, which will be transported into the flap inner cavities by pipes. The frequency of the synthetic jets is from 30 to 50 Hz, and the maximum velocity from the slot is higher than 60 m/s.

The wing/flap cavities are connected to surface slots [4], and the slots parameters are shown in Table 13.1. There is a single flap on the wing at about from 11 to 67.7% of the model span. The wing slot center is at about 40% chord-wise position and lies from 24 to 64% of the span of the half model with the slot length of 0.7 m. The slot on the flap is at about 25% flap-chord position and lies from 22.5 to 51.4% of the span of the half model with the slot length of 0.5 m, and the outer part can be sealed with 0.3 m remained for synthetic jets flow control. The angle of the jets from slots is 30° to the tangential direction.

Table 13.1 Parameters of wing/flap slots

		Wing slot	Flap slot
Chordwise position of the slot center (from the leading edge of wing or flap, divided by chord length) (%)		40	24.5
Slot length	Absolute length (m)	0.7	0.5
	Relative length (divided by half span) (%)	40.3	28.8
Spanwise position of slot beginning (divided by half span) (%)		24	22.5
Slot width (mm)		0.5	0.3
Slot angle (tangential to local surface)		30°	30°

13.2.4 Forces Measurements

Loads measurements are conducted by an external six components strain gauge balance (8DB09-01) for half model. Balance data are acquired by VXI system, and is used to compute aerodynamic coefficients.

The thrust of the flow control system are obtained by conducting wind tunnel runs with flow control system working while the free stream speed is 0. The results from actual flow control runs subtracted the thrust of the flow control system are the clean aerodynamic forces. All the coefficients used in this paper are based on wind axis.

13.2.5 PIV Visualization

The average flow distribution is visualized by PIV method. The light sheet is a longitudinal plane which is paralleling to the symmetrical plane and at the 40% half-span position (0.69 m to the symmetrical plane, while the local chord length is 0.3692 m with flap deflection angle of 0°). PIVCAM 13-8 CCD camera (resolution ratio 1280*1280) located right above the light sheet. Pressure atomization olive seeding machine located at the first diffusion section downstream the test section, the tracer particles flow through the whole wind tunnel circuit to arrive the test section. The image capture ratio is 8 frames per second, and 50 pairs of pictures are obtained and averaged on each state.

13.2.6 Shear Stress Measurement

Shear stress measurements are conducted by a MEMS system. The output of voltage ΔU varies with different shear stress, and the formation is $1/\Delta U = a^*\tau^n + b$, in which a, b, n are constants for the same sensor, $a > 0$, $n < 0$. There is no correction for the shear stress sensor because only differences are used. The stress sensors are situated behind the surface slots.

13.3 Results and Discussion

13.3.1 Wing Jets Flow Control

Results of the model with wing jets flow control at the take-off and landing states are shown in Figs. 13.2 and 13.3, with the wind speed from 30 to 60 m/s and the flap angle of 20°. Wing jets can obviously increase the maximum lift coefficient (maximum 40.8%) and stall angle (maximum 4.34°). With the increase of the flux of the wing

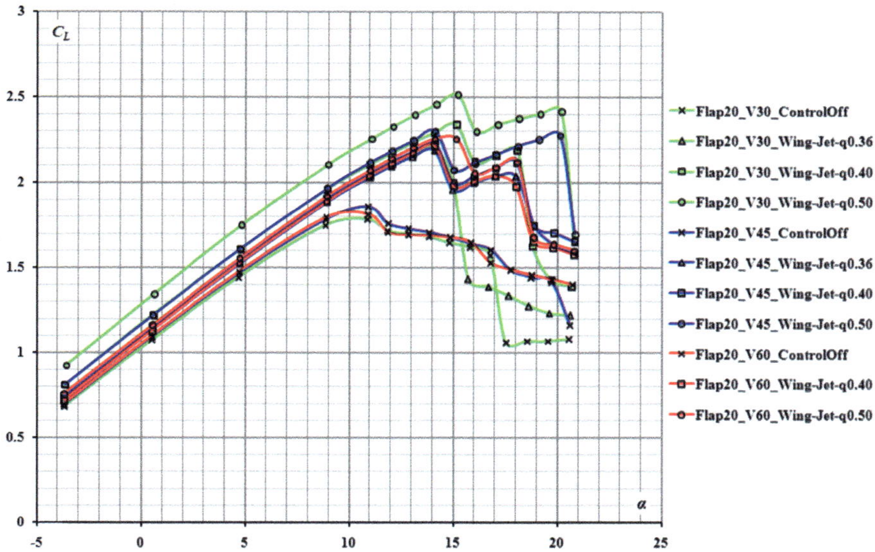

Fig. 13.2 Lift coefficient of the model with wing jets flow control

jets, the maximum lift coefficient increases. The separation is well controlled at $q = 0.50$ kg/m^3 at $\alpha = 18°$, $V_\infty = 30$ m/s. The effect of the flow control decreases with the increase of the wind speed.

Shear stress results of the model with wing jets flow control at take-off and landing state are shown in Fig. 13.4, where ΔU monotonous increasing with τ, and "$\Delta U_{\text{on}} - \Delta U_{\text{off}}$" means the shear stress change between flow control on and off. The results show that while the separation flow is effectively controlled, the shear stress is increasing and the lift coefficient is increased.

13.3.2 Flap Jets Flow Control

Results of the model with flap jets flow control at take-off and landing state are shown in Fig. 13.5, with the wind speed from 30 to 60 m/s and the flap angle of 20°. Flap jets can increase the maximum lift coefficient (maximum 19.0%) and stall angle (maximum 2.16°). With the increase of the flux of the flap jets, the maximum lift coefficient increases. The effect of the flow control decreases with the increase of the wind speed.

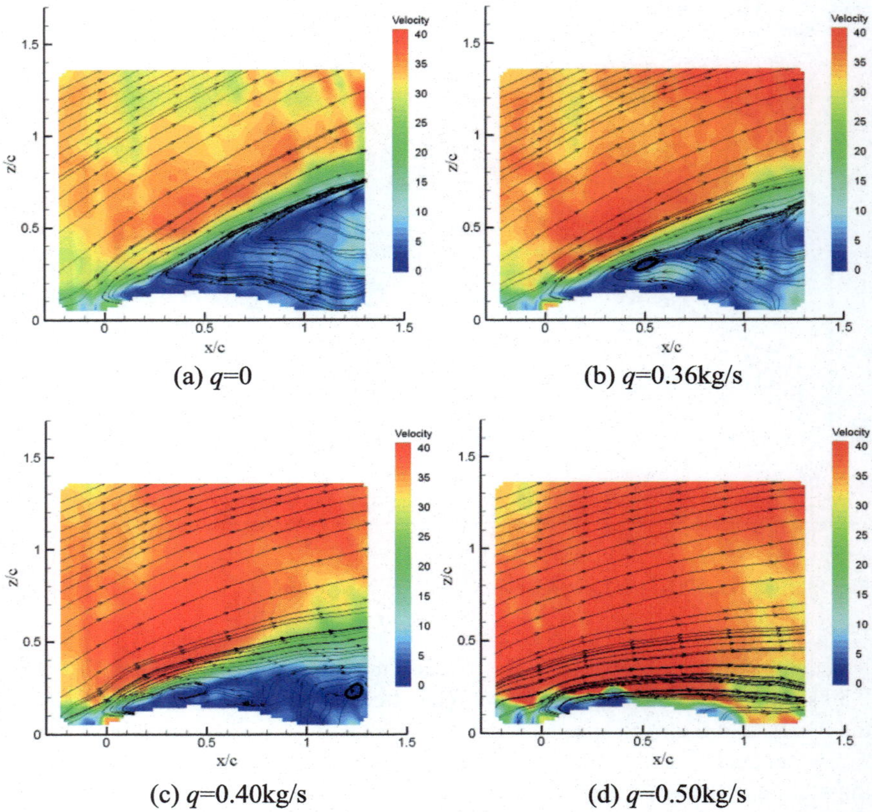

Fig. 13.3 PIV results of the model with wing jets flow control ($\alpha = 18°$, $V_{\infty} = 30$ m/s)

13.3.3 Flap Synthetic Jets Flow Control

Results of the model with flap synthetic jets flow control at take-off and landing states are shown in Figs. 13.6 and 13.7, with the wind speed from 20 to 30 m/s and the flap angle of 20°. Flap synthetic jets can increase the maximum lift coefficient by 0.87% with the slot length of 0.5 m and wind speed $V_{\infty} = 30$ m/s, compared to 0.77% with the slot length of 0.3 m and wind speed $V_{\infty} = 20$ m/s. However, there is no sign of stall angle increase using flap synthetic jets control methods. The separation is not obviously changed at $\alpha = 14°$, $V_{\infty} = 20$ m/s.

Compared with flap jets, the effect of synthetic jets is smaller, but it needs less energy input and don't need air supply. Hence, the evaluation of two control methods need take both the control effect and energy cost into account.

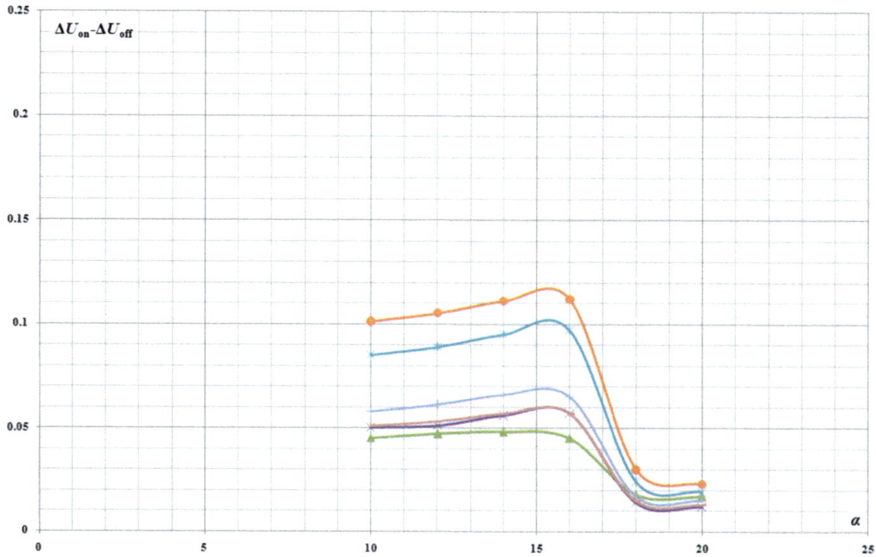

Fig. 13.4 Shear stress results of the model with wing jets flow control ($V_\infty = 30$ m/s)

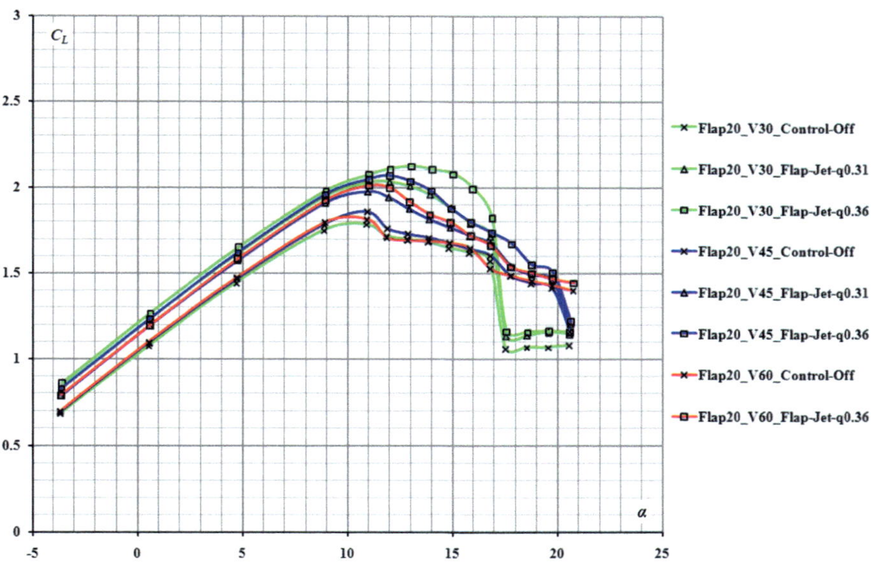

Fig. 13.5 Lift coefficient of the model with flap jets flow control

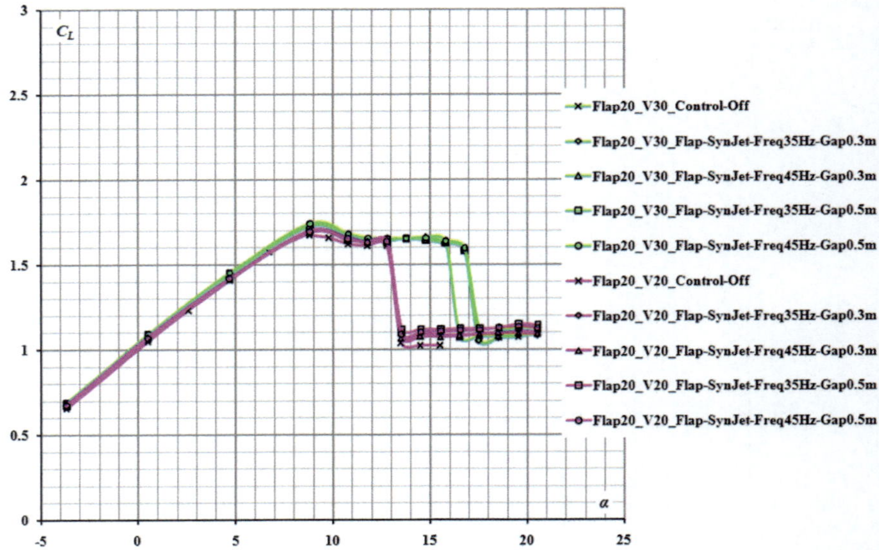

Fig. 13.6 Lift coefficient of the model with flap synthetic jets flow control

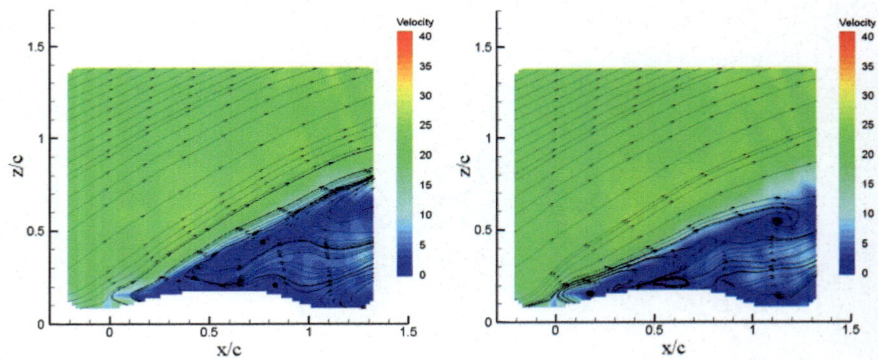

(a) f=0 (b) f=45Hz

Fig. 13.7 PIV results of the model with synthetic jets flow control ($\alpha = 14°$, $V_\infty = 20$ m/s)

13.4 Conclusions

(1) Flow control methods such as jets/synthetic jets are applied on a high lift configuration wing-body half model, and loads/PIV/shear stress measurements are conducted in FL-8 low speed wind tunnel of AVIC ARI.

(2) The maximum lift coefficient of the taking-off/landing configuration can be increased 40.8%, and the stall angle of attack can be increased 4.34° with wing

jets flow control; the maximum lift coefficient can be increased 19.0% with flap jets flow control, and 0.87% with flap synthetic jets flow control, while the energy expense is much smaller.

(3) Increasing shear stress is beneficial to separation control, thus beneficial to lift promotion.

(4) With the increase of the flux of the jets, the maximum lift coefficient increases. The effect of the flow control decreases with the increase of the wind speed.

References

1. Liu G, Sun N (2013) Wind tunnel test investigation of typical flow control devices on a high lift configuration. AVIC Aerodynamics Research Institute, 12
2. Zhang P, Wang J, Feng L (2008) Progress on synthetic jets technology and its application. Sci China (Sci E Technol Sci) 38(3)
3. Luo Z, Xia Z (2005) Advances in synthetic jet technology and applications in flow control. Adv Mech 35(2)
4. Tong Z, Liu P, Duan H (2010) Numerical simulation of the flap blowing and suction control for two dimension multi-element airfoil. Aircr Des 30(4)

Chapter 14
Effective Synthetic Jet Control for Separation Control on BFS

Xiao Ming, Xingyu Ma, Reinhard Geisler and Peng Li

Abstract The flow separation behind a two-dimensional backward facing step is controlled by synthetic jet which is different from traditional ones. The synthetic jet is generated from a 2 mm horizontal slot which is at up corner of the step and facing downstream direction. In order to investigate the effectiveness of the synthetic jet at different actuation frequencies (0, 35, 50 and 100 Hz), 2C-2C particle image velocimetry and fluorescence oil-film was used to indicate the flow pattern on the surface and measure quantitatively the skin friction downstream of the step. The velocity vector field as well as global skin friction clearly indicates reduction of the reattachment length under proper actuation. The comparisons among frequencies indicate that the reattachment length depends on the actuation frequencies. The most effective reduction of reattachment length as much as 43.7% is achieved at the frequency of 100 Hz, corresponding to a Strouhal number $St_h \approx 0.3$, based on the free-stream velocity and step height. The Reynolds shear stress is considerably increased and large-scale vortices are analyzed. The skin friction downstream is increased by 20% at the frequency of 35 Hz ($St_h \approx 0.026$). In the conclusion, the streaming mechanism of a synthetic jet is briefly discussed.

Keywords Flow control · Synthetic jet · Backward facing step · Particle image velocimetry · Fluorescence oil-film

X. Ming (✉) · P. Li
College of Aerospace Engineering, Nanjing University of Aeronautics and Astronautics, Yudao ST 29, Nanjing, China
e-mail: mingam@nuaa.edu.cn; mingam@126.com

X. Ma · R. Geisler
German Aerospace Center, Institute of Aerodynamics and Flow Technology, Göttingen, Germany

X. Ma
Department of Physics, University of Göttingen, Göttingen, Germany

N. Qin et al. (eds.), *Advances in Effective Flow Separation Control for Aircraft Drag Reduction*, Computational Methods in Applied Sciences 52,
https://doi.org/10.1007/978-3-030-29688-9_14

14.1 Introduction

Flow separation control is directly related to more efficient air transportation and less emission into the environment. The aerospace industry has been striving for better aerodynamic designs, yet it still needs improvement to achieve the target set out in the ACARE (The Advisory Council of Aeronautical Research in Europe) 2020 vision for 50% reduction in aircraft emissions. Among various aerodynamic designs, active separation flow control techniques play an important role in the improvement of aerodynamic efficiency by applying artificial actuators in flow field, which provides an additional dimension for further improvement of aircraft performance, in particular, during taking-off and landing. As "game-changing" techniques, active flow control shows significant potential of further improvement in current highly optimized aircraft designs. In FP7, the EU-China joint project "MARS" is aiming at effective flow separation control by active flow control techniques.

As a fundamental experimental and numerical test case, a backward facing step (BFS) has a very simple geometry and provides a separated shear layer from a fixed step edge. Depending on the ratio of the incoming boundary layer thickness to the step height, various separated flow characteristics can be generated. The separating and reattaching flow as well as a recirculation region depends on incoming free-stream velocity, the laminar/turbulent flow states and the step geometry. Numerous investigations have been carried out on characteristics of BFS flows.

Synthetic jets for active flow separation control have been investigated extensively. In the 1980s, Ming et al. [1] pioneered the study on synthetic jets as a flow control method. By using a loudspeaker and a nozzle, he observed that the velocity at the nozzle outlet exhibits a significant peak under a particular frequency, which is referred to as "acoustic streaming". It was applied to flow around a circular cylinder to reduce the separated region or to manipulate the direction of the wake. Later, Glezer et al. [2, 3] explored further ideas of synthetic jets for flow control and made the idea popular.

In the present study, a streamwise synthetic jet is applied to turbulent BFS flow is experimentally investigated. Particle image velocimetry (PIV) and fluorescence oil-film are employed in the measurements of the flow velocity field and global skin friction with and without control and measure the. In the conclusion, the streaming mechanism of the synthetic jet is briefly discussed.

14.2 Experiments

14.2.1 Flow Facilities and Synthetic Jet

The wind tunnel tests were carried out in the straight-flow wind tunnel in the University of Aeronautics and Astronautics (NUAA) and in the closed-loop wind tunnel in the German Aerospace Center (DLR), as shown in Figs. 14.1 and 14.2. The wind tunnel in NUAA has 4 m total length, 0.5 m height and 0.3 m width. The Reynolds

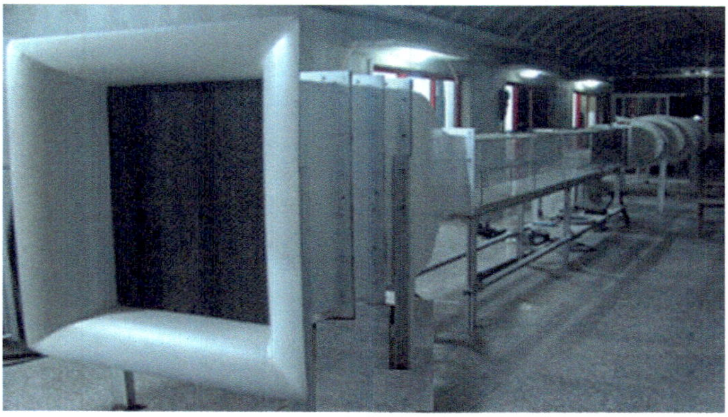

Fig. 14.1 Photo of the straight-flow subsonic wind tunnel in NUAA

Fig. 14.2 Photo of the closed-loop subsonic wind tunnel in DLR

number is $Re_h = 8.0 \times 10^4$, based on the free-stream velocity of $U_1 = 40$ m/s and the step height h = 30 mm. On the other hand, the wind tunnel in DLR has an open test section of 0.7 m height and 1.05 m width. The Reynolds number is $Re_h = 2.0 \times 10^4$, based on the free-stream velocity of $U_2 = 10$ m/s and the step height h = 30 mm. The incoming boundary layer was artificially tripped upstream of the step to generate a turbulent boundary layer. A two-dimensional coordinate system has its origin point at the corner on the wall, a horizontal X-axis and a vertical Y-axis.

The actuator consists of a square chamber box, a thin horizontal slot, a pipe and a piston. The slot is 2 mm wide close to the step edge and facing to the downstream direction (Fig. 14.3). The square box has a cross-section of 30 × 30 mm^2. The frequency of the reciprocating piston can be controlled by the servo motor. Synthetic

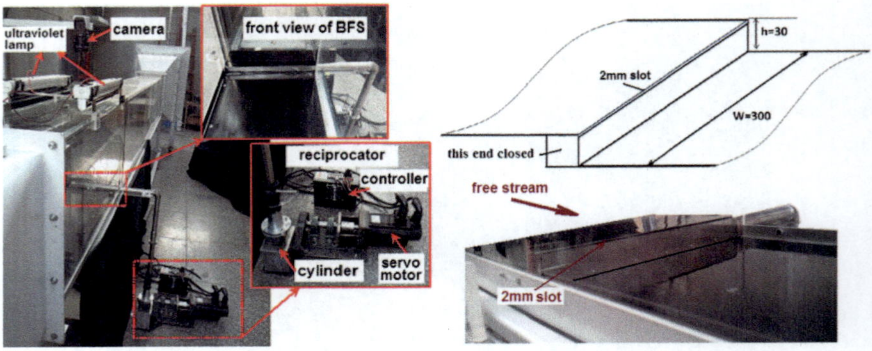

Fig. 14.3 Photo and schematic of the synthetic jet

jet is generated through the thin slot at the upper corner of the step. Three actuation frequencies of f = 0, 35 and 50 Hz were tested in NUAA by using fluorescence oil-film, while one frequency f = 100 Hz was tested in DLR by using PIV. The corresponding Strouhal number, based on the free-stream velocity and step height, are $St_h = 0, 0.026, 0.0375$ and 0.30, respectively.

14.2.2 Particle Image Velocimetry Measurement

Digital 2D-2C PIV has been extensively used in fluid flow measurements for more than two decades [4]. The PIV measurement was carried out in DLR. In the present study, the BFS flow was measured in a single field of view by using a two-dimensional horizontal-vertical laser light sheet and one side-view camera, as shown in Fig. 14.4. The laser sheet was 1 mm thick and delivered 30 mJ per pulse. The incoming flow was homogeneously seeded by oil droplets. Double-frame particle images were recorded

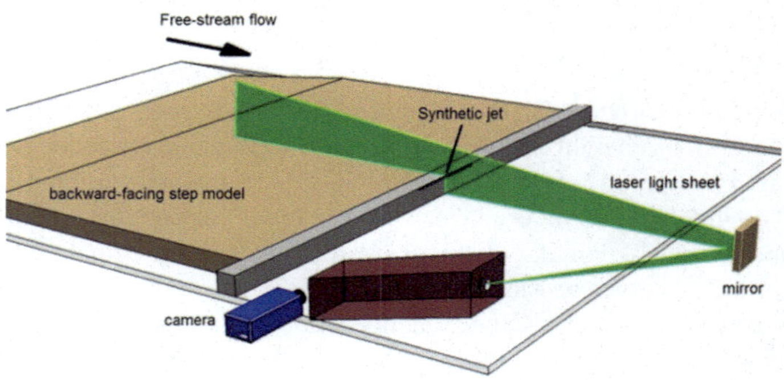

Fig. 14.4 Schematic of 2D-2C PIV system

by a pco.4000 camera (11 M pixel, 14 bit) with a Nikon lens (85 mm, f/4) mounted in the side view. Viewing distortion was corrected by a two-dimensional calibration process before recording. In the first step, 4000 non-phase-locked double-frame images were recorded for the flows with and without control. In the second step, 1000 phase-locked double-frame images at 12 equal-interval phases were recorded in the controlled case. The double-frame particle images were evaluated by a multi-grid cross-correlation method with image deformation and a final interrogation window size of 16 × 16 pixel at 75% overlap, resulting in a vector spacing of 0.30 in the physical coordinate.

14.2.3 Fluorescence Oil-Film Measurement

As fluid flows past a thin film of viscous liquid, for instance silicone oil, on a plane surface. The oil-film moves in response to the surface skin friction, the gravity force and surface tension due to curvature of a surface. The thickness of the film varies as a function of position and time. Given a steady flow condition, Squire [5] and Tanner [6] provided two different forms of the thin oil-film equation, respectively, and the equivalence of the two forms can be demonstrated by metric transformation. Liu et al. [7] mixed silicone oil with luminescent molecules which emit radiation at a longer wavelength due to the Stokes shift when illuminated by a UV light source. The detailed physics and mathematics method is given by Liu et al. [8]. In order to accelerate convergence and improve the accuracy of solution, the pyramid iterative evolution method (PIEM) is proposed. There are two critical steps in the optimization algorithm. First, the original image of fluorescence oil-film is decomposed into a series of image pyramids for accelerating convergence [9]. The second critical step is the iterative evolution of the original image of fluorescence oil-film. This step can improve the accuracy of solution. As we know, the intensity of fluorescence oil-film will vary in response to these wall stresses, thus we can use the solution of the measurement model to simulate the evolution process, and in turn to evaluate the accuracy of solution. If the desired accuracy of solution is not reached, the solution process can be done repeatedly based on the recent image of similar evolution. Figure 14.5 shows the schematic of fluorescence oil-film technique. For skin friction measurement the downstream wall of the step is polished and with black colour. Silicon oil will be applied on the smooth surface. Luminescent molecules will be dropped and fluorescence oil will be made, so that a longer wavelength radiation can be emitted.

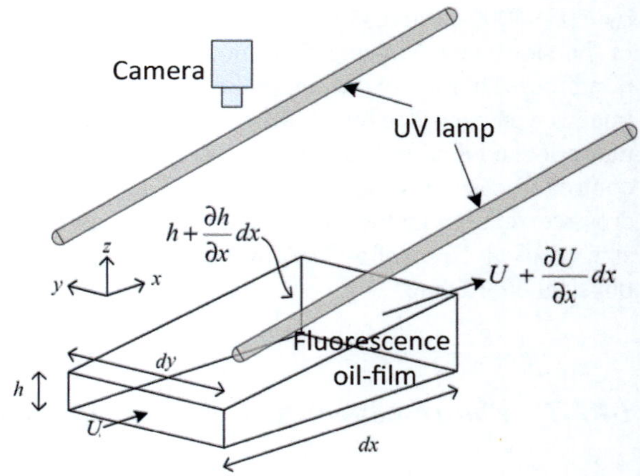

Fig. 14.5 Schematic of fluorescence oil-film technique

14.3 Results

14.3.1 Time-Averaged Flow Field and Reynolds Shear Stress

Time-averaged velocity vector fields of the clean case at f = 0 Hz and controlled case at f = 100 Hz are shown in Fig. 14.6. In the clean case, the main and the secondary recirculation regions are identified behind the step and the reattachment length is X/h = 7.1. The comparison shows the streamlines are drawn downward closer to the wall. The main recirculation region is clearly suppressed and its center position moves upstream and closer to the wall. The reattachment length is reduced to X/h = 4.0, resulting in a reduction rate of R = 43.7%. As a result the recirculation region is smaller than that of the clean case [10].

Reynolds shear stress is an essential quantity representing momentum transfer in turbulent shear flows. Based on the triple decomposition [11], the Reynolds shear stress can be decomposed as:

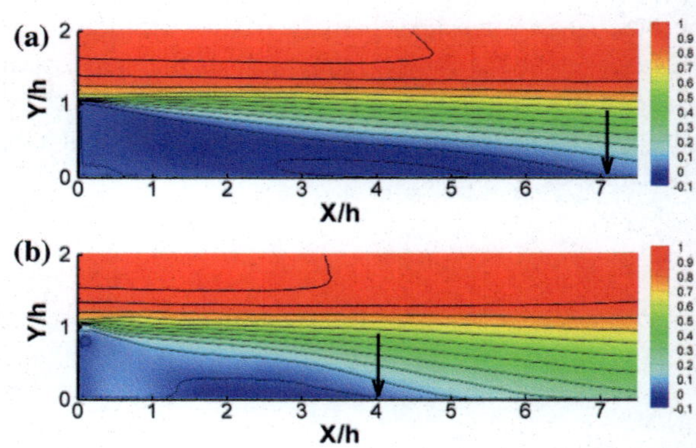

Fig. 14.6 Time-averaged velocity contour fields at f = 0 Hz (**a**) and f = 100 Hz (**b**). The black arrows indicate the reattachment points

$$\frac{\tau_{ij}}{-\rho} = \overline{(\tilde{u}+u')\cdot(\tilde{v}+v')} = \overline{\tilde{u}\cdot\tilde{v}} + \overline{u'\cdot\tilde{v}} + \overline{\tilde{u}\cdot v'} + \overline{u'\cdot v'} \tag{14.1}$$

Given the non-correlation condition between the periodic velocity components random fluctuating ones:

$$\overline{u'\cdot\tilde{v}} = 0 \quad \text{and} \quad \overline{\tilde{u}\cdot v'} = 0 \tag{14.2}$$

So the Reynolds shear stress is decomposed as:

$$\frac{\tau_{ij}}{-\rho} = \overline{\tilde{u}\cdot\tilde{v}} + \overline{u'\cdot v'} \tag{14.3}$$

On the right-hand side the first part is the contribution of the coherent structures, and the second part is the contribution of the incoherent turbulence.

The total Reynolds shear stresses of f = 0 and 100 Hz are compared in Fig. 14.7, which shows two isolated high-shear regions within the shear layer due to the actuations. The one close to the synthetic jet is due to the spanwise vortices rolling-up in the initial part of the shear layer. The other one further downstream is due to the vortex break down resulting in an increase of incoherent Reynolds shear stress. Therefore, the rolling-up and breakdown processes of vortices play an important role in the momentum transfer within the shear layer. Furthermore, the incoherent Reynolds shear stress can be decomposed into phase-averaged parts, based on the phase-locked PIV data. It can be seen in Fig. 14.8 that the interface between the shear layer and the free-stream flow is evolving along the spanwise vortices. Both the vortex rolling-up and break down processes lead to the considerable increases of Reynolds shear stress. The high-momentum fluid flow is engulfed into the turbulent shear layer due to the Biot-Savart induction [12], which produces more turbulent

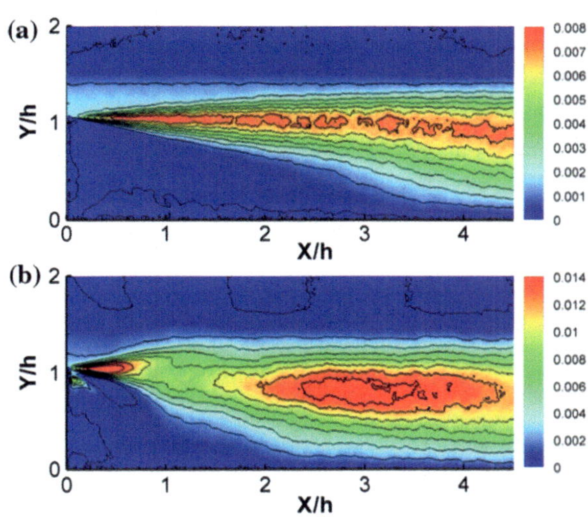

Fig. 14.7 Comparison of total Reynolds shear stress. **a** f = 0 Hz; **b** f = 100 Hz

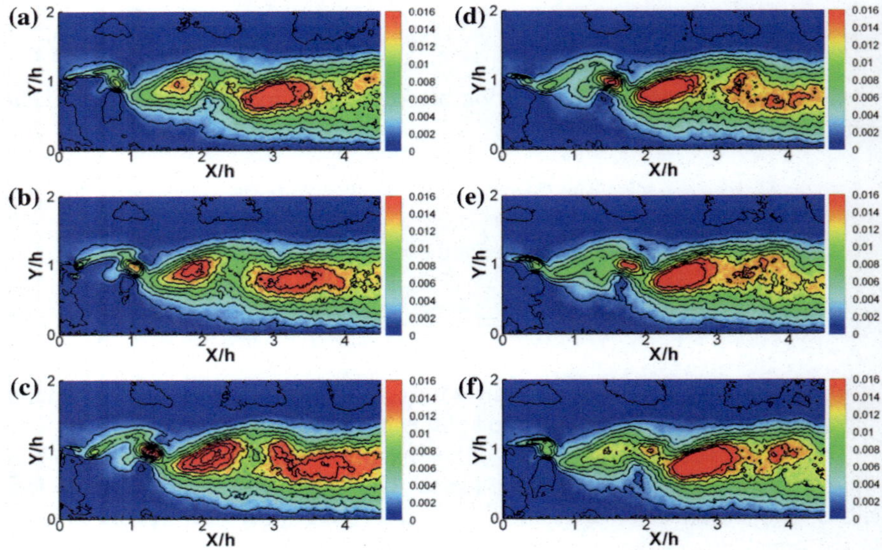

Fig. 14.8 Phase-averaged incoherent Reynolds shear stress at f = 100 Hz. **a–f**: 0, π/3, 2π/3, π, 4π/3 and 5π/3

fluctuations within the shear layer than turbulent mixing process due to viscosity alone.

14.3.2 POD Analysis

In order to reveal the spatial structures of the coherent structures, the snapshot POD method by Sirovich [13] is applied to the initial part of the separated shear layer. Detailed mathematical algorithm of the snapshot POD method is presented by Meyer et al. [14]. It decomposes the ensemble flow field into a linear combination of spatial orthogonal modes as:

$$U = \overline{U} + \sum_{i=1}^{N} \Phi_i \bullet a_i^T \tag{14.4}$$

Φi is the i-th mode and ai is the corresponding coefficient. Because the POD modes are ranked based on descending turbulent kinetic energy, it thereby ensures that energetic flow structures can be represented by the first few modes.

The POD eigenvalue distributions of f = 0 and 100 Hz are plotted in Fig. 14.9. For both cases the first two modes contain the most turbulent kinetic energy and the other modes decay logarithmically. The comparison shows the energetic fluctuating motions of the controlled case at f = 100 Hz are greater than those of the clean case at f = 0 Hz. Additionally, the first two modes of the controlled case contain nearly

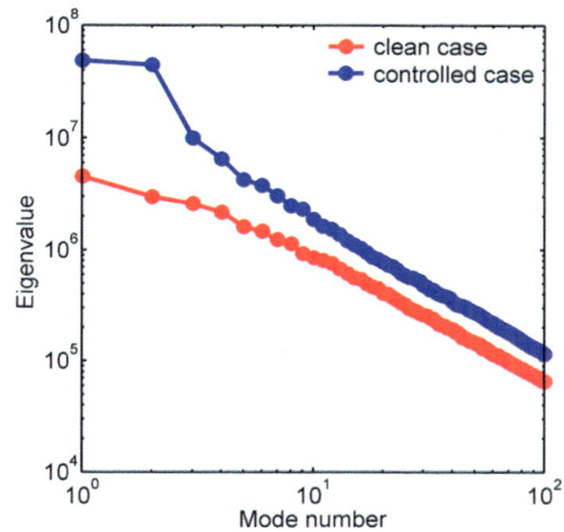

Fig. 14.9 POD eigenvalue distributions of f = 0 Hz (in red) and f = 100 Hz (in blue). Only the first 100 eigenvalues are plotted for clarity

equivalent energy, whose coherence is discussed in detailed by Ma et al. [10]. A phase angle φ between the two POD modes is defined by Perrin et al. [15]:

$$\phi = arctan\left(\frac{\sqrt{\lambda_i} \cdot a_j}{\sqrt{\lambda_j} \cdot a_i}\right) \tag{14.5}$$

Phase-averaged reconstructions are shown in Fig. 14.10, showing counter-rotating vortices at different phases. The regular patterns indicate coherent vortices represented by the first two modes which are mutually orthogonal. In other words, the reconstructed regular patterns are equivalent to the interference of two oscillating modes. An estimated wavelength of the vortices is ΔX/h ≈ 1.0, which agrees well with the spatial scale in the phase-averaged incoherent Reynolds shear stress in Fig. 14.8. These vortices show the distinct frequencies and coherent features, which is totally different from the wide frequency bandwidth and multi-scale structures as the nature behavior of the clean BFS flow [16].

14.3.3 Global Skin Friction

The global skin friction distributions at f = 0, 35 and 50 Hz downstream of the step are obtained by fluorescence oil-film measurement. Figure 14.11 shows that the reattachment length is moved significantly upstream due to the synthetic jet actuations. By changing the actuation frequency, flow reattachment varies as well. Comparison of skin friction distributions in Fig. 14.12 shows that the skin friction is considerably increased by approximately 20% by the synthetic jet actuation, which leads to an effective reduction of flow separation [17]. The increase of skin friction agrees well with the increased Reynolds shear stress in the PIV measurement.

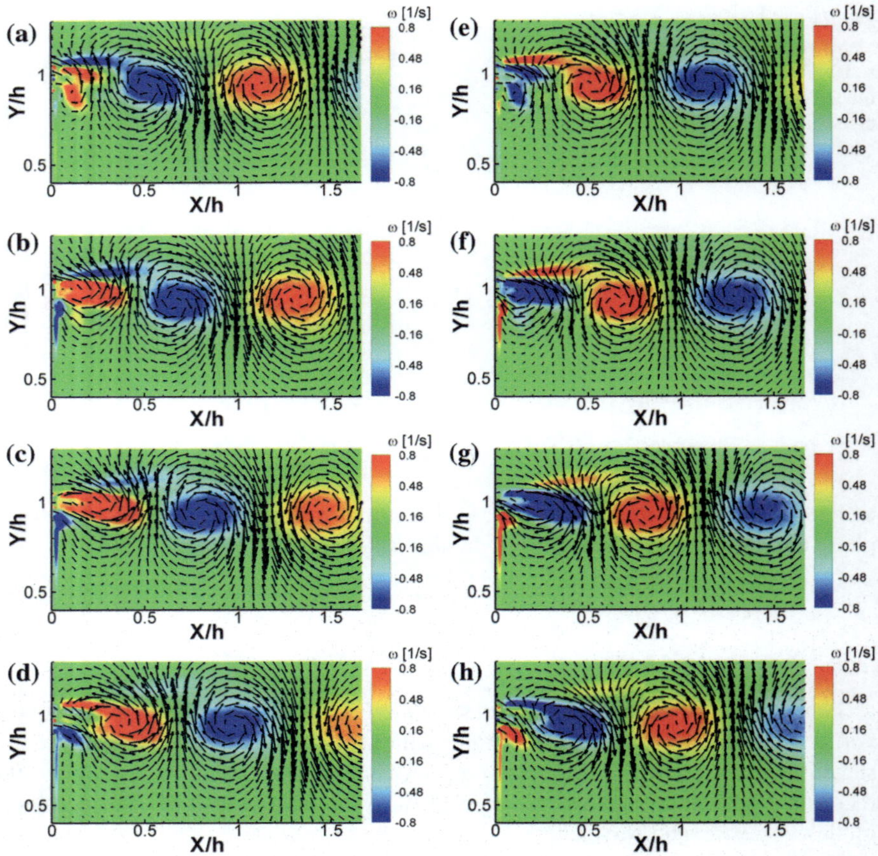

Fig. 14.10 Phase-averaged reconstruction by the first two POD modes at f = 100 Hz. **a–h**: 0, π/4, π/2, 3π/4, π, 5π/4, 3π/2 and 7π/4. The color indicates spanwise vorticity

14.4 Conclusions and Discussion

The synthetic jet is applied in turbulent BFS flow. 2D-2C PIV and fluorescence oil-film techniques are used to measure the velocity vector field and skin friction on the reattaching surface. Periodic actuations can generate spanwise vortices in the turbulent shear layer. Phase-averaged flow fields provide insight into the roll-up and pairing processes. The coherent vorticity are extracted by POD and represented by predominant POD modes. The PIV and Fluorescence oil-film measurements show considerable increases of the Reynolds shear stress within the separated shear layer as well as increased skin friction downstream of the reattachment area due to the

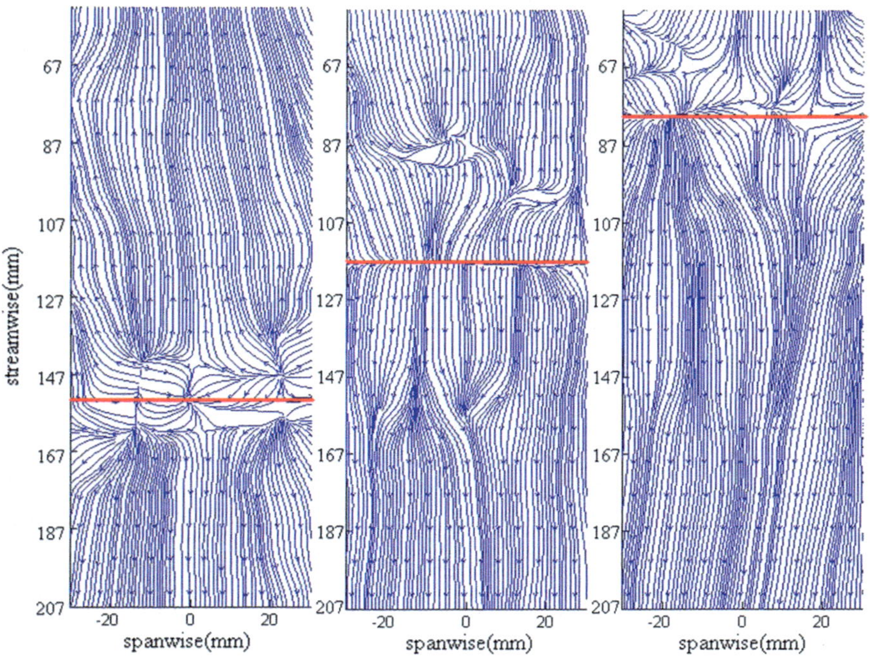

Fig. 14.11 Surface flow patterns downstream of the BFS at f = 0 Hz (left), 35 Hz (middle) and 50 Hz (right). The free-stream direction is from top to bottom. The red lines indicate reattachment areas

actuation. The streamwise synthetic jet results in an effective flow separation control in the backward-facing step flow.

Synthetic jets have been a very popular active flow control technique and they have been successfully used among various flow control applications. However, the physical mechanism of a synthetic jet in flow separation control is still not completely understood. The formation of a synthetic jet is attributed to the streaming effect, as shown in Fig. 14.13. In the pulling phase, fluid is sucked from the neighboring area and the fluid momentum is converted into fluid pressure, while in the pushing phase the air is pushed out of the chamber. The mechanic energy of the vibrating diaphragm is converted into flow momentum. At resonant frequency, depending on the geometry of the actuator, the streaming reach a maximum and the control is most effective.

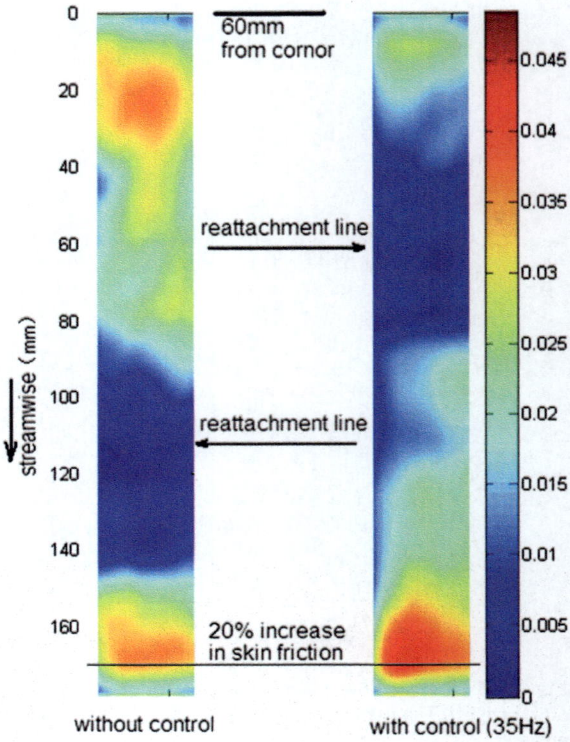

Fig. 14.12 Skin friction comparison between f = 0 Hz (left) and f = 35 Hz (right)

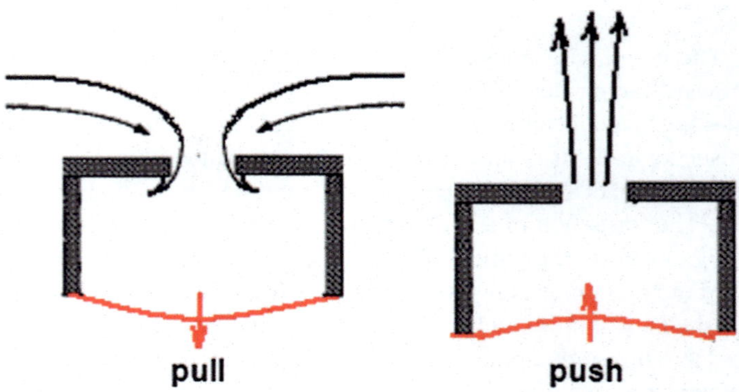

Fig. 14.13 Streaming mechanism of a synthetic jet

Acknowledgements The authors would like to thank the EU-China co-funded "MARS" project for supporting the active flow control research. We are also grateful to the Nanjing University of Aeronautics and Astronautics and German Aerospace Center for great support the experiments.

References

1. Ming X, Dai CH (1991) A new phenomenon of acoustic streaming. Acta Mech Sinica 7(3). In: Proceedings of the international conference on fluid dynamics measurement and its applications, Beijing China, Oct 1989
2. Glezer A (1990) The formation of vortex rings. Phys Fluid 31(12)
3. Glezer A, Amitay M (2002) Synthetic jet. J Fluid Mech 34
4. Raffel M, Willert CE, Wereley ST, Kompenhans J (2007) Particle image velocimetry: a practical guide. Springer, Berlin
5. Squire LC (1961) The motion of a thin oil sheet under the boundary layer on a body. J Fluid Mech 11(2):161–179
6. Tanner LH, Blows LG (1976) A study of the motion of oil films on surfaces in air flow, with application to the measurement of skin friction. J Phys E Sci Instrum 9(3):194–202
7. Liu T, Sullivan JP (1998) Luminescent oil-film skin friction meter. AIAA J 36(8):1460–1465
8. Liu T, Woodiga S, Montefort J, Conn KJ, Shen L (2009) Global skin friction diagnostics in separated flows using luminescent oil. J Flow Vis Image Process 16(1):19–39
9. Winter KG (1977) An outline of the techniques available for measurement of skin friction in turbulent boundary layer. Progr Aerospace Sci 18:1–57
10. Ma X, Geisler R, Agocs J, Schröder A (2015) Investigation of coherent structures generated by acoustic tube in turbulent flow separation control. Exp Fluids 56:46
11. Hussain F, Reynolds WC (1970) The mechanics of an organized wave in turbulent shear layer. J Fluid Mech 41:241–258
12. Hussain F (1986) Coherent structures and turbulence. J Fluid Mech 173:303–356
13. Sirovich L (1987) Turbulence and the dynamics of coherent structures. Part 1: coherent structure. Q Appl Math 45(3):561–571
14. Meyer KE, Pedersen JM, Özcan O (2007) A turbulent jet in crossflow analysed with proper orthogonal decomposition. J Fluid Mech 583:199–227
15. Perrin R, Braza M, Cid E, Cazin S, Barthet A, Sevrain A, Mockett C, Thiele F (2007) Obtaining phase averaged turbulence properties in the near wake of a circular cylinder at high Reynolds number using POD. Exp Fluids 43:341–355
16. Bhattacharjee S, Scheelke B, Troutt TR (1986) Modification of vortex interactions in a reattaching separated flow. AIAA J 24(4):623–629
17. Fukagata K, Iwamoto K, Kasagi N (2002) Contribution of Reynolds stress distribution to the skin friction in wall-bounded flows. Phys Fluids 14(11):L73–L76

Chapter 15
Novel Spanwise Vortex Generator for Separation Control on BFS: Experiment and Simulation

Palma Gonzalez, Ning Qin and Xiao Ming

Abstract This paper is on the effects of a novel span vortex generator on the periodic components of the turbulent shear layers and the Reynolds stresses. The turbulence modelling approach is 3D simulations IDDES, a hybrid RANS/LES technique. The geometry for the study is taken from the experimental configurations for the case. The case comprises a turbulent flow over a backward facing step (BFS), where separation is induced after the step edge. The results from the simulations are compared to the experimental data with and without control. Spanwise vortex generators consist of a strip of magnets placed along the span of the BFS upstream of step and the device oscillates at a given frequency of 280 Hz and amplitude of 0.002m. Turbulent structures, Reynolds stresses, skin friction distributions and velocities are analysed and compared to the experimental measurements. A remarkable effect of the device is observed especially in the reattachment length which is considerably reduced. Experimental measurements for the baseline case were available and a comparison with the available data is performed.

Keywords Span-wise vortex generator · Experiment · Hybrid RANS-LES · Separation control · BFS

15.1 Introduction

A flow over a backward facing step was chosen to study the flow control with span-wise vortex generators, SVG. The experimental work was carried out by Nanjing University of Aeronautics and Astronautics. This study was performed within the MARS project.

P. Gonzalez · N. Qin
Department of Mechanical Engineering, University of Sheffield, Sheffield S1 3JD, UK
e-mail: n.qin@sheffield.ac.uk

X. Ming (✉)
Department of Aerodynamics, Nanjing University of Aeronautics and Astronautics, Nanjing 210008, China
e-mail: mingam@126.com

N. Qin et al. (eds.), *Advances in Effective Flow Separation Control for Aircraft Drag Reduction*, Computational Methods in Applied Sciences 52,
https://doi.org/10.1007/978-3-030-29688-9_15

First of all, a baseline simulation was performed and then, SVG were configured and a controlled case simulation was carried out using DGDES code. 3D SVG simulations provide further understanding on the effects of SVG in a BFS flow.

A classic vortex generator, VG (Figs. 15.1 and 15.2) is an aerodynamic device which consists of a small vane placed on top of a surface of an airplane fuselage or any other vehicle such as cars and they can be also installed on a propeller blade of a wind turbine [19]. Basically, when the air is flowing over the surface where the VG is installed, this device forces the creation of vortices which will delay the separation of the flow over the surface and hence, delaying the aerodynamic stall of the wing. The effects of the vortex generators in the flow over a backward facing step has been also studied both experimental and computationally [24].

Flow control using steady streamwise vortex generators is well known in the aeronautical field [1, 3, 4, 12–15, 20, 22] and have been widely investigated. In general in a turbulent flow, there is a large loss of energy associated with the separation of the boundary layer. For such reason, flow separation control in fluid mechanics has been and it still is nowadays extremely important for the aerospace industry. VGs are used to control boundary layer separation at a wide range of flow speeds (low speed see Jenkins et al. [11]. Supersonic flow: McCormick [15]; Mounts and Barber [17]). VGs were analysed to establish the relation between the parameters of the device and

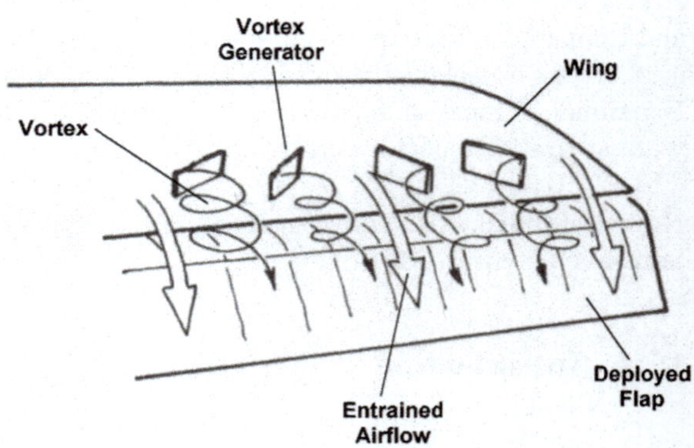

Fig. 15.1 Illustration of VGs on a wing

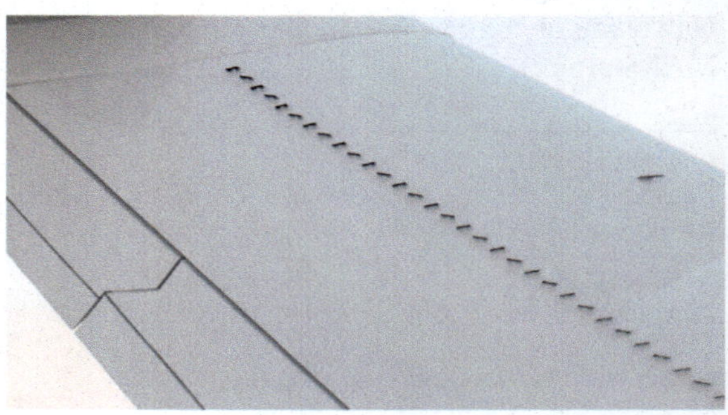

Fig. 15.2 Vortex generators on the vertical fin of a Boeing 727-100

the resulting flow characteristics [1, 3, 4, 22], the aircraft performance improvement and also noise control [8] and distortion reduction [2, 7].

Similar to the vortex generators, different types of passive control have been studied such as cavities or rods [10] and blockages in a channel [5, 6]. In 2003, Neumann and Wengle [18] found that a minimum distance between the step edge and a control fence—a small obstruction upstream of the step inducing turbulence in the flow—is required to achieve the maximum reduction of the recirculation bubble downstream of the step. Yokoyama et al. [23], found the formation of a small vortex after a rectangular bump placed before the step edge. Once the vortex went beyond the step edge, it broke into faster and slower vortices causing a compression effect which led to a reattachment reduction.

Several studies focused on researching for more effective flow control beyond fixed vortex generators were carried out by Mochizuki et al. [16]. An experimental investigation of the effects caused by a streamwise vortex generator with periodic perturbation on Reynolds stresses in a turbulent jet wall was performed. In the study, the device is an oscillating vortex generator of a delta wing shape. The triple decomposition mentioned in the introduction of this thesis was analysed [21] and together with the averaging of the flow, periodic variations in the strength were observed as a direct consequence of the periodic oscillation of the VG. As a conclusion of such study, effects on the oscillating vortex generators were captured at different Reynolds stress components.

The novelty of the present research relies on the fact that the periodical perturbations were introduced in the flow by a flat surface oscillating vertically to the flow direction—with a 2 mm gap between the surface and the bottom wall—in order to analyse the effects of the spanwise vortex generator in the reattachment length as well as Reynolds stresses. In this study, it was assumed that the spanwise vortex generator was as rectangular strip located above the edge of the step, with a 2 mm gap between the SVG and the wall, like the experiment configuration.

In this paper, the spanwise vortex generator, Fig. 15.3, is simulated according to Nanjing University of Aeronautics and Astronautics experimental case.

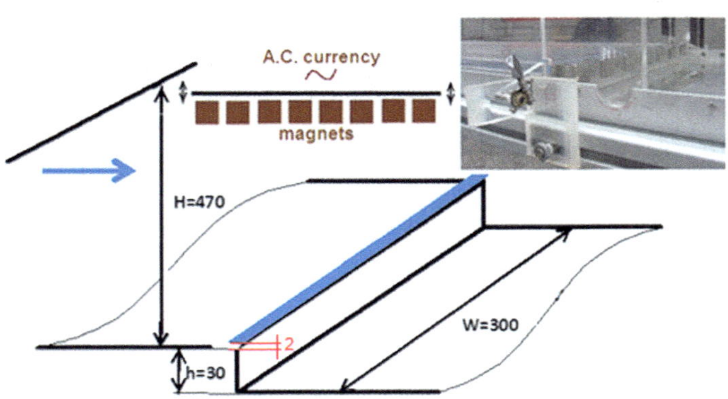

Fig. 15.3 Spanwise Vortex Generator (SVG) setup in NUAA wind-tunnel

15.2 Geometry and Computational Domain

All the experimental work for this study was carried out at NUAA facilities. Figure 15.4 shows a picture of the wind-tunnel where the experiments took place.

The wind tunnel had a total height of 0.47 m and a span width of 0.30 m. The step height was 0.03 m. Experiments were carried out at a free stream velocity of 19.8 m/s and the Reynolds number based on step height was 39,000. In this case, the boundary layer thickness at step edge location was around 2H according to the experimental readings.

A diagram of the SVG configuration can be seen in previous Fig. 15.3. Experimentally, the SVG consisted of a metal tape stretched over a strip of magnets placed along the span of the wind tunnel. The magnets had a width of 0.005 m and were located at 0.002 m away from the wall. A.C. current is applied to the metal tape generating a Lorentz force that moves the tape up and down. The frequency and magnitude of the movement is controlled by the applied current. The experimental frequency of 280 Hz produced oscillation of the strip and the amplitude was 0.002 m.

In order to perform the simulations, it was assumed that the SVG performed as a flat surface 0.005 m long located just before the step edge at 0.002 mm over the wall with a frequency of oscillation as the experimental, 280 Hz and maximum amplitude of 0.002 m. See Fig. 15.5.

Regarding the computational mesh, and 8 million cells provided with very reliable results. Figure 15.6 shows the computational mesh and an instantaneous capture at 0.29 s of a detail of the step region when SVG were activated in simulations.

Fig. 15.4 Wind-tunnel facilities at Nanjing University of Aeronautics and Astronautics

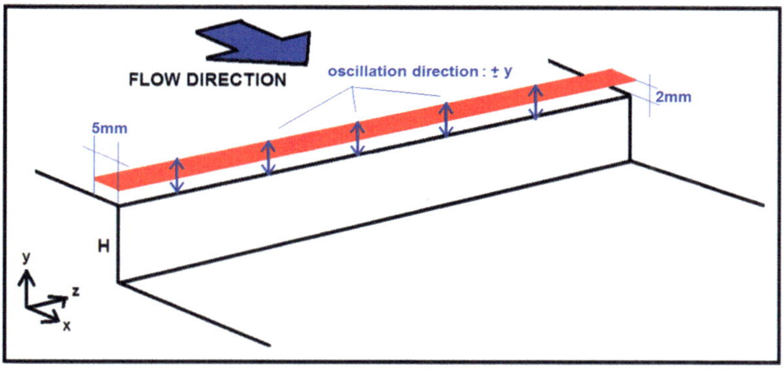

Fig. 15.5 Schematics of the simulated SVG

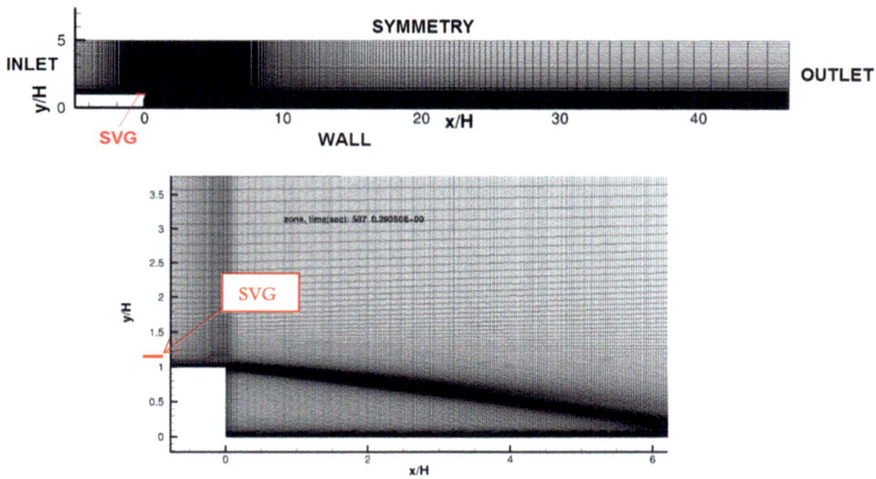

Fig. 15.6 Computational mesh and detail of step region when SVG are in operation

15.3 Implementation of Spanwise Vortex Generators in DGDES

When the experimental device SVG was described in previous sections, it was said the SVG were located 0.002 m over the wall moving up and down with 0.002 m amplitude at a frequency of 280 Hz. In DGDES, the configuration was the same. A 0.005 m long by the span length long surface was defined along the spanwise and it was placed 0.002 m over the wall. The oscillation in time corresponds to a sinusoid wave of maximum amplitude of 0.002 m, the SVG amplitude; therefore its position is given by,

$$y_{SVG} = A_0 \sin(2\pi f t + \varphi) \tag{15.1}$$

where A_0 is the maximum amplitude of the SVG, f is the frequency of the actuator, t is the physical time and φ is the initial phase angle if at some point an exploration of the deployment effects of the SVG from different locations wanted to be investigated.

So the velocity of the SVG is obtained by deriving its y position, Eq. (15.1), by time,

$$V_{SVG} = \frac{dy_{SVG}}{dt} = 2\pi f A_0 \cos(2\pi f t + \varphi) \tag{15.2}$$

An example of the actuation of the SVG was already displayed in Fig. 5.5 where a detail of the grid at the step region was shown and it can be appreciated how the SVG defined region is moving upwards at that exact time position.

15.4 Results and Discussion: Comparison with Experimental Database

15.4.1 Analysis of Coherent Structures

An analysis of the turbulent structures was carried out to evaluate the effects of the spanwise vortex generators. Q-criterion [9] was used to identify the aforementioned structures. In Fig. 15.7, Q-criterion of the baseline—left column—is depicted versus

(a)

(b)

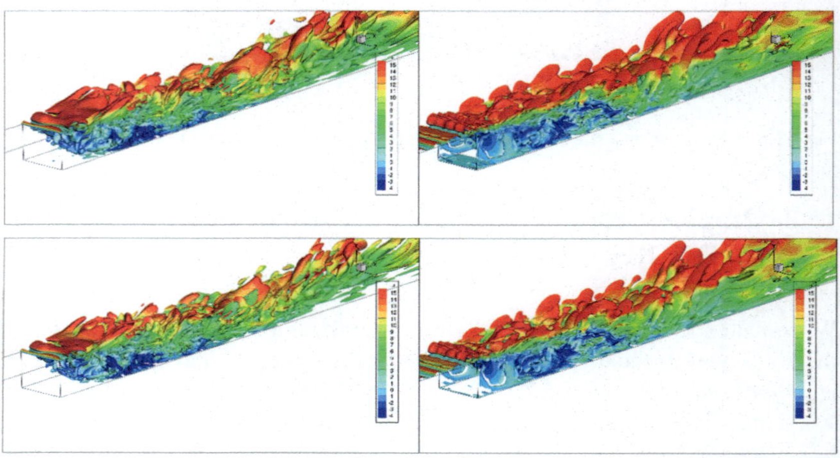

Fig. 15.7 Q-criterion of baseline—left column—versus SVG case at **a** 100,000 and **b** 200,000. Iso-surfaces coloured by streamwise velocity

the controlled case—right column—for three different values of Q: (a) Q = 100,000 and (b) Q = 200,000. The figures are coloured by the streamwise velocity values.

For (a) Q = 100,000, for the baseline case obvious hairpin structures can be seen downstream of step. There is a bigger destruction of turbulent structures on the uncontrolled case, especially near the step which means the device introduces large instabilities in the mean flow hence there are more turbulent structures in the flow field. The effects of SVG are appreciated for the controlled case: a pseudo-homogeneous layer of mushroom eddies is formed downstream of step. The lower parts of these structures move also towards the wall breaking into much smaller structures. From the reattachment area and downwards, bigger structures are present for the controlled case. Finally at (b) Q = 200,000 hairpin-like structures are clearly observed for the uncontrolled case. Smaller structures are formed downstream of the reattachment area, unlike in the controlled case, where again a layer of turbulent structures—mushroom eddies and hairpin further downstream—is appreciated at y = H. An enhanced mixing region is observed in the recirculation area is observed in the controlled case compared to the baseline flow.

15.4.2 Reattachment Region Studies

In this section the streamlines of both simulations are depicted and compared, Fig. 15.8.

It is observed that the actuation of SVG reduces the recirculation bubble size and also the secondary bubble is smaller for the controlled case. For the baseline case, the flow attaches to the wall at 7H, although there is a series of smaller embedded eddies after the reattachment point. The controlled case flow reattaches to the bottom wall at 6H, which is a reduction of 12.5%. The experimental reduction of the reattachment length was 16%, indicating some discrepancy between simulation and experiment.

With regards to the skin friction distribution, experimental measurements were taken by means of the oil film technique. The experimental partner warned the computational side that this technique is a relative quantitative measurement as calibration is needed for absolute value of skin friction. When looking at the next graph next, one must bear in mind that the comparison between the skin friction coefficients between the experiments and the simulations is only relative and qualitative, not quantitative.

In Fig. 15.9 the shear stress contours are shown. It can be seen that there is an overall increase of this Reynolds component within the separation area, confirming again that there exists and improvement of the flow mixing, in particular near the reattachment region. Turbulent shear stress is also increased at the SVG location.

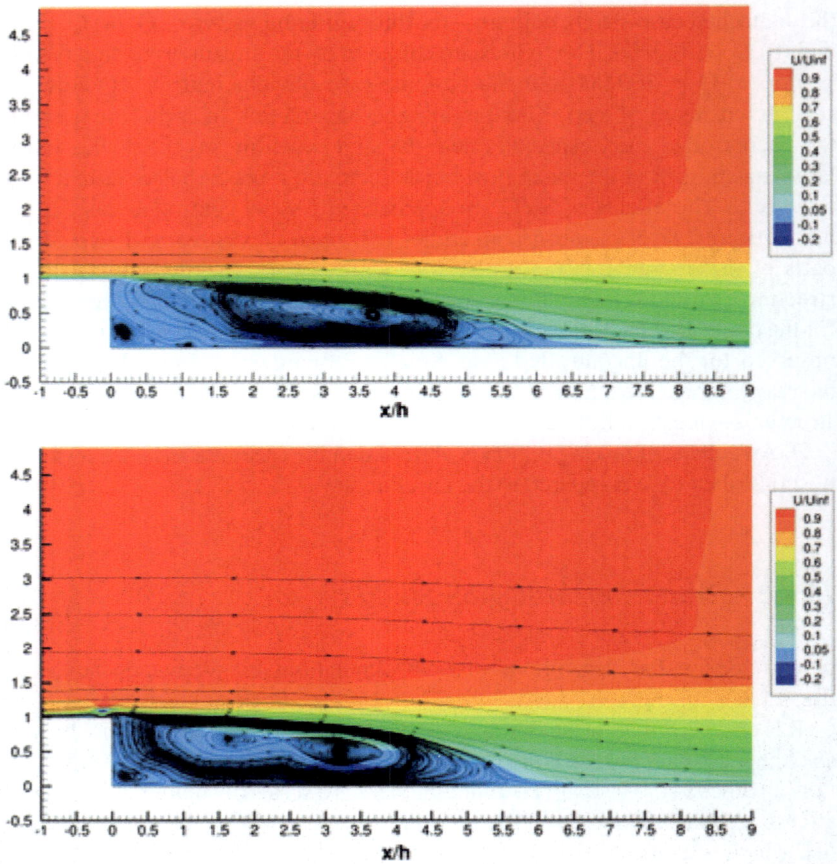

Fig. 15.8 Streamlines of flow field: baseline –top—and controlled case –bottom

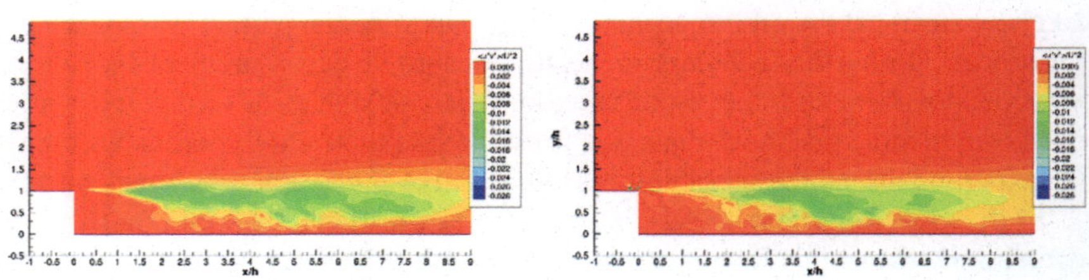

Fig. 15.9 Time and spanwise averaged Reynolds shear stress $<u'v'>/U_0^2$ contour: Baseline compared to SVG actuation

15.5 Conclusions

A novel device known as spanwise vortex generators has been investigated in this paper as a first attempt. For all simulations, an 8 M mesh was used. According to experiments a reduction of 16% of the reattachment length was achieved by the application of this device upstream of the step and simulations provided a reduction of 12.5%, at an oscillation frequency of 280 Hz and an amplitude of 0.002m. Turbulent structures and Reynolds stresses captured the effects of the oscillation of the SVG captured such effect downstream of the step and even further downstream near the end of the computational domain. The installation of this device has shown to generate a highly turbulent flow and the produced turbulent structures remain in the domain further downstream of the step.

Spanwise vortex generator induces an organised motion into and interacts with the separated shear layer from the upper corner of the step. The reattachment point moves upstream and the skin friction downstream increases.

References

1. Allan BG, Yao C-S, Lin JC (2002) Numerical simulation of vortex generator vanes and jets. In: 1st AIAA flow control conference, St. Louis, MO, 24–27 June 2002. AIAA paper 3160
2. Anabtawi AJ, Blackwelder RF, Lissaman PBS, Liebeck RH (1999) An experimental investigation of boundary layer ingestion in a diffusing S-duct with and without passive active control. In: 37th AIAA aerospace sciences meeting and exhibit, Reno, NV, 11–14 Jan 1999. AIAA paper 90-0739
3. Ashill PR, Fulker JL, Hackett KC (2001) Studies of flow induced by sub boundary layer vortex generators (SBVGs). In: 39th AIAA aerospace sciences meeting and exhibit, Reno, NV, 8–11 Jan 2001. AIAA paper 0887
4. Ashill PR, Fulker JL, Hackett KC (2002) Research at DERA on sub boundary layer vortex generators (SBVGs). In: 40th AIAA aerospace sciences meeting and exhibit, Reno, NV, 14-17 Jan 2002. AIAA paper 0968
5. Griffith MD, Hourigan K, Thompson MC (2004) Numerically modelling blockage effects on the flow between flat plates. In: 15th Australasian fluid mechanics conference, The University of Sydney, Sydney, Australia, 13–17 Dec 2004
6. Griffith MD, Thompson MC, Leweke T, Hourigan K, Panderson W (2007) Wake behaviour and instability of flow through a partially blocked channel. J Fluid Mech 582:319–340
7. Hamstra JW, Miller DN, Yagle PJ, Truax PP, Anderson BH, Wendt BJ (2000) Active flow control technology demonstration. In: 11.2, 22nd international congress of the aeronautical sciences, Harrogate, UK, 27 Aug–1 Sep 2000. ICAS-2000-6
8. Holmes AE, Hickey PK, Murphy WR, Hilton DA (1987) The application of sub-boundary layer vortex generators to reduce canopy Mach rumble interior noise on the Gulfstream III. In: AIAA 25th aerospace sciences meeting, Reno, NV, 12–15 Jan 1987. AIAA paper 87-0084
9. Hunt JCR, Wray AA, Moin P (1988) Eddies, stream, and convergence zones in turbulent flows. In: Center for turbulence research report CTR-S88, pp 193–208
10. Isomoto K, Honami S (1989) The effect of inlet turbulence intensity on the reattachment process over a backward facing step. J Fluids Eng 111:87–92. Transactions of the ASME
11. Jenkins L, Gorton SA, Anders S (2002) Flow control device evaluation for an internal flow with an adverse pressure gradient. AIAA paper 0266

12. Kerho M, Hutcherson S, Blackwelder RF, Liebeck RH (1993) Vortex generators used to control laminar separation bubbles. J Aircr 30(3):315–319
13. Lin JC, Howard FG, Bushnell DM, Selby GV (1990) Investigation of several passive and active methods for turbulent flow separation control. In: AIAA 21st fluid dynamics, plasma dynamics and lasers conference, Seattle, WA, 18–20 June 1990. AIAA paper 90-1598
14. Lin JC, Selby GV, Howard FG (1991) Exploratory study of vortex-generating devices for turbulent flow separation control. In: AIAA 29th aerospace sciences meeting, Reno, NV, 7–10 Jan 1991. AIAA paper 91-0042
15. McCormick DC (1992) Shock-boundary layer interaction control with low-profile vortex generators and passive cavity. In: 30th AIAA aerospace sciences meeting and exhibit, Reno, NV, 6–9 Jan 2012. AIAA paper 92-0064
16. Mochizuki S, Yamada S, Osaka H (2006) Reynolds stress field in a turbulent wall jet induced by streamwise vortex with periodic perturbation. Exp Fluids 40:372–382
17. Mounts JS, Barber TJ (1991) Numerical analysis of shock-induced separation alleviation using vortex generators. In: 30th AIAA aerospace sciences meeting and exhibit, Reno, NV, 6–9 Jan 1992. AIAA paper 92-0751
18. Neumann J, Wengle H (2003) DNS and LES of passively controlled turbulent backward-facing step flow. Flow Turbul Combust 71:297–310
19. Peppler IL (1996) From the ground up, 27th Revised edn. Aviation Publishers Co. Limited, Ottawa Ontario, p 23
20. Rao DM, Kariya TT (1988) Boundary-layer submerged vortex generators for separation control—an exploratory study. AIAA paper 88-3546-CP
21. Reynolds WC, Hussain AKMF (1972) The mechanics of an organized wave in turbulent shear flow, part 3: theoretical models and comparison with experiments. J Fluid Mech 54(2):263–288
22. Yao C-S, Lin JC, Allan BG (2002) Flowfield measurement of device-induced embedded streamwise vortex on a flat plate. In: 1st flow control conference, St. Louis, MO, 24–27 June 2002. AIAA paper 3162
23. Yokoyama H, Tuskamoto Y, Kato C, Iida A (2007) Self-sustained oscillations with acoustic feedback in flows over a backward-facing step with a small upstream step. Phys Fluids 19:106104
24. Zulkefli NF, Tai EN, Mujeebu MA, Abdullah MZ, Ahmad KA (2009) Numerical and experimental investigations of passive flow control devices on a backward facing step. Int J Eng Technol 6(2):21–29

Chapter 16
Modelling and Numerical Simulation for Flow Control

Ruyun Hu, Liang Wang and Song Fu

Abstract This study deals with numerical prediction through Improved-Delayed-Detached-Eddy Simulation of the separating-reattaching flow over a backward-facing step and the manipulated flow with harmonic actuation for investigating the mechanism of reattaching acceleration. The predicted base flow agrees well with the experimental results in terms of both the time-averaged sense and the spectra content. Mode decomposition analyses, including Proper Orthogonal Decomposition and Dynamic Mode Decomposition are applied to the base flow data to extract the salient coherent structures, that is, shear layer mode and shedding mode. Then the base flow is excited with slot-shaped harmonic actuators with different excitation amplitudes and frequencies corresponding to those of the salient coherent structures. The flow data are analyzed with Triple Decomposition Technique to recognize the excited flow due to the actuation. We find that the reattachment is accelerated under the excitation of shear layer mode, while a significant promotion obtained under the excitation of vortex pairing. A better understanding of the separating-reattaching flow is presented and the mechanism of reattaching acceleration is finally proposed.

Keywords Backward-facing step · Flow control · Detached-Eddy simulation · Dynamic mode decomposition

---

R. Hu
School of Mechanical Engineering, Shanghai Jiao Tong University,
800 Dongchuan Road, Shanghai 200240, China
e-mail: hury@mail.sjtu.edu.cn

L. Wang · S. Fu (✉)
School of Aerospace Engineering, Tsinghua University,
Haidian District, Beijing 100084, China
e-mail: fs-dem@tsinghua.edu.cn

L. Wang
e-mail: wangliang12@mail.tsinghua.edu.cn

N. Qin et al. (eds.), *Advances in Effective Flow Separation Control for Aircraft Drag Reduction*, Computational Methods in Applied Sciences 52,
https://doi.org/10.1007/978-3-030-29688-9_16

16.1 Introduction

Massively separated turbulent flows play an important role in engineering design since they occur in many aerodynamic applications, such as aircraft body, landing gear and combustor. Usually the flows are highly three-dimensional (3D) and mainly unsteady with complex flow physics. The characteristic frequencies are well-known but the interactions between salient coherent structures still remain to be understood. The interactions between coherent structures and pressure field are the major source of noise, which is a major concern of the greener aviation. In fundamental fluid dynamics, several typical prototypes are extensively investigated, such as flow over airfoil, blunt plate, cavity and backward-facing step (BFS). Among them, the flow over a backward-facing step is the simplest in terms of geometry but presents highly complex unsteady flow physics, which provides an ideal application for numerical simulation. On the aspect of flow control, BFS flow separates at a fixed location, which is helpful to focusing on investigating on the reattaching process. Bubble length is governed by the reattaching process of the separated shear layer. Reduction of it is a major concern of flow control on BFS flow, but with poor understanding of the mechanism. Both the base flow and the excited flow are highly unsteady, which is beyond the capability of the traditional unsteady Reynolds-Averaged Navier-Stokes (RANS) equations. Direct-Numerical-Simulation (DNS) or Large-Eddy-Simulation (LES) are capable of this kind of flow, but huge computational resource is required for the Reynolds number concerned here. Recently developed hybrid RANS-LES method (HRLM) is a trade-off of the computational resource and the accuracy, which is applicable for the cases concerned. Therefore, we here simulated harmonic actuation of the BFS flow with Improved-Delayed-Detached-Eddy-Simulation (IDDES), which is zonal-type HRLM, to reveal the controlling mechanism.

In the past, both the BFS flow and its control have widely been studied experimentally. Eaton and Johnston [1] gave a comprehensive review of the available experimental data of BFS flow and outlined the dependance of the time-averaged reattachment location on the geometry and inflow conditions. They also mentioned that the behavior of the separated shear layer layer resembles that of the planar mixing layer, which rolls-up due to Kelvin-Helmholtz (K-H) instability following by vortex pairing events downstream [2]. However, there exist different opinions on the coherent motions approaching reattachment. One is that coherent structures in the shear layer are split into two parts, one part recirculates into the bubble and the other convects downstream [3, 4]. Another is that part of the coherent structures sweeps upstream and the other convects downstream [5–7]. Beyond the reattachment, there still exist large scale spanwise coherent structures without vortex pairing [8, 9]. The characteristic frequency, $St_h = fU_0/h$, in this region ranges from 0.06 to 0.10 [10, 11]. Proper Orthogonal Decomposition (POD) with Linear Stochastic Estimation (LSE) was applied to extract the coherent structures of BFS flow by Hudy et al. [12]. They concluded that BFS flow contains two distinct modes: *shear layer mode* and *wake mode*. Regarding the control of the BFS flow, a variety of actuators were applied, finding that the effective excitation frequencies in terms of bubble length reduction are lower than $St_{h,ex} = 0.8$ [11, 13, 14]. Furthermore, the most effective excitation frequency is around 0.20 [14–16].

On the other hand, only few numerical studies of the BFS flow exist in the literature, especially the high Reynolds number case concerned here. Armaly et al. [17] first performed a 2D simulation of laminar BFS flow. Neto et al. [18] conducted DNS and LES of a 3D turbulent BFS flow, observing that intense longitudinal vortices exist between spanwise vortices in the shear layer. A detailed DNS of the BFS flow measured by Jovic and Driver [19] gave a characteristic frequency of 0.06, falling in the range as aforementioned [20]. The authors are aware of a DNS of the BFS flow with the highest Reynolds number of 9000 by Kopera et al. [21]. The velocity and wall pressure fluctuations are resolved by Detached-Eddy-Simulation (DES) at a Reynolds number of $37,000$, which shows the capability of DES method on high Reynolds number flow [22]. Regarding the perturbed flow, Dejoan and Leschziner [23] first conducted a LES of a perturbed turbulent BFS flow at a Reynolds number of $3,700$, concluding that the high level of sensitivity to the perturbation is due to a strong interaction between *shear layer mode* and *shedding mode*. DES, a zonal-type of HRLM, was first proposed based on the S-A turbulence model [24] by Spalart et al. [25] for massively separated flows, and then developed by Spalart et al. [26] to tackle the "grid-induced separation (GIS)" problem, named Delayed DES (DDES) and extended to the SST turbulence model [27] by Strelets [28] and Menter et al. [29]. For the consideration of internal flow, a combination of DDES and Wall Modelled LES (WMLES) was proposed by Shur et al. [30] to resolve the "modelled-stress depletion (MSD)" problem, which is called Improved DDES (IDDES).

On the actuators applied to the BFS flow, most of them take effects. For passive control, a fence installed upstream of the step reduced the bubble length by generating higher level of turbulence [31, 32]. For active control, an audio speaker installed on the upper wall of the wind tunnel near the step increased upstream turbulent intensity and modified the vortex interaction processes within the separation bubble, resulting in a significant reduction of bubble length [13]. And an oscillating flap located at the step regularized vortical structure formation, enhancing the mixing and entrainment of the excited shear layer to reduce the reattachment length [11]. Most recently, a plasma actuator was applied to both reducing the reattachment length and the wall pressure fluctuations [33]. Besides, miscellaneous actuators were investigated [16, 34–39]. Among them, slot synthetic jet is a simple and effective one for 2D separated flows, like BFS flow and flow over a ramp [14, 15, 40–42].

In the present study, a harmonic actuation, analog to a slot synthetic jet is selected with a consideration of numerical modelling. We use IDDES with the SST baseline turbulence model to simulate the BFS base flow and the excited flows. Then POD and DMD techniques are applied to analyze the 3D huge flow data to extract the salient coherent structures to provide a guidance of the choosing the exciting frequencies. The excited flows are analyzed using Triple Decomposition Technique (TDT) to demonstrate the exciting effects. Finally, mechanism of bubble length reduction is revealed by the streamwise distribution of vorticity thickness. The following sections are organized as: Section 16.2 is IDDES formulation and the baseline turbulence model; Section 16.3 is numerical setup and parameters for harmonic actuators; Section 16.4 is results and discussion; and Sect. 16.5 comes with a conclusion.

16.2 The IDDES Formulation and the Baseline Turbulence Model

16.2.1 SST Turbulence Model

Considering the performance of varieties of turbulence models in BFS flow [44], the incompressible mass and momentum conservation equations are solved along with the two-equation SST model in the present study. The SST model reads as follows:

$$\left.\begin{aligned}\frac{\partial k}{\partial t}+\frac{\partial\left(ku_j\right)}{\partial x_j}&=P_k+\mathit{Diff}_k-D_k\\ \frac{\partial \omega}{\partial t}+\frac{\partial\left(\omega u_j\right)}{\partial x_j}&=P_\omega+\mathit{Diff}_\omega-D_\omega+(1-F_1)\,CD_{k\omega}\end{aligned}\right\}, \tag{16.1}$$

where the production, diffusion and dissipation terms in the $k-$equation read,

$$\left.\begin{aligned}P_k&=\left(2\nu_{turb}S_{ij}-\frac{2}{3}k\delta_{ij}\right)S_{ij}\\ \mathit{Diff}_k&=\frac{\partial}{\partial x_j}\left((\mu+\sigma_k\nu_{turb})\frac{\partial k}{\partial x_j}\right)\\ D_k&=\beta^* k\omega\end{aligned}\right\}, \tag{16.2}$$

and the production, diffusion, dissipation and cross-diffusion terms in the $\omega-$equation read,

$$\left.\begin{aligned}P_\omega&=\gamma\frac{\omega}{k}P_k\\ \mathit{Diff}_\omega&=\frac{\partial}{\partial x_j}\left((\mu+\sigma_\omega\nu_{turb})\frac{\partial \omega}{\partial x_j}\right)\\ D_\omega&=\beta\omega^2\\ CD_{k\omega}&=\frac{2\sigma_{\omega 2}}{\omega}\frac{\partial k}{\partial x_j}\frac{\partial \omega}{\partial x_j}\end{aligned}\right\}. \tag{16.3}$$

The SST-based turbulent viscosity is defined,

$$\nu_{turb,SST}=\min\left(\frac{k}{\omega},\frac{a_1k}{SF_2}\right). \tag{16.4}$$

The readers are referred to the articles of Menter et al. [27, 29] for more details.

16.2.2 IDDES Formulation

DES approach was first proposed by Spalart et al. [25] based on the SA turbulence model [24] for combining the strength of LES in the separated region and RANS near the wall. Under the inspiration of Spalart's work, Strelets [28] developed the DES approach based on the SST turbulence model by substituting the dissipation term in Eq.(16.2) with

$$D_k = k^{3/2}/l_{DES}, \tag{16.5}$$

where $l_{DES} = \min(l_{RANS}, l_{LES})$, $l_{RANS} = k^{1/2}/(\beta^* \omega)$ is the length scale of SST model, $l_{LES} = C_{DES}\Delta$ is the length scale of Smagorinsky model.

The GIS and the MSD problems occur because the grid scale, Δ, plays a decisive role on the DES length scale. Spalart et al. [26] dealt with the GIS problem by introducing the flow field information to the formulation of DES length scale, resulting in the DDES method. Shur et al. [30] designed an "elevation function" to compensate Reynolds stress in the buffer layer between LES region and RANS region to battle the MSD problem, and combined the strength of DDES and WMLES, proposing IDDES approach for internal flow. The length scale, l_{DES}, is then substituted with the length scale of IDDES,

$$l_{IDDES} = \tilde{f}_d (1 + f_e) l_{RANS} + \left(1 - \tilde{f}_d\right) l_{LES}. \tag{16.6}$$

The dissipation term then reads,

$$D_k = k^{3/2}/l_{IDDES} \tag{16.7}$$

The readers are referred to the articles of Shur et al. [30] for more details.

16.3 Actuator Model and Numerical Setup

The current simulation refers to the experiment of Driver et al. [9] in a low-speed wind tunnel, the size of whose inlet section is 1,000, 151 and 101 mm in the streamwise, spanwise and wall-normal directions, respectively. The step height, h, is 12.7 mm. The aspect ratio (the ratio of the spanwise size of wind tunnel to step height) is 12, which is large enough to ensure a quasi-2D flow in the mid-span of the wind tunnel [43]. The Reynolds number based on the free-stream velocity, $U_0 = 44.2$ m/s, and the step height, h, is $Re_h = 3.7 \times 10^4$. The boundary layer thickness is $\delta = 1.5h$ at $4h$ upstream of step.

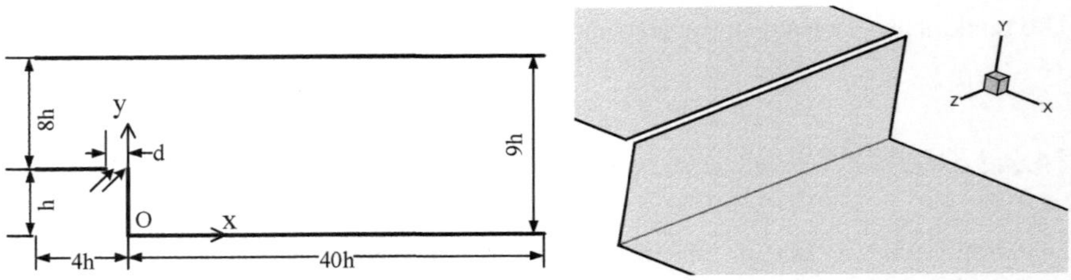

(a) Dimensions (b) Perspective view

Fig. 16.1 Schematic of the slot harmonic actuator

16.3.1 Actuator Model

The slot harmonic actuator is located at the step edge, as shown in Fig. 16.1 with a width, d, of $0.1h$. The jet is prescribed by a velocity profile, which is uniformly distributed in spanwise direction and harmonically oscillates in time.

$$u_{ex} = U_{ex} f\,(\xi, \eta) \sin\left(2\pi f_{ex} t + \varphi\right), \tag{16.8}$$

where $\varphi = 0$ is the initial phase angle, f_{ex} is the excitation frequency, $U_{ex} = VR \cdot U_0$ is the amplitude and VR is velocity ratio.

The velocity profile is a laminar-like profile. An additional coefficient 1.5 is applied to ensure that the mass flow rate equal the one of the uniformly distributed profile. The profile read,

$$f\,(\xi, \eta) = 1.5[1.0 - 4.0\,(\xi + 0.5)^2], \tag{16.9}$$

where $\xi = x/d \in [-1, 0]\eta = z/d$. The jet angle is 45° deviating from the streamwise direction.

16.3.2 Computational Domain and Boundary Conditions

We consider the full height of the wind tunnel, which is $8h$ upstream of step, so a non-slip boundary condition is applied to the upper and lower wall. The dimensions of 2D geometry are shown in Fig. 16.1a. The inlet is located at $x/h = -4h$, where the time-averaged velocity is available and the turbulence information is not. Hence, we conducted a 2D RANS of the wind tunnel and extracted the time-averaged velocity and turbulence information at the corresponding location to the inlet, feeding the 3D simulation. The outlet is located at $x/h = 40h$, which is downstream enough to apply a convective boundary condition. Considering that the BFS flow is quasi-2D, the spanwise boundary condition is set as periodic condition. And the length of spanwise computational domain is $4h$, which is large enough to weaken the spanwise correlation [20].

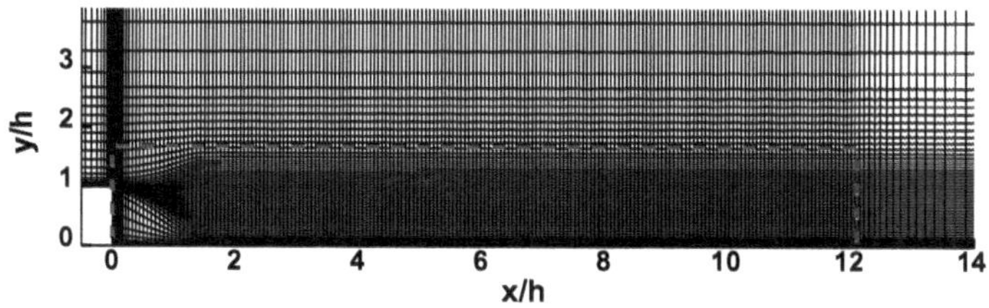

Fig. 16.2 Zoom-in show of mesh in focus region (marked by dash line)

16.3.3 Grid Generation

A mesh refinement study has been undertaken to check its influence on the statistical properties. The mesh that gives rather grid independent results, consists in two subdomains of $(N_x \times N_y \times N_z) = (80 \times 110 \times 160)$ and $(N_x \times N_y \times N_z) = (584 \times 170 \times 160)$ grid cells along the streamwise, spanwise and wall-normal directions, respectively. The grid points are uniformly distributed along the spanwise direction with a cell size of $\Delta z = 0.025h$. The first cell height in wall unit near the wall is below unit, with the wall unit calculated from the skin friction coefficients at a far downstream location. Cell height in the boundary layer grows in the geometrical progression manner with a ratio of 1.05. The grid points in the major part of the focus region are uniformly distributed along the streamwise and wall-normal direction, resulting in cubic cells in this region. The mesh at the slot is further refined and the grid points are uniformly distributed with a cell size of $0.0025h$. A zoom-in show of the mesh in the focus region is presented in Fig. 16.2.

16.3.4 Numerical Implementation

The incompressible mass and momentum conservation equations, along with the SST model equations are solved by SIMPLE method in the framework of finite volume method. The velocity and the pressure variables are co-located on the cell center, and the "momentum interpolation" [45] technique is applied to tackle the "checkerboard problem". The convection term are discretized by TVD scheme, and the viscous term by a second order central difference scheme. An adaptive dissipation coefficient [28] is adopted to resolve turbulent fluctuations in the LES region and to avoid the numerical oscillation in the RANS region. A second order implicit iteration is applied to the equations. The time-step of each iteration is $\Delta t = 0.0384h/U_0$ for the baseline case.

16.4 Results and Discussions

Except for the base flow, two groups of excited flow are simulated. One group is designed to investigate the amplitude effects of the slot harmonic excitation. The excitation frequency is fixed at $St_{h,ex} = 0.2$, which is the most effective excitation frequency in the literature, and velocity ratio $VR = 0.01, 0.05, 0.10, 0.20, 0.30$. Another is for the frequency effects. The velocity ratio is fixed at $VR = 0.01$, which ensures that the jet does not form, and the excitation frequencies $St_{h,ex} = 0.069, 0.112, 0.224, 0.471$.

Figure 16.3 presents a snapshot of λ_2-criterion of base flow. Iso-surfaces of λ_2 are colored with streamwise velocity. The separated shear layer rolls-up due to Kelvin-Helmholtz instability. Streamwise vortices occur between consecutive rollers downstream owing to the vortex dynamics. Then the large scale vortices breakdown further downstream and imping on the wall. And finally, a new boundary layer develops. The entrainment of the large scale spanwise coherent structures in the separated shear layer is evident from Fig. 16.4.

The base flow data are averaged for 13 flow-through times (1 flow-through time equals $44.0h/U_0$) after a transient period. The velocity and pressure are sampled at

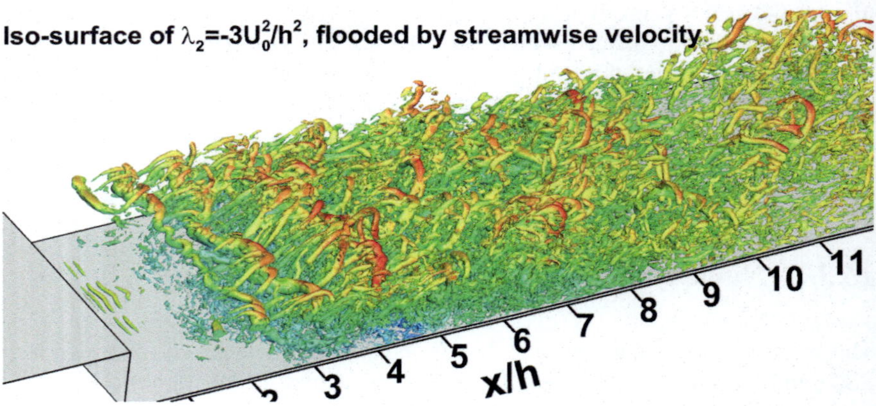

Fig. 16.3 Snapshot of an iso-surface of the λ_2 criterion colored with streamwise velocity

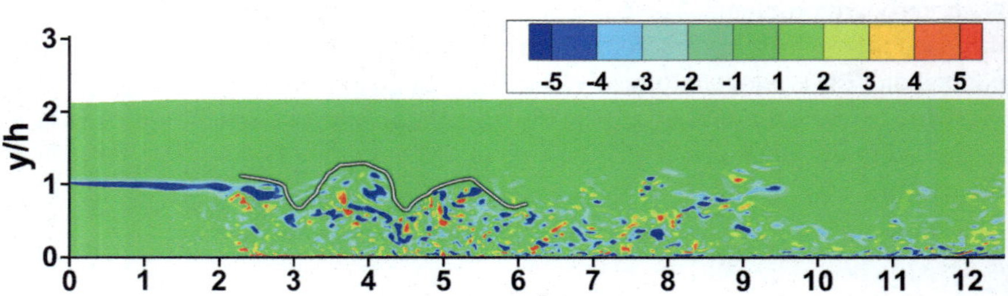

Fig. 16.4 Contours of non-dimensional spanwise vorticity on the plane of mid-span

a lot of locations (including locations near the lower wall, around reattachment, in the bubble and the shear layer) every time-step. The length of the available data is about 14,000. Meanwhile, the 3D flow data, including velocity and pressure in the focus region are collected every 10 time-steps with the data length of 601, thus the time span of them are $6010\Delta t \approx 231.0h/U_0$. The time-step for the excited flow is $1/(200St_{h,ex})$, and the flow data are averaged for 20 cycles after a transient period. Hence, the length of sampled data is 4000. The phase-averaged data are collected at four different phases of the excitation cycle, that is, $\varphi = 0.1\pi, 0.6\pi, 1.1\pi, 1.6\pi$.

We first validated the IDDES results of the base flow and analysed the spatial and temporal dynamics of the flow in the focus region. We then presented the amplitude effects and the frequency effects of the excited flow by TDT. We finally shed light on the mechanism for the bubble length reduction through an observation of the distribution of momentum thickness.

16.4.1 Base Flow

16.4.1.1 Averaged Flow and Statistical Quantities

Figure 16.5 presents the time-and spanwise-averaged streamlines of the base flow, which is consist of a large bubble and a secondary corner vortex at the step. Comparison of the time-averaged reattachment location with the experimental result is necessary to show the capability of the numerical approach. Several criterion were available for determining the reattachment location, i.e. streamwise velocity, skin friction coefficient, forward flow fraction. In this study, the skin friction coefficient is adopted to determine the reattachment location, as shown in Fig. 16.6. The solid line gives the skin friction coefficient distribution along the streamwise direction, which shows a good agreement with the experimental data. The reattachment location is where $C_f = 0$. A close-look on the curve gives $x_r/h = 6.19$, which shows an excellent agreement with the experimental result, $x_{r,exp}/h = 6.26 \pm 0.10$.

The wall-normal profiles of statistical quantities are compared with the experimental data at several streamwise locations as shown in Fig. 16.7. In the early stage of the shear layer, the simulation gives a steeper velocity gradient but much lower turbulent kinetic energy and shear stress, which resembles that of a laminar separation. This may be caused by the lack of turbulence content at the inlet. However, the turbulence content grows rapidly and the peaks of the profiles match that of the experimental data at $x/h = 2.0$. The peak location is further away from the wall than the experimental one further downstream. And the peak values are slightly larger. Finally in the relaxation region, an excellent agreement is obtained.

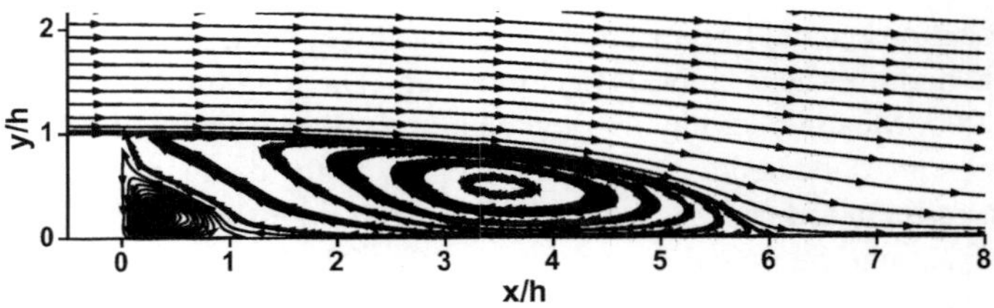

Fig. 16.5 Time-and spanwise-averaged streamlines of base flow

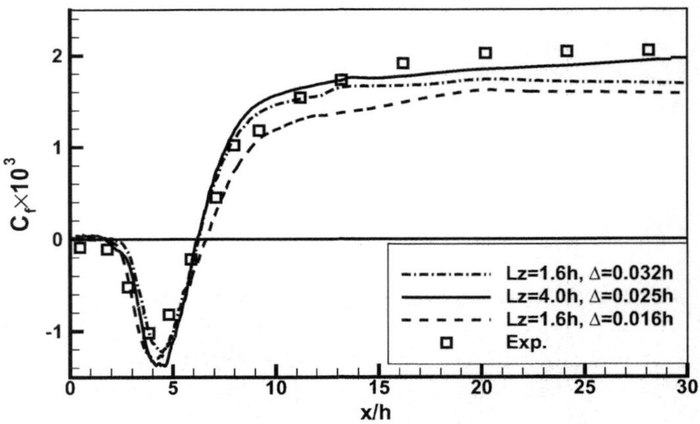

Fig. 16.6 Skin friction coefficients of base flow

16.4.1.2 Dynamic Mode Analyses

POD analysis

The POD approach was first introduced to the fluid dynamic community to extract coherent structures by Lumley [46]. POD decomposes the flow field into time-averaged field and superimpose of a series of spatially orthogonal modes, with a time-varying coefficient. A snapshot POD algorithm was developed by Sirovich [47] and Meyer et al. [48], making it ease of use. The POD modes ranks from large to small according to mode energy. We perform POD analysis on the pressure field in the focus region using 600 snapshots. Figure 16.8 presents the distribution of relative energy of the first 100 modes. The first 14 modes contain a large fraction of energy.

Several significant modes are shown in Fig. 16.9. The first three modes represent shedding motion of coherent structures, the dominant frequencies of which are given in Fig. 16.10a, $St_h = 0.067$ and 0.112. While, the last three modes represent the coherent motion in the shear layer, the dominant frequency of which is given in Fig. 16.10b, $St_h = 0.224$.

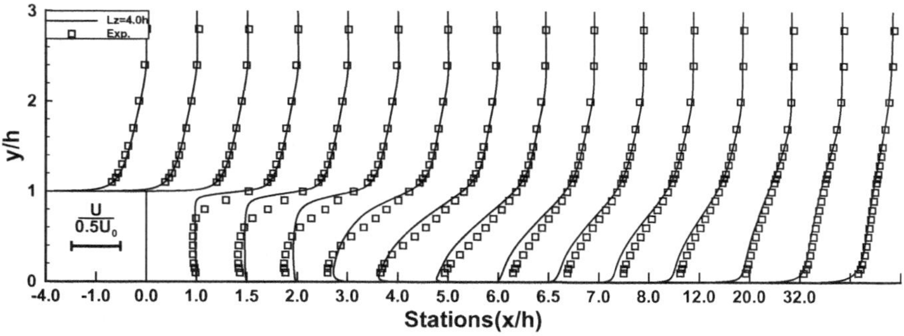

(a) Streamwise velocity

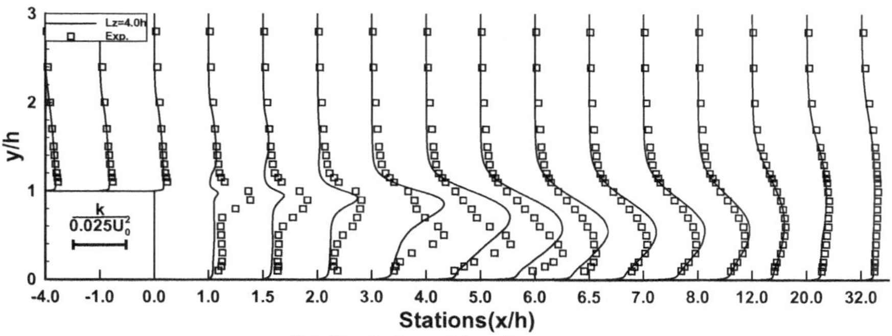

(b) Turbulent kinetic energy

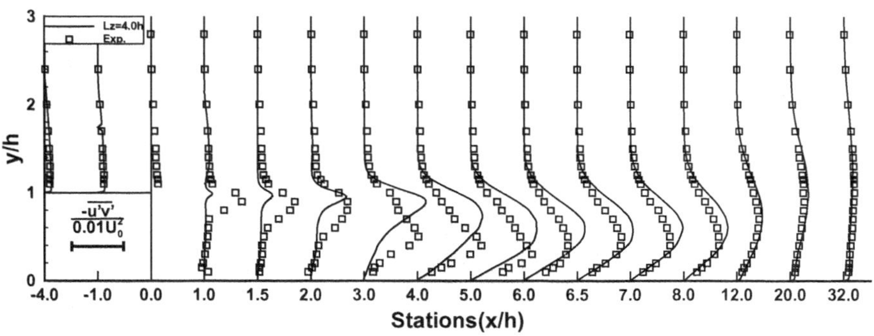

(c) Shear stress

Fig. 16.7 Wall-normal profiles of statistical quantities at several streamwise locations of base flow

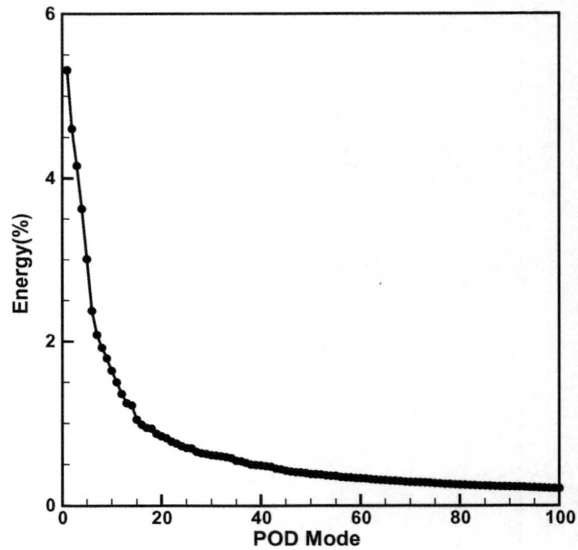

Fig. 16.8 Distribution of pressure POD mode energy

DMD analysis

The POD analysis doesn't recognize higher frequency motions in the shear layer, but the DMD analysis does. DMD was first proposed by Schmid and Sesterhenn [49]. And a connection was established with dynamic system by Rowley et al. [50]. The DMD analysis could be regarded as an algorithm for Koopman modes of dynamic system. A numerical stable algorithm was given by Schmid [51]. The DMD modes are both spatially and temporally orthogonal, so the dynamic important modes could be recognized by DMD without frequency aliasing.

Several dynamic important DMD modes are given in Fig. 16.11. The shedding modes and shear layer modes are both recognized, and a low-frequency mode (mode 1, $St_h = 0.024$) is also recognized. Mode 2 and mode 3 are shedding modes, while mode 4 to mode 6 are shear layer modes. Mode 4 represents larger scale coherent structures produced by vortex pairing, the characteristic frequency of which is $St_h = 0.225$. While, mode 5 represents smaller scale coherent structures produced by K-H instability, the characteristic frequency of which is $St_h = 0.471$. Consider the dynamic importance and energy contribution of the coherent motions, mode 2 to mode 5 are excited by a slot harmonic actuator, which will be presented in Sect. 16.4.3.

16.4.2 Amplitude Effects

By literature survey, we find that the most effective excitation frequency is around 0.20. In order to investigate the amplitude effects, the excitation frequency was fixed

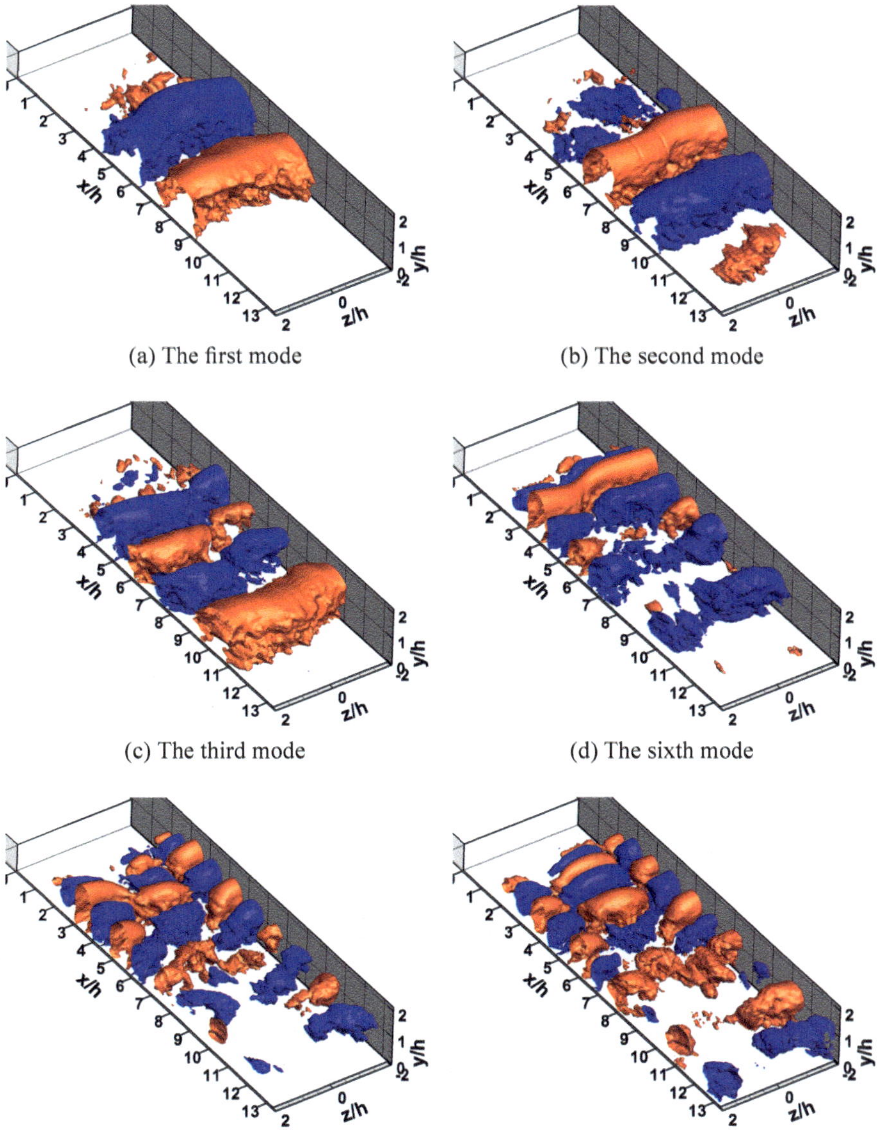

(a) The first mode (b) The second mode

(c) The third mode (d) The sixth mode

(e) The seventh mode (f) The eighth mode

Fig. 16.9 POD modes of pressure field. Iso-surface of $p' = -0.001$ is colored with blue, iso-surface of $p' = 0.001$ is colored with orange

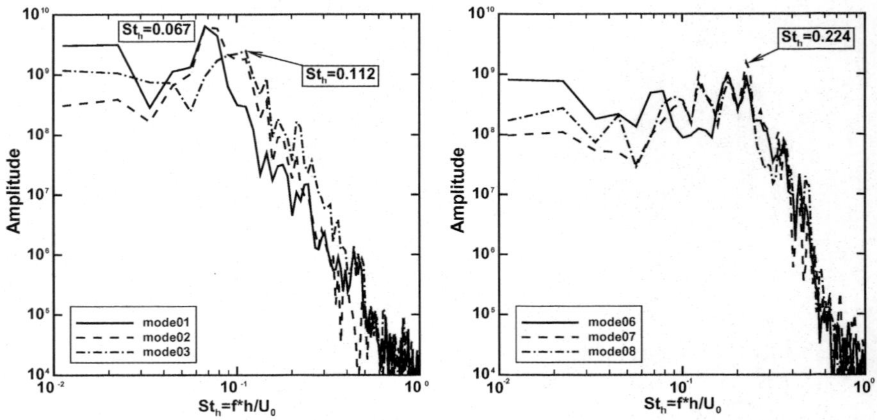

(a) Spectra of the first to the third modes (b) Spectra of the sixth to the eighth modes

Fig. 16.10 Spectra of the POD coefficients of pressure field

at $St_{h,ex} = 0.20$, and five cases for $VR = 0.01, 0.05, 0.10, 0.20, 0.30$ are included. Triple decomposition technique is applied here to explain the excitation effects, which is proposed by Hussain and Reynolds [52].

$$f(x, y, z, t) = \bar{f}(x, y, z) + \tilde{f}(x, y, z, t) + f'(x, y, z, t), \tag{16.10}$$

where, $\bar{f}(x, y, z)$ is time-average,
$\tilde{f}(x, y, z, t)$ is periodic component, $\tilde{f} = \langle f \rangle - \bar{f}$,
and $f'(x, y, z, t)$ is stochastic turbulence, $f' = f - \langle f \rangle$, $\langle\ \rangle$ stands for phase-average.

Figure 16.12 presents the contour of spanwise-averaged periodic component of non-dimensional spanwise vorticity ($\tilde{\omega}_z h/U_0$) for phase $\varphi = 1.1\pi$ for the five different velocity ratio cases. The jet does not form for case $VR = 0.01$. For cases $VR = 0.05$ and 0.10, jets form but the spanwise vorticity near the slot does not roll-up. For cases $VR = 0.20$ and 0.30, jets form and the spanwise vorticity rolls-up rapidly. The results agree well with the criterion proposed by Holman et al. [53].

Figure 16.13 gives the spanwise-averaged streamlines on phases $\varphi = 0.1\pi, 0.6\pi, 1.1\pi, 1.6\pi$ for the five different velocity ratio cases. There exist two consecutive spanwise vortices in the bubble for all five cases. The scale of the spanwise vortices increase and then decrease evolving downstream, finally, evanish at the reattachment. The location of the first spanwise vortex shifts upstream along with the increasing VR because the excitation only takes advantage of the intrinsic instability of the shear layer when the jet does not form, and additional strong shear takes effects when the jet forms. The curvature of the streamlines increase along with the increasing VR. The larger curvature enhances then entrainment of the spanwise vortices, leading to faster spreading of the shear layer, resulting in earlier reattachment of the shear layer. The bubble length reductions for different velocity ratios are illustrated by

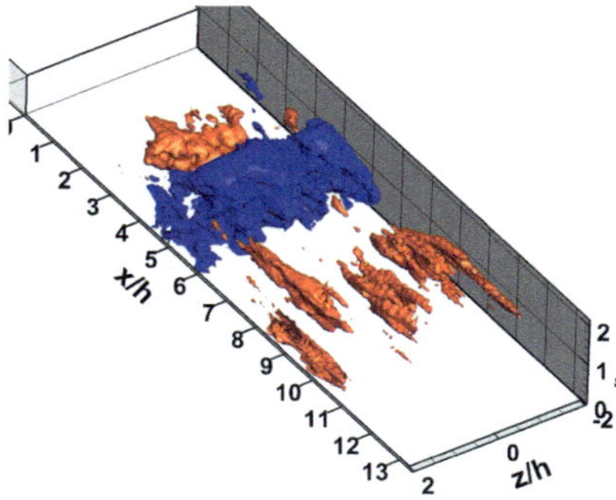

(a) Mode 1, $St_h = 0.024$

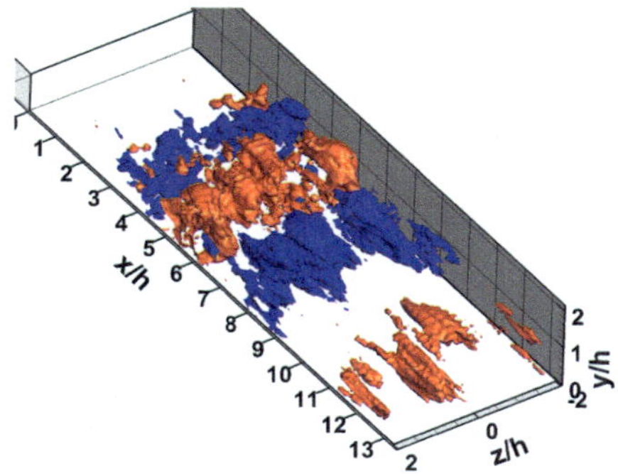

(b) Mode 2, $St_h = 0.069$

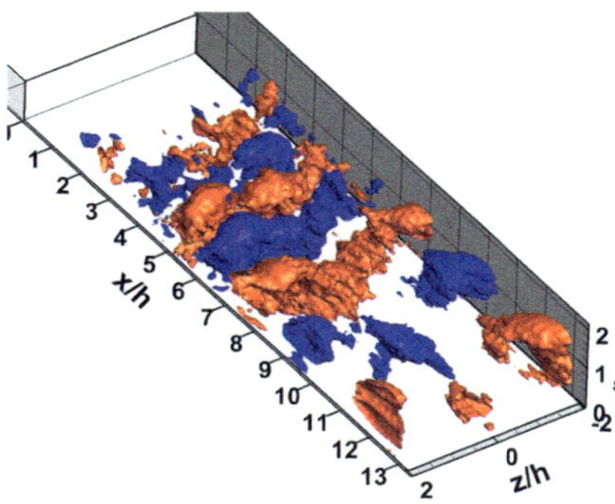

(c) Mode 3, $St_h = 0.114$

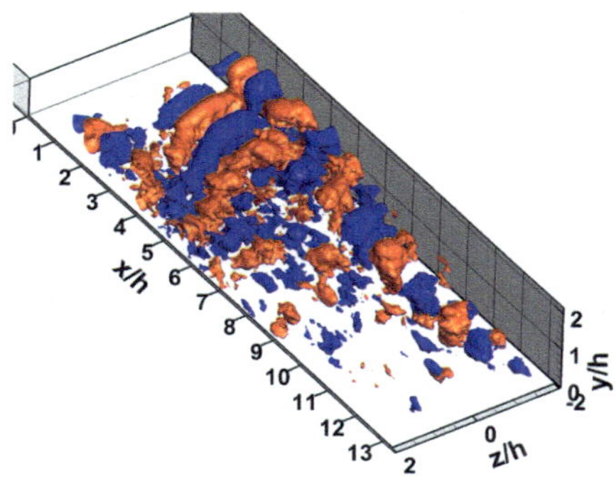

(d) Mode 4, $St_h = 0.225$

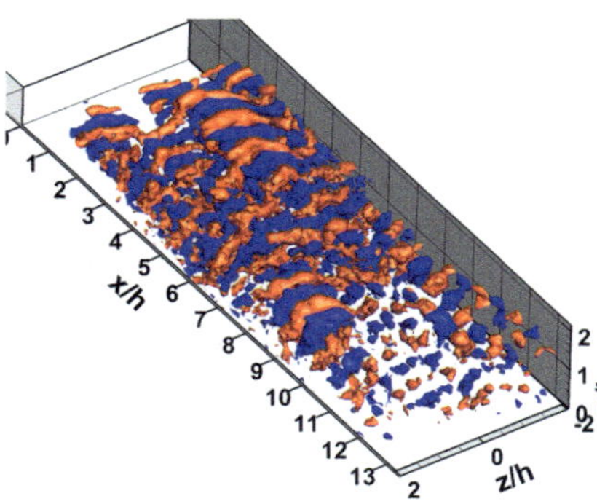

(e) Mode 5, $St_h = 0.471$

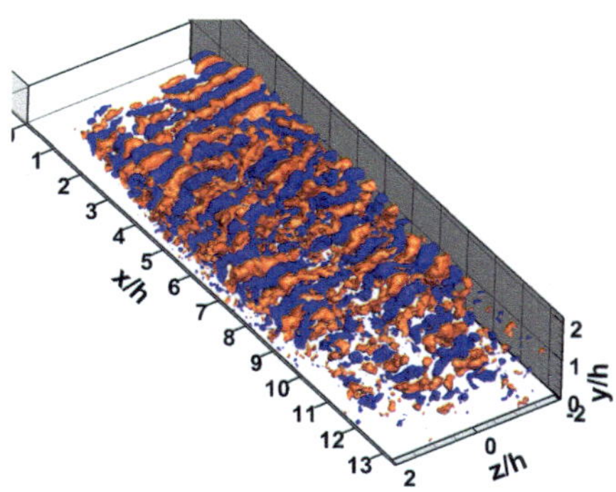

(f) Mode 6, $St_h = 0.642$

Fig. 16.11 DMD modes of pressure field. Iso-surface of $p = -0.001$ is colored with blue, iso-surface of $p = 0.001$ is colored with orange

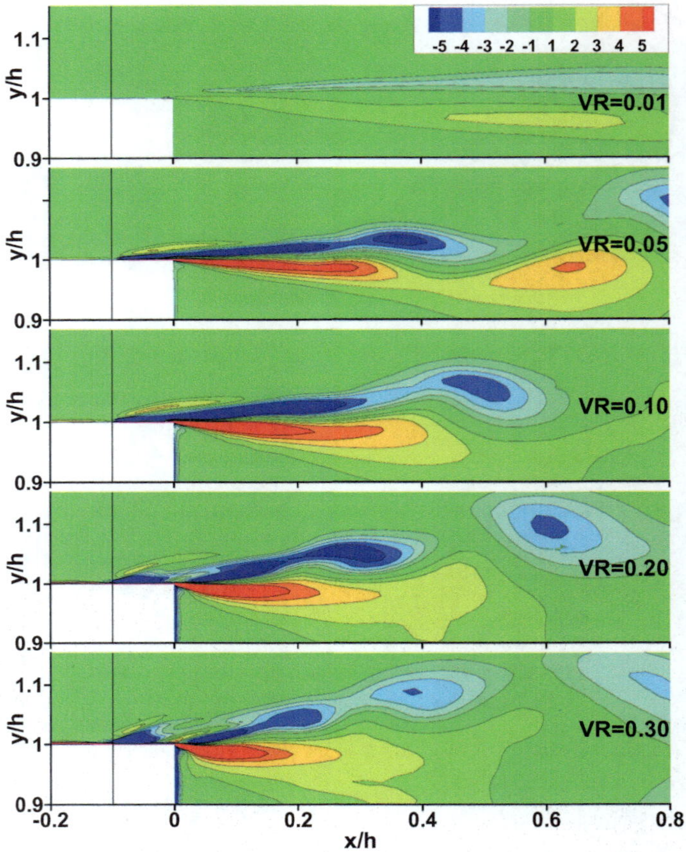

Fig. 16.12 Spanwise-averaged periodic component of non-dimensional spanwise vorticity for phase $\varphi = 1.1\pi$ for the five different VR cases

the solid line in Fig. 16.14. The reattachment location moves upstream rapidly when jet forms, but the bubble length reduction ratio decrease when both jet forms and spanwise vortex rolls-up. In general, the formation of the jet plays an important role on the reattachment.

16.4.3 Frequency Effects

It has been shown that the jet does not form for case both $VR = 0.01$ and $St_{h,ex} = 0.20$ in the previous subsection. And the excitation only takes advantage of the intrinsic instability of the shear layer. For the sake of utilizing the intrinsic instability of the

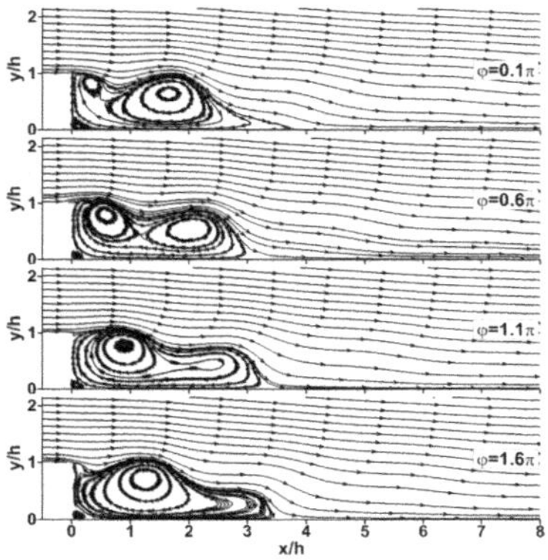

(a) $VR = 0.30$

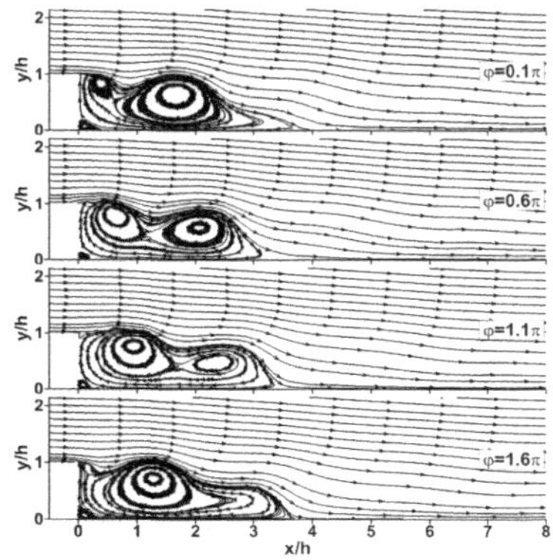

(b) $VR = 0.20$

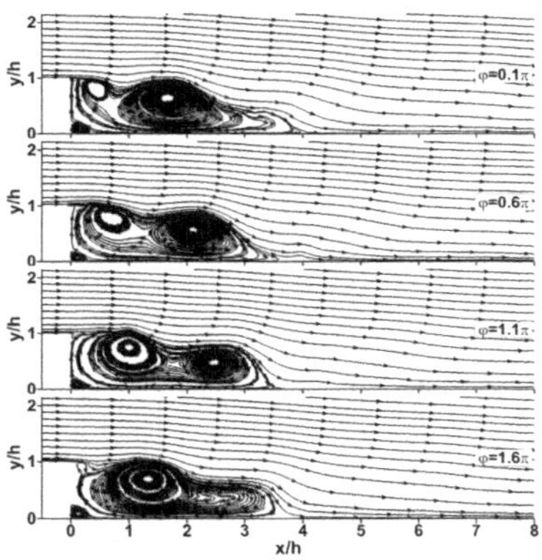

(c) $VR = 0.10$

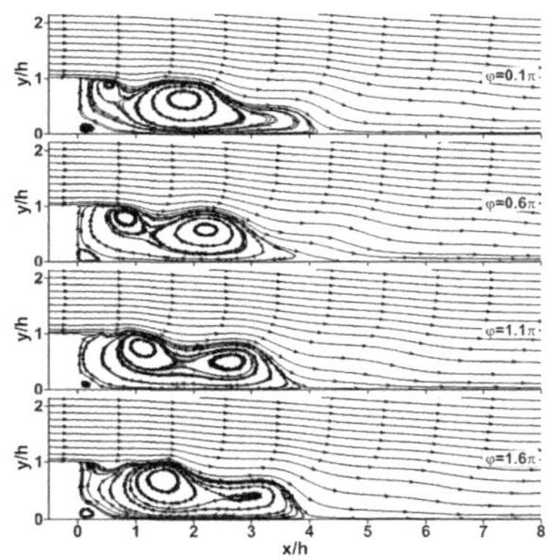

(d) $VR = 0.05$

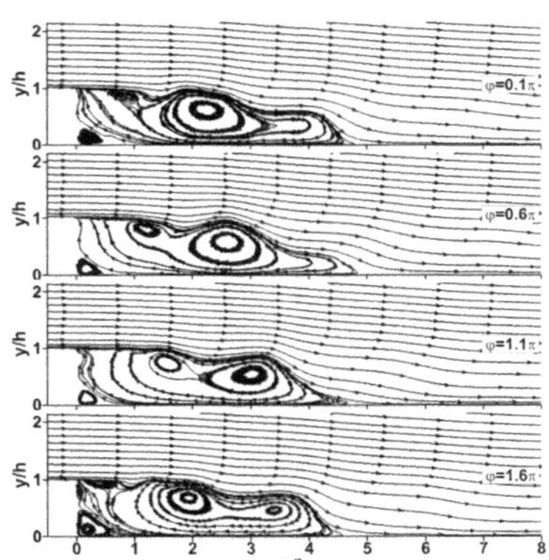

(e) $VR = 0.01$

Fig. 16.13 Spanwise-averaged streamlines on phases $\varphi = 0.1\pi, 0.6\pi, 1.1\pi, 1.6\pi$

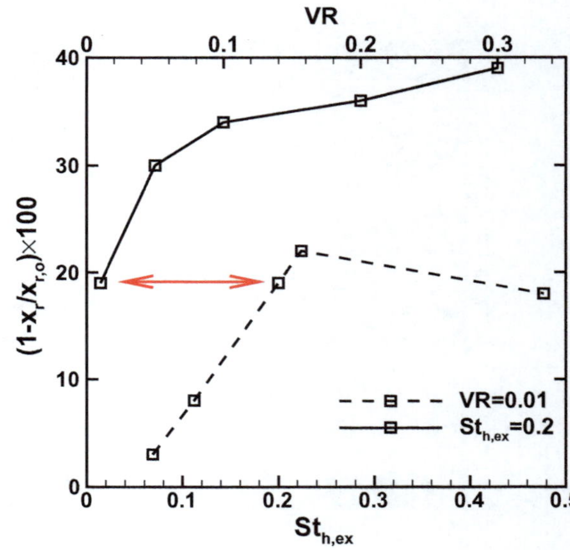

Fig. 16.14 Bubble length reduction for different velocity ratios and excitation frequencies

base flow, four excitation frequencies are considered, that is, $St_{h,ex} = 0.069$, 0.112, 0.224, 0.471. And the excitation velocity ratio is fixed at $VR = 0.01$, for which the jet does not form for all four cases. The excitation frequencies considered here are related to the instability modes of the shear layer, that is to say, $St_{h,ex} = 0.069$ and 0.112 correspond to shedding mode, $St_{h,ex} = 0.224$ to vortex pairing and $St_{h,ex} = 0.471$ to K-H instability, respectively.

Figure 16.15 shows the contours of spanwise-averaged periodic component of non-dimensional pressure field ($\tilde{p}/(0.5\rho U_0^2)$) on phases $\varphi = 0.1\pi, 0.6\pi, 1.1\pi, 1.6\pi$ for four different excitation frequencies. Figure 16.15a, b present the ones excited under $St_{h,ex} = 0.069$ and 0.112 (the shedding modes). The excited flows are dominated by vortex shedding. In other words, the shedding mode is excited. While, vortex pairing and K-H instability are also excited as suggested in Fig. 16.15c, d. The intrinsic instabilities of the base flow resonance with the excitation, prevailing in the excited flow.

Figure 16.16 shows the spanwise-averaged streamlines on phases $\varphi = 0.1\pi, 0.6\pi$, 1.1π, 1.6π for four different excitation frequencies. The roller in the shear layer is smeared, and the separation bubble expands and shrinks during an excitation cycle as depicted in Fig. 16.16a, b. Meanwhile, the reattachment location oscillates with the separation bubble. However, there exists two consecutive spanwise vortices in the separation bubble for the case exciting vortex pairing as shown in Fig. 16.16c, which increases the curvature of the streamlines, resulting in enhancing the entrainment of the shear layer flow. The rollers convect downstream and evanish at reattachment location which keeps almost stationary during the excitation cycle. Besides, the rollers dominate in the shear layer as shown in Fig. 16.16d. And the reattachment location keeps stationary, also. In summary, the reattachment is accelerated for all

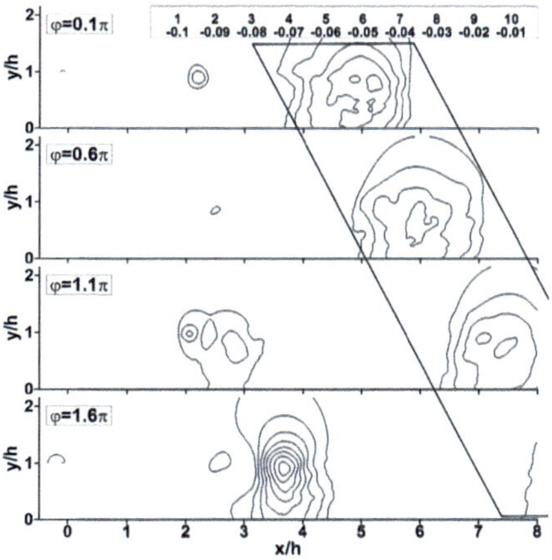

(a) $St_{h,ex} = 0.069$

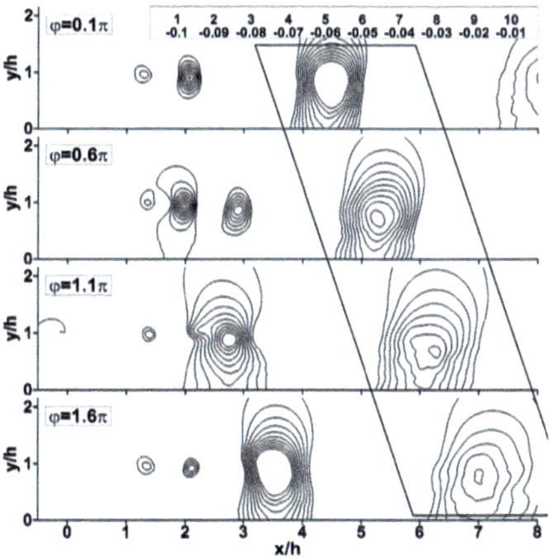

(b) $St_{h,ex} = 0.112$

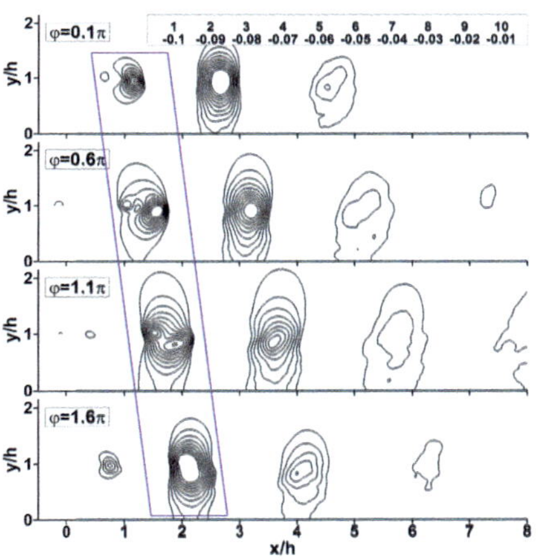

(c) $St_{h,ex} = 0.224$

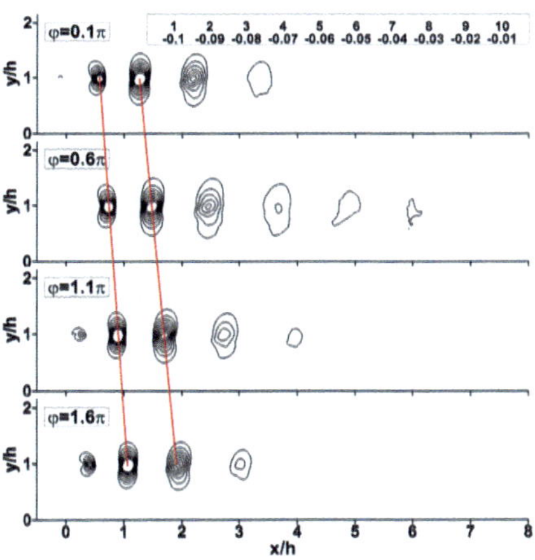

(d) $St_{h,ex} = 0.471$

Fig. 16.15 Spanwise-averaged periodic component of non-dimensional pressure field on phases $\varphi = 0.1\pi, 0.6\pi, 1.1\pi, 1.6\pi$ for four different excitation frequencies

four cases whatever exciting the shedding modes, vortex pairing or K-H instability as depicted by the dash line in Fig. 16.14. We obtain a maximum bubble length reduction for exciting vortex pairing.

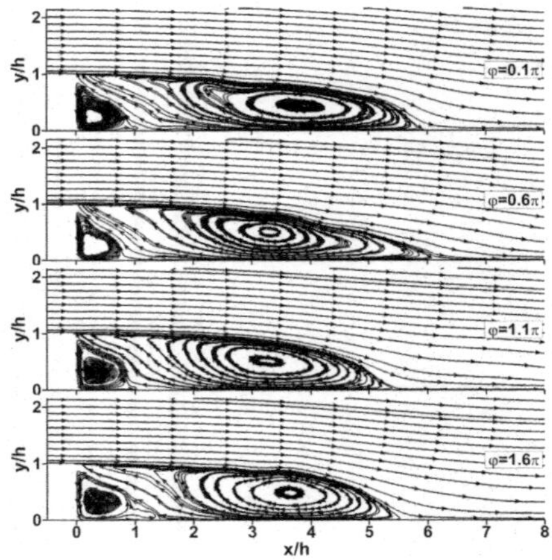

(a) $St_{h,ex} = 0.069$

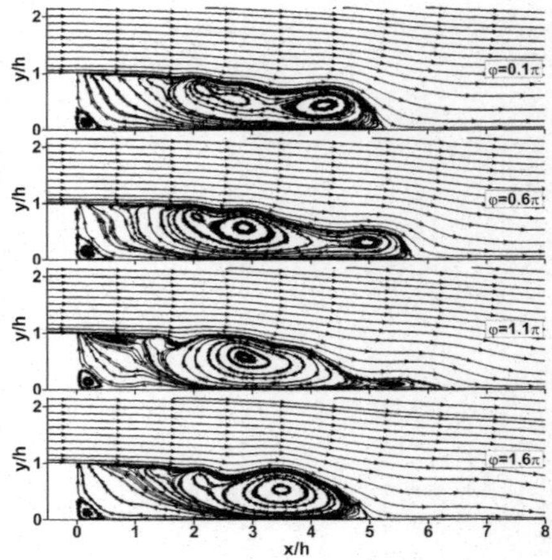

(b) $St_{h,ex} = 0.112$

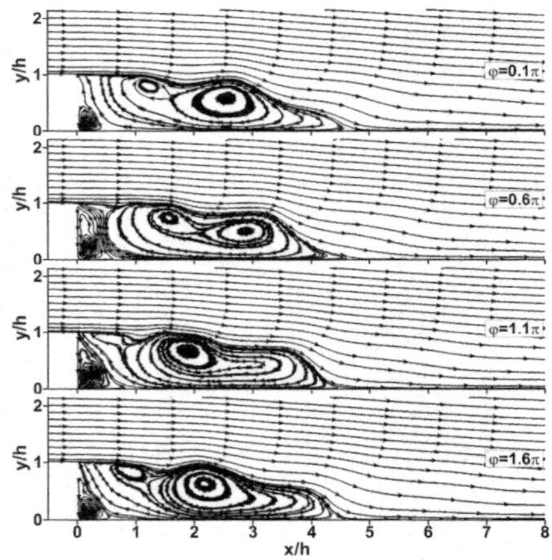

(c) $St_{h,ex} = 0.224$

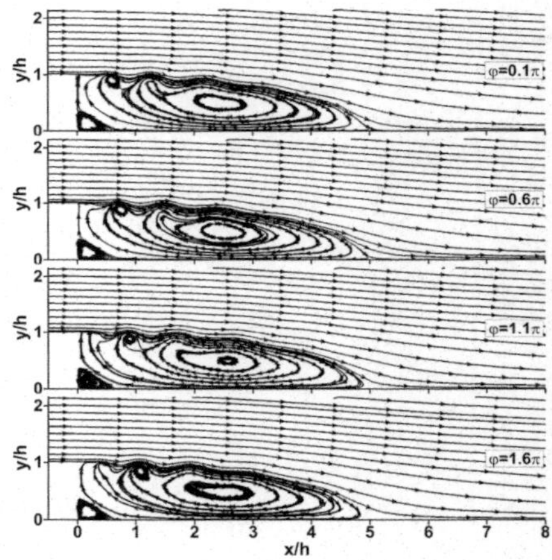

(d) $St_{h,ex} = 0.471$

Fig. 16.16 Spanwise-averaged streamlines on phases $\varphi = 0.1\pi, 0.6\pi, 1.1\pi, 1.6\pi$ for four different excitation frequencies

16.4.4 Mechanism for Bubble Length Reduction

As declared in Sect. 16.4.3, vortex shedding dominates in the excited flow if exciting the shedding mode. The excitation frequency dominates in the reattachment and the relaxation region as depicted in Fig. 16.17b and d, and the development of the shear layer is not significantly altered as shown in Fig. 16.17a and c. Consequently, the distribution of the vorticity thickness is also not significantly changed compared

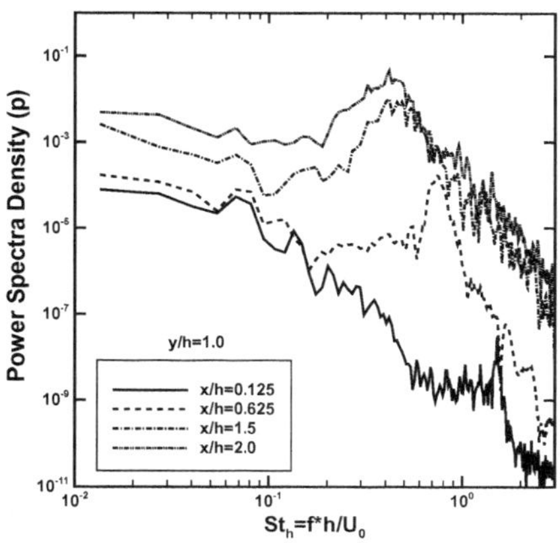

(a) $St_{h,ex} = 0.069$, shear layer region

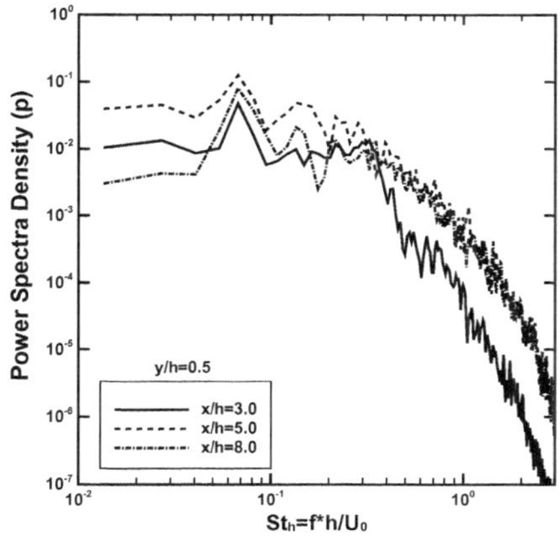

(b) $St_{h,ex} = 0.069$, reattachment region & relaxation region

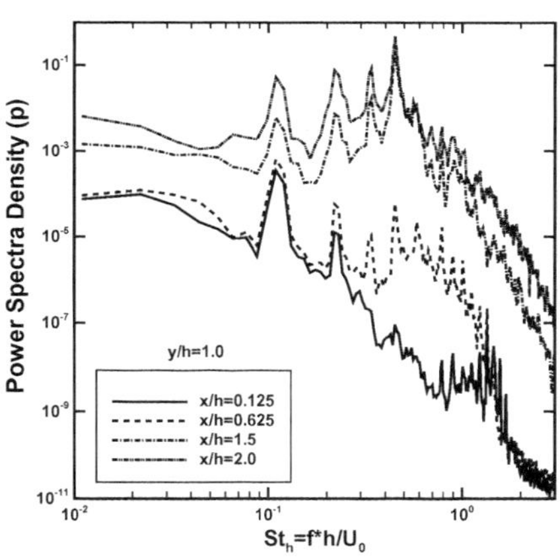

(c) $St_{h,ex} = 0.112$, shear layer region

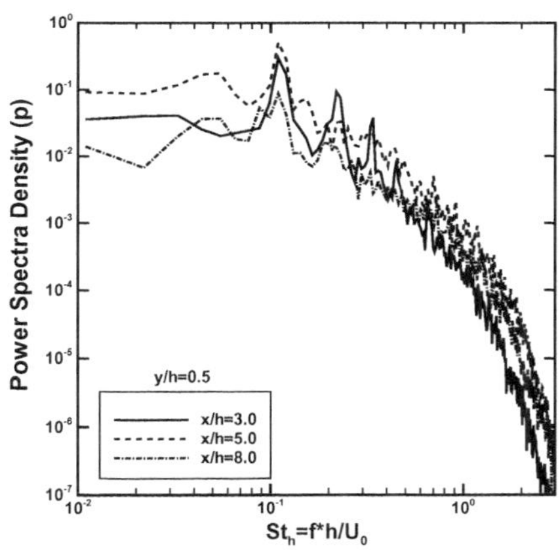

(d) $St_{h,ex} = 0.112$, reattachment region & relaxation region

Fig. 16.17 Power spectra density of pressure for $St_{h,ex} = 0.069$ and 0.112

with base flow as depicted by the triangle symbols in Fig. 16.19, especially the one excited by $St_{h,ex} = 0.069$.

If exciting vortex pairing, vortex pairing process dominates in the separated bubble as depicted by solid line in Fig. 16.18b, but shedding process dominates downstream as depicted by dash line and dash-dot line in Fig. 16.18b. The exciting frequency dominates in the shear layer, and the rolling-up of the it is accelerated as depicted in Fig. 16.18a. As a result, the vorticity thickness grows rapidly between $x/h = 1.0$ and 2.0, the growth rate of which is larger than that of the base flow where vortex pairing occurs, followed by a linear growth as the base flow does, as depicted by solid line

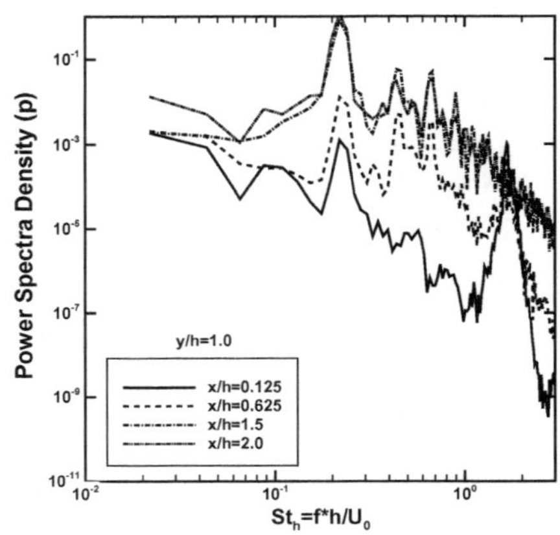

(a) $St_{h,ex} = 0.224$, shear layer region

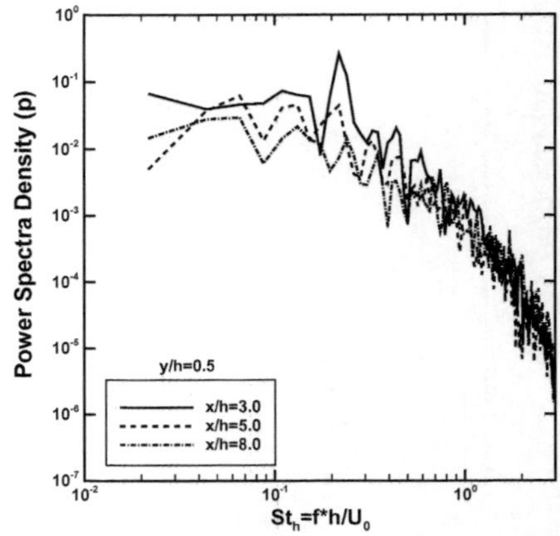

(b) $St_{h,ex} = 0.224$, reattachment region & relaxation region

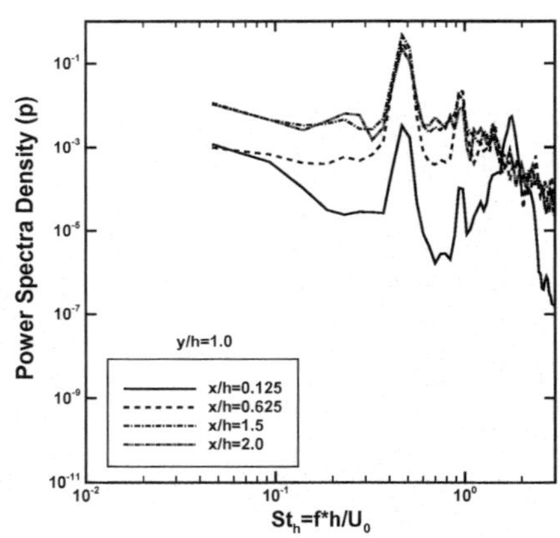

(c) $St_{h,ex} = 0.471$, shear layer region

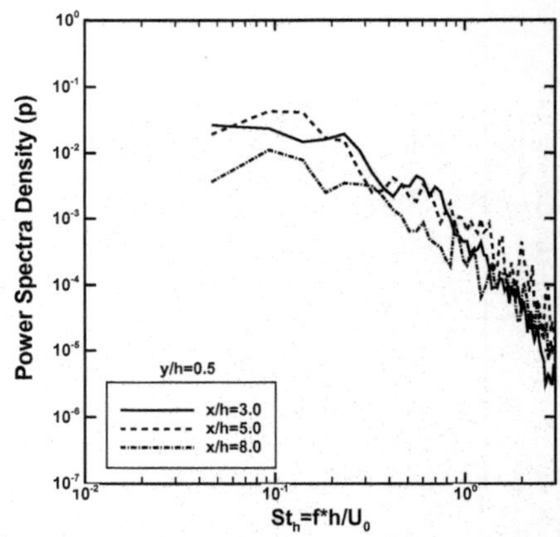

(d) $St_{h,ex} = 0.471$, reattachment region & relaxation region

Fig. 16.18 Power spectra density of pressure for $St_{h,ex} = 0.224$ and 0.471

with upside-down triangle in Fig. 16.19. It is because that the vortex pairing process shift upstream in the excited flow, and it is more organized. In another words, the vortex pairing process enhanced if exciting vortex pairing.

If exciting K-H instability, rolling-up of the shear layer is accelerated, and the dominant frequency of the shear layer is extremely close to the exciting frequency as given in Fig. 16.18c. Consequently, vorticity thickness grows rapidly in the early stage of the shear layer, followed by a linear growth as depicted by dash line with upside-down triangle in Fig. 16.19. Quickly, it restores the behavior of the base flow. Besides, the reattachment region and the relaxation region are dominated by the shedding process as given in Fig. 16.18d.

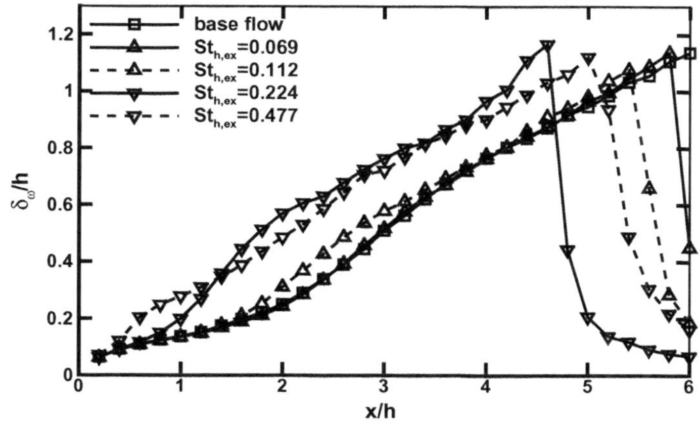

Fig. 16.19 Vorticity thickness distribution along streamwise direction

16.5 Conclusions

We conducted IDDES of the BFS base flow and the excited ones. The predicted base flow agrees well with the experimental results for both the time-averaged one and the spectra content. Base flow contains two distinct modes: shear layer mode and shedding mode, the former including Kelvin-Helmholtz instability ($St_h = 0.471$) and vortex pairing ($St_h = 0.224$) and the later representing large scale spanwise vortices shedding from the separated bubble ($St_h = 0.069$ and 0.112).

The amplitude effects and the frequency effects were investigated. Bubble length reduction increases with increasing velocity ratio, and the formation of jet plays an important role. If the jet does not form, i.e. $VR = 0.01$, the exciting frequency plays an outstanding role. Taking advantage of the intrinsic instability of the shear layer of base flow, bubble length are largely reduced. And we obtained a maximum reduction of the bubble length if exciting vortex pairing. Vortex pairing results in larger scale spanwise vortices, hence the entrainment of the shear layer is enhanced. Consequently, the development of the shear layer is accelerated and the shear layer reattaches earlier. As a result, bubble length reduction is obtained.

References

1. Eaton JK, Johnston JP (1981) A review of research on subsonic turbulent flow reattachment. AIAA J 19(9):1093–1100
2. Brown GL, Roshko A (1974) On density effects and large structure in turbulent mixing layers. J Fluid Mech 64(04):775–816
3. Bradshaw P, Wong F (1972) The reattachment and relaxation of a turbulent Shear layer. J Fluid Mech 52(01):113–135

4. Hasan M (1992) The flow over a backward-facing step under controlled perturbation: laminar separation. J Fluid Mech 238:73–96
5. McGuinness MD (1978) Flow with a separation bubble: steady and Unsteady Aspects. University of Cambridge
6. Kim J, Kline S, Johnston J (1978) Investigation of separation and reattachment of a turbulent shear layer: flow over a backward-facing step. Report MD-37, thermosciences division, department of mechanical engineering
7. Chandrsuda C (1975) A reattaching turbulent Shear layer in incompressible flow. Imperial Coll Sci Technol
8. Troutt T, Scheelke B, Norman T (1984) Organized structures in a reattaching separated flow field. J Fluid Mech 143:413–427
9. Driver DM, Seegmiller HL, Marvin JG (1987) Time-dependent behavior of a reattaching shear layer. AIAA J 25(7):914–919
10. Eaton JK, Johnston JP (1982) Low frequency unsteadyness of a reattaching turbulent Shear layer. In: Proceedings of turbulent Shear flows 3, Springer, pp 162–170
11. Roos FW, Kegelman JT (1986) Control of coherent structures in reattaching laminar and turbulent Shear layers. AIAA J 24(12):1956–1963
12. Hudy LM, Naguib A, Humphreys WM (2007) Stochastic estimation of a separated-flow field using wall-pressure-array measurements. Phys Fluids (1994-present) 19(2):024103
13. Bhattacherjee S, Scheelke B, Troutt T (1986) Modification of vortex interactions in a reattaching separated flow. AIAA J 24(4):623–629
14. Chun KB, Sung H (1996) Control of turbulent separated flow over a backward-facing step by local forcing. Exp Fluids 21(6):417–426
15. Yoshioka S, Obi S, Masuda S (2001) Turbulence statistics of periodically perturbed separated flow over backward-facing step. Int J Heat Fluid Flow 22(4):393–401
16. Kapiris PG, Mathioulakis DS (2014) Experimental study of vortical structures in a periodically perturbed flow over a backward-facing step. Int J Heat and Fluid Flow 47:101–112
17. Armaly BF, Durst F, Pereira J et al (1983) Experimental and theoretical investigation of backward-facing step flow. J Fluid Mech 127:473–496
18. Neto AS, Grand D, Métais O et al (1993) A numerical investigation of the coherent vortices in turbulence behind a backward-facing step. J Fluid Mech 256:1–25
19. Jovic S, Driver DM (1994) Backward-facing step measurements at low Reynolds number $Re_h = 5000$
20. Le H, Moin P, Kim J (1997) Direct numerical simulation of turbulent flow over a backward-facing step. J Fluid Mech 330:349–374
21. Kopera MA, Cantwell C, Kerr RM, et al (2014) Direct numerical simulation of turbulent flow over a backward-facing step. University of Warwick
22. Dietiker JF, Hoffmann KA (2009) Predicting wall pressure fluctuation over a backward-facing step using detached eddy simulation. J Aircr 46(6):2115–2120
23. Dejoan A, Leschziner MA (2004) Large eddy simulation of periodically perturbed separated flow over a backward-facing step. Int J Heat Fluid Flow 25(4):581–92
24. Spalart PR, Allmaras SR (1992) A one equation turbulence model for aerodinamic flows. AIAA J 94
25. Spalart P, Jou W, Strelets M et al (1997) Comments on the feasibility of LES for wings, and on a hybrid RANS/LES approach. Adv DNS/LES 1:4–8
26. Spalart PR, Deck S, Shur M et al (2006) A new version of detached-eddy simulation, resistant to ambiguous grid densities. Theor Comput Fluid Dyn 20(3):181–195
27. Menter FR (1994) Two-equation eddy-viscosity turbulence models for engineering applications. AIAA J 32(8):1598–1605
28. Strelets M (2001) Detached eddy simulation of massively separated flows. In: Proceedings of AIAA, aerospace sciences meeting and exhibit, 39th, Reno, NV
29. Menter F, Kuntz M, Langtry R (2003) Ten years of industrial experience with the SST turbulence model. Turbul Heat Mass Trans 4(1):625–632

30. Shur ML, Spalart PR, Strelets MK et al (2008) A hybrid RANS-LES approach with delayed-DES and wall-modelled LES capabilities. Int J Heat Fluid Flow 29(6):1638–1649
31. Neumann J, Wengle H (2003) DNS and LES of passively controlled turbulent backward-facing step flow. Flow Turbul Combust 71(1–4):297–310
32. Neumann J, Wengle H (2004) Coherent structures in controlled separated flow over sharp-edged and rounded steps. J Turbul 5(22):14
33. Benard N, Pons-Prats J, Periaux J et al (2016) Turbulent separated Shear flow control by surface plasma actuator: experimental optimization by genetic algorithm approach. Exp Fluids 57(2):1–17
34. Heenan AF, Morrison JF (1998) Passive control of backstep flow. Exp Thermal Fluid Sci 16(1):122–132
35. Kang S, Choi H (2002) Suboptimal feedback control of turbulent flow over a backward-facing step. J Fluid Mech 463:201–227
36. Lai JC, Yue J, Platzer MF (2002) Control of backward-facing step flow using a flapping foil. Exp Fluids 32(1):44–54
37. Morioka T, Honami S (2004) Dynamic characteristics in a control system of backward facing step flow by vortex generator jets. AIAA paper 2125
38. Uruba V, Jonáš P, Mazur O (2007) Control of a channel-flow behind a backward-facing step by suction/blowing. Int J Heat Fluid Flow 28(4):665–72
39. Henning L, King R (2007) Robust multivariable closed-loop control of a turbulent backward-facing step flow. J Aircr 44(1):201–8
40. Dandois J, Garnier E, Sagaut P (2007) Numerical simulation of active separation control by a synthetic jet. J Fluid Mech 574:25–58
41. Avdis A, Lardeau S, Leschziner M (2009) Large eddy simulation of separated flow over a two-dimensional hump with and without control by means of a synthetic slot-jet. Flow Turbul Combust 83(3):343–70
42. Leschziner MA, Lardeau S (2011) Simulation of slot and round synthetic jets in the context of boundary-layer separation control. Philos Trans R Soc Lond A Math Phys Eng Sci 369(1940):1495–1512
43. De Brederode V, Bradshaw P (1972) Three-dimensional flow in nominally two-dimensional separation bubbles: flow behind a rearward facing step. Dept Aeronaut Imperial Coll Sci Technol
44. Shur M, Strelets M, Zaikov L, et al (1995) Comparative numerical testing of one-and two-equation turbulence models for flows with separation and reattachement AIAA paper
45. Rhie C, Chow W (1983) Numerical study of the turbulent flow past an airfoil with trailing edge separation. AIAA J 21(11):1525–1532
46. Lumley JL (1967) The structure of inhomogeneous turbulent flows. Atmos Turbul Radio Wave Propag 166–178
47. Sirovich L (1987) Turbulence and the dynamics of coherent structures. Part I: coherent structures. Q Appl Math 45(3):561–571
48. Meyer KE, Pedersen JM, Özcan O (2007) A turbulent jet in crossflow analysed with proper orthogonal decomposition. J Fluid Mech 583:199–227
49. Schmid P, Sesterhenn J (2008) Dynamic mode decomposition of numerical and experimental data. In: Proceedings of 61st annual meeting of the APS division of fluid dynamics
50. Rowley CW, Mezić I, Bagheri S et al (2009) Spectral analysis of nonlinear flows. J Fluid Mech 641:115–127
51. Schmid PJ (2010) Dynamic mode decomposition of numerical and experimental data. J Fluid Mech 656:5–28
52. Hussain AKMF, Reynolds WC (1970) The mechanics of an organized wave in turbulent Shear flow. J Fluid Mech 41(2):241–258
53. Holman R, Utturkar Y, Mittal R et al (2005) Formation criterion for synthetic jets. AIAA J 43(10):2110–2116

Chapter 17
Effects of Flow Control Devices on an Airfoil

Lan Xie, Yao Zheng, Jianfeng Zou, Tingwei Ji, Xing Shi and Fangfang Xie

Abstract A set of active flow control methods on controlling flow separation and drag reduction over a NACA0015 airfoil were numerically studied by solving three-dimensional incompressible unsteady Reynolds-averaged Navier-Stokes equations. Four achievements have been made: (1) Through the comparison of several turbulence models, DES k-epsilon model was adopted since it can well capture the unsteady separation around the airfoil. (2) The numerical results of a NACA0015 airfoil without/with the blowing/suction and pulsed jets as flow control devices have been proved to be very consistent with the experimental results. (3) The trends of Reynolds stresses were analyzed to reveal the flow-control mechanism over airfoil under various active flow control devices. (4) A fluidic mechanism for pulsed jets and blowing/suction devices can significantly improve the control performance, and optimization could be conducted to obtain better results in the future.

Keywords Turbulence models · Flow control · Drag reduction

17.1 Introduction

Flow separation and drag reduction are directly related to the issues such as aviation transport efficiency and emissions to the atmosphere. In the year of 2020, the aviation industry expects to reach the goal of 50% in emission reduction, which drives the development on more efficient flow control methods. The cooperation between China and Europe makes an effort on research in this field.

In the entire MARS project, we are responsible for a part of numerical simulation on the flow separation and analysis of the drag reduction control system. DES and LES were selected as the turbulence models to analyse the characteristics of oscillating surface, blowing/suction and pulsed jets. Then a relationship between the flow control and Reynolds stress will be established. According to the content of the

L. Xie · Y. Zheng (✉) · J. Zou · T. Ji · X. Shi · F. Xie
School of Aeronautics and Astronautics, Center for Engineering and Scientific Computation, Zhejiang University, Hangzhou 310027, Zhejiang, People's Republic of China
e-mail: yao.zheng@zju.edu.cn

N. Qin et al. (eds.), *Advances in Effective Flow Separation Control for Aircraft Drag Reduction*, Computational Methods in Applied Sciences 52,
https://doi.org/10.1007/978-3-030-29688-9_17

project, a series of numerical simulation researches must be required. The main contents include further development of the algorithm of fluid-structure interaction and related modules based on the existing computational fluid dynamics codes, research in control characteristics of oscillating surface, blowing/suction and fluidic vortex generators (FVG), using comprehensive and theoretical analysis to establish a relationship between the flow control and Reynolds stress, fundamental research in active control of smart structure design and layout.

Our work is to carry out researches using our developed parallel CFD solver which is based on the LES/DES technique. We focused on methodologies for describing Reynolds stresses, mining data, and obtaining a better resolution for the boundary layer separation finally. Detailed routine is described as following:

(1) Numerical studies with traditional control methods, such as blow/suck and oscillating jets. Verification and validation studies have been conducted to testify and further improve the capability of our CFD solver on capturing boundary layer separation.
(2) Development of moving mesh solvers and strategies for oscillating control devices by implementing the fluid-structure interaction simulation methodology, as well as proposing appropriate highly efficient unsteady treatments.
(3) Numerical studies on intelligent vortex generators applied to the close loop control process. Intelligent structures, including sensors, actuators and the close-loop process, represented by numerical models, will be introduced into our CFD solver thereafter.

17.2 Numerical Methods

It's well known that the results from LES (Large Eddy Simulation) are more accurate than RANS (Reynolds Averaged Navier-Stokes), especially in the simulation of large eddy structure in large scale flow separation. While dealing with high Reynolds number flow, LES has a particularly high requirement of the grid density, which makes the computing expense extremely huge. It's almost impossible to use LES in engineering application. In constract, RANS can accurately simulate the boundary layer structure with a well-defined wall treatment approach, but the simulation on large scale flow separation is out of its ability; therefore, there have been extensively studies on improving and optimizing LES to deal with turbulence flow in the near-wall region.

Siauw et al. [1, 2] analyzed the influence of flow control on NACA0015 by experiments. The experiment conditions are: angle of attack is 11°, Reynolds number based on chord length and incoming flow is approximately 1×10^6. They visualized the flow field through heater in spanwise direction and observed the evolvement rule of longitudinal vortex generated from the jet hole. 44 jet holes were arranged at 30%c of the airfoil profile along the spanwise direction. And several pressure sensors were installed in the area from 30%c to trailing edge of the airfoil surface for lift enhancement and drag reduction by feedback flow control. Geissler [3] introduced the numerical simulation through the modifications of boundary conditions corresponding to flow control modes, like steady blowing and suction, synthetic jet, moving surface and virtual slip caused by some special materials. He also investigated on the dynamic stall airfoil problem, and compared the changes of lift coefficient, resistance coefficient, pitch moment and skin-friction coefficient before and after the flow control. Hwang [4, 5] summarized the contribution of micro blowing/suction control forms to reduce skin friction through experimental measure, and Li's [6] numerical studies verified the control characteristics, and further researches on mixing of porous wall flow had been discussed.

Above all, previous researchers had carried out a large number of numerical simulation and experimental works on blowing/suction flow control method. Nonetheless, there is a lack of comprehensive studies on flow control effect of lift enhancement and drag reduction over an airfoil as well as separation control. Therefore, based on previous experience, in this paper we will perform a series of numerical simulation on the blowing/suction flow control techniques on the two-dimensional NACA0015 airfoil. The DES k-epsilon turbulence model [7] was utilized to obtain the best control mode under different AOAs. Furthermore, we conduct a basic flow field analysis on the two-dimensional and three-dimensional NACA0015 airfoil and discussed the effect of blowing/suction control method by analyzing the Reynolds stress.

DES was proposed by Spalart et al. [8] in 1997, which combines the techniques both in RANS and LES. Firstly, Spalart proposed a DES (Detached-Eddy Simulations) model by using one-equation Spalart-Allmaras turbulence model in the near-wall region for a better resolution of turbulent separation and using LES method in area far from wall. After that, Menter [9] and Strelets [10] further developed hybrid RANS/LES method based on a two-equation turbulence model. It combined the advantages of RANS and LES, using RANS in near-wall region and LES in the main flow. In this way, we can control the computing resources and obtain accurate solutions for the boundary layer structure and large scale turbulence separation vortex structure.

In DES method, the turbulence model with eddy viscosity hypothesis is applied in flow attached area, which means RANS with Reynolds average hypothesis is used to simulate the turbulent flow in the boundary layers. Therefore, it will maximize its advantage in processing wall shear layer. In the area away from wall, turbulent closed mode automatically turns into a sub-grid viscosity model [11]. If the grid density meets the requirement, it is equivalent to simulate turbulent flow with LES. It shows that DES combines the advantages of RANS and LES, which saves a lot of time and works out more accurate results. Currently, one-equation Spalart-Allmaras and Menter SST model are relatively common used in the application of DES method. Based on the concept of DES method, Li and Ren [7] developed a new DES combined with the k-epsilon mode and LES. This k-epsilon based DES method had been widely used to achieve good performances in boundary layer (adhesion area) and main flow area (separation area) respectively, which can predict the flow characteristics more accurately.

In this paper we adopted above mentioned DES k-epsilon turbulence model to conduct numerical simulations on flow control study for the NACA0015 airfoil.

17.2.1 Standard k-Epsilon Turbulence Model

Equation of k-epsilon turbulence model is written as:

k equation:

$$\frac{\partial(\rho k)}{\partial t}+\frac{\partial\left(\rho u_j k\right)}{\partial x_j}=\frac{\partial}{\partial x_j}\left(\mu+\frac{\mu_T}{\sigma_k}\right)\frac{\partial k}{\partial x_j}+G_k-\rho\varepsilon \tag{17.1}$$

ε equation:

$$\frac{\partial(\rho\varepsilon)}{\partial t}+\frac{\partial\left(\rho u_j \varepsilon\right)}{\partial x_j}=\frac{\partial}{\partial x_j}\left(\mu+\frac{\mu_T}{\sigma_\varepsilon}\right)\frac{\partial \varepsilon}{\partial x_j}+\frac{\varepsilon}{k}(C_1 G_k-C_2\rho\varepsilon) \tag{17.2}$$

where G_k is the generation rate of turbulence kinetic energy k, and defined as:

$$G_k=\mu_T S=\mu_T\left(\frac{\partial u_i}{\partial x_j}+\frac{\partial u_j}{\partial x_i}\right)\frac{\partial u_i}{\partial x_j},\quad S=\left(\frac{\partial u_i}{\partial x_j}+\frac{\partial u_j}{\partial x_i}\right)\frac{\partial u_i}{\partial x_j} \tag{17.3}$$

Turbulent eddy viscosity coefficient μ_T is defined as:

$$\mu_T=\rho C_\mu\frac{k^2}{\varepsilon} \tag{17.4}$$

where C_1, C_2, C_μ, σ_k and σ_τ are constants in the model.

17.2.2 DES Method Based on k-Epsilon Model

According to the dimensional analysis, the length scale of energy-containing eddies in turbulence flow is clarified as:

$$l_{k-\varepsilon}=\frac{k^{3/2}}{\varepsilon} \tag{17.5}$$

Replace the ε in k Eq. (17.1) by (17.5), introduce length scale to the k equation, the result is:

$$\frac{\partial(\rho k)}{\partial t}+\frac{\partial\left(\rho u_j k\right)}{\partial x_j}=\frac{\partial}{\partial x_j}\left(\mu+\frac{\mu_T}{\sigma_k}\right)\frac{\partial k}{\partial x_j}+G_k-\rho\frac{k^{3/2}}{l_{k-\varepsilon}} \tag{17.6}$$

Modify the length scale based on the idea of DES as follow:

$$\tilde{l} = \min(l_{k-\varepsilon}, C_{DES}\Delta) \tag{17.7}$$

where C_{DES} is a constant of DES model, equal to 0.65 [4], $\Delta = \max(\Delta x, \Delta y, \Delta z)$ is the maximum of local computational grid cell size. Replace the l in (17.6), k equation due to DES method can be rewritten as:

$$\frac{\partial(\rho k)}{\partial t} + \frac{\partial(\rho u_j k)}{\partial x_j} = \frac{\partial}{\partial x_j}\left(\mu + \frac{\mu_T}{\sigma_k}\right)\frac{\partial k}{\partial x_j} + G_k - \rho\frac{k^{3/2}}{\tilde{l}} \tag{17.8}$$

Meanwhile, the turbulent eddy viscosity coefficient should be redefined as:

$$\mu_T = \rho C_\mu k^{1/2}\tilde{l} \tag{17.9}$$

ε is rather larger near the wall, while turbulence kinetic energy k is limited, so that $l_{k-\varepsilon}$ is really smaller than the maximum of cell size of computational grid, in other words $l_{k-\varepsilon} \le C_{DES}\Delta$, and it leads to a result that the real computational model is the standard k-epsilon model of DES k-epsilon near the wall; when it is far away from the wall, the ε decreases rapidly along with the mushrooming $l_{k-\varepsilon}$, and eventually to a big value which is much larger than the maximum of cell size of computational grid, ($l_{k-\varepsilon} > C_{DES}\Delta$), considering the partial equilibrium of equality between the G_k and dissipative term $\rho\varepsilon$ $\left(G_k = \mu_T S = \rho\frac{k^{3/2}}{\tilde{l}}\right)$ and $\tilde{l} = C_{DES}\Delta$, (17.9) can be written as:

$$\mu_T = \rho\left(C_\mu^{3/4}C_{DES}\Delta\right)^2\sqrt{S} \propto \sqrt{S}\Delta^2 \tag{17.10}$$

17.2.3 Test Cases on DES k-Epsilon Turbulence Model

In order to test the accuracy of DES k-epsilon turbulence model, several simulations have been conducted under AOA = 11° over a NACA0015 airfoil with DES S-A, DES k-epsilon, DES $k-\omega$ and IDDES turbulence models. According to the results shown in Fig. 17.1, the contour of vorticity magnitude in the downstream clarifies that DES k-epsilon and IDDES turbulence models can well predict the vortex shedding in the wake of NACA0015 airfoil, while DES S-A and DES $k-\omega$ could hardly describe the flow characteristics. Moreover, the streamlines in Fig. 17.1 also demonstrate the same conclusion, as well as the separation points, which have been listed in Table 17.1. By comparing the separation points, except DES $k-\omega$, all turbulence models obtain the separation flow with small errors. Above all, DES k-epsilon model is mostly close to the experimental value, while DES $k-\omega$ gets no separation.

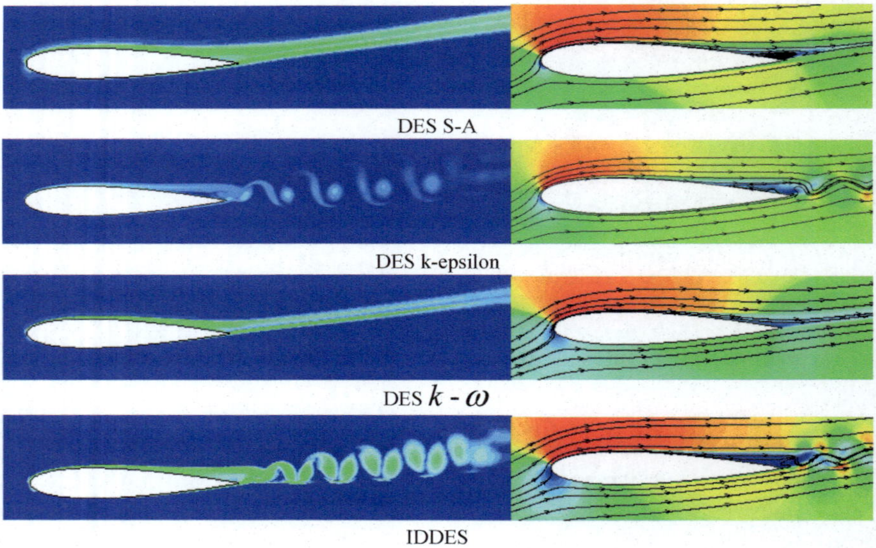

Fig. 17.1 Vorticity magnitude (left) and streamline (right) with different turbulence models

Table 17.1 Separation points, lift and drag coefficients for the NACA0015 test case (without flow control)

Model	Separation point (%c)	Error (%)	Cl	Error (%)	Cd
Experiments of Siauw et al. [1, 2]	70		0.912		
DES S-A	79	12.9	0.9281	1.8	0.0288
DES k-epsilon	71	1.4	0.9151	0.3	0.0301
DES $k - \omega$	N/A	N/A	1.0113	10.9	0.0274
IDDES	78	11.4	0.9064	0.6	0.0326

For the lift and drag coefficients of the NACA0015 test case without flow control, all turbulence models approximately give the results as expected. Especially, DES k-epsilon and IDDES can give the best lift coefficient in comparison with the experimental measurements of Siauw in Fig. 17.2, which presented the variation of lift coefficient along with the angle of attack. The simulation of DES k-epsilon verified a great agreement with experimental results with the separation point at 71%c and lift coefficient of 0.9151.

In the present study, DES k-epsilon, which has made a good performance in simulation of flow characteristic over a NACA0015 airfoil, is chosen to deal with the further numerical simulation of NACA0015 airfoil cases with blowing/suction control devices.

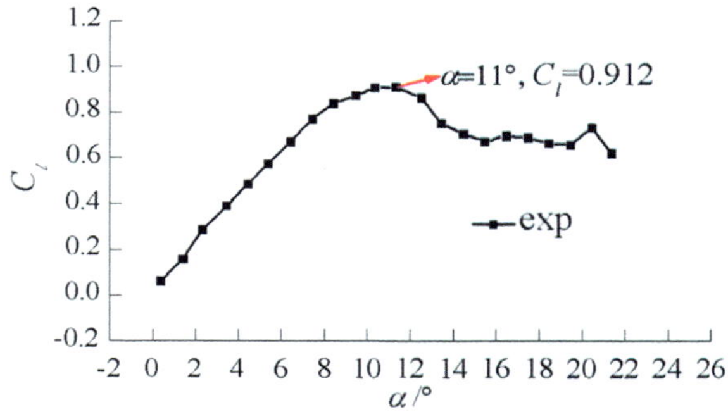

Fig. 17.2 Lift coefficient along with the angle of attack in experiment

17.3 Mechanism Studies of 2-D Blowing/Suction Flow Control

17.3.1 Computational Parameters and Boundary Conditions

Considering a NACA0015 airfoil, the flow mechanism of separation flow control with blowing/suction approaches will be investigated. According to the experiment configuration by Gilarranz et al. [8], the inflow velocity of 35 m/s and the control slots width of 0.53%c are set in present simulations, and the Reynolds number based on chord is approximate to 8.96×10^5. The flow control positions are located at 12, 30 and 70%c, respectively. The time step is $\Delta t = 5.357 \times 10^{-5} s$, then the dimensionless time is $\Delta t^+ = 5 \times 10^{-3}$ so that every flow period is guaranteed to contain enough time steps. The flow is initialized by uniform inflow, lift and drag coefficients are averaged after a quantity of time steps with 30 iterations per time step when the periodicity of the flow is achieved.

The computational domain is shown in Fig. 17.3, the velocity inlet and pressure outlet conditions are separately utilized in inlet and outlet boundaries, while no slip wall condition is enforced on the airfoil surface.

The structured mesh used in present study contains 79,000 nodes and the airfoil surface is resolved with 472 points (Fig. 17.4). The local mesh refinements are applied near the jet location. The minimum grid size near the wall is $1 \times 10^{-5}c$, and the value of y^+ approximate to 0.3.

17.3.2 Comparison of Flow Control Effects with Blowing and Suction Approaches

According to the researches in [12], jet angle of 25° to the wing surface shows a best control effect, and the control velocity is fixed as 56.6 m/s. In order to compare the different aspects of blowing and suction flow control, two sets of numerical computational parameters are listed in Table 17.2. In the first part, suction approach has been reproduced with velocity of −56.6 m/s; while in the second part, the velocity is 56.6 m/s in the blowing approach.

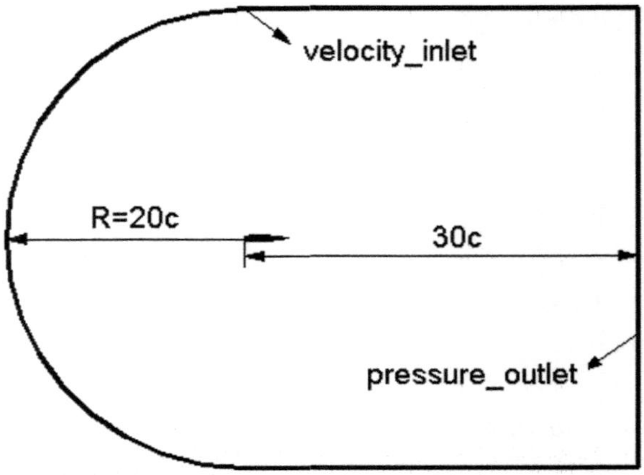

Fig. 17.3 NAC0015 airfoil computational domain

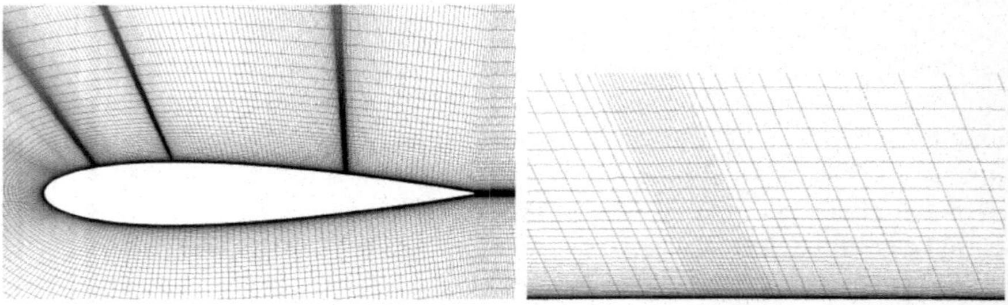

Fig. 17.4 Computational grid for NACA0015 airfoil and mesh near control device

Table 17.2 Calculation design of different control forms and parameters

Control forms	Control location (%c)	AOA	Velocity (m/s)
Suction	12	10°–24°	−56.6
	30	10°–24°	−56.6
	70	10°–24°	−56.6
Blowing	12	10°–24°	56.6
	30	10°–24°	56.6
	70	10°–24°	56.6

The variety of AOA is in the range of 10°–24°, while the control location is at 12, 30, and 70%c, respectively. The lift and drag coefficients under different control forms are shown in Figs. 17.5, 17.6 and 17.7.

The suction control approach shows a great potential in lift enhancement and drag reduction, when the control location is at 12%c. When AOA is less than 18°, blowing approach seems to be weak in flow control, as it tends to reduce the lift and increase the drag; when AOA is larger than 18°, due to the separation point moving upstream, blowing approach could reach the goal of lift enhancement and drag reduction as well as suction.

The suction control approach is also better in lift enhancement and drag reduction, when the control location is at 30%c. When AOA is less than 16°, there is no change on the lift and drag as blowing approach is concerned; when AOA is larger than 16°, due to the separation point moving upstream along with the increasing of AOA, blowing approach could reach the goal of lift enhancement and drag reduction; when AOA reaches to 24°, separation point moves to the leading edge, which is far from the control location, so that it's difficult to control the flow.

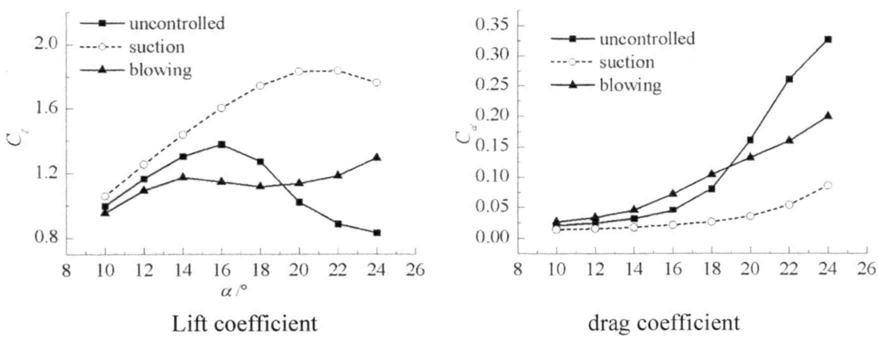

Fig. 17.5 Install different control location at 12%c

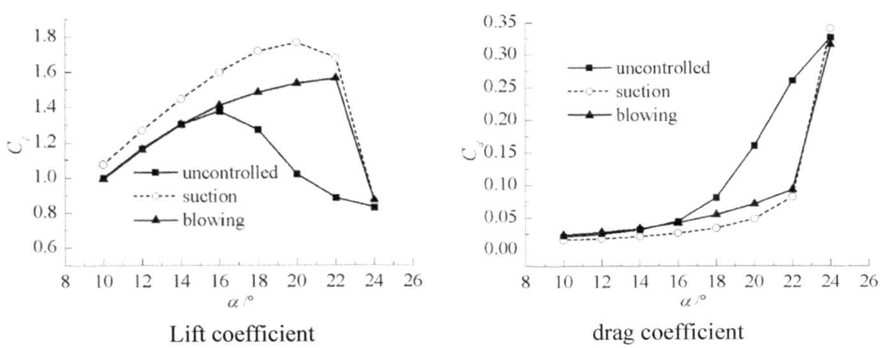

Fig. 17.6 Install different control location at 30%c

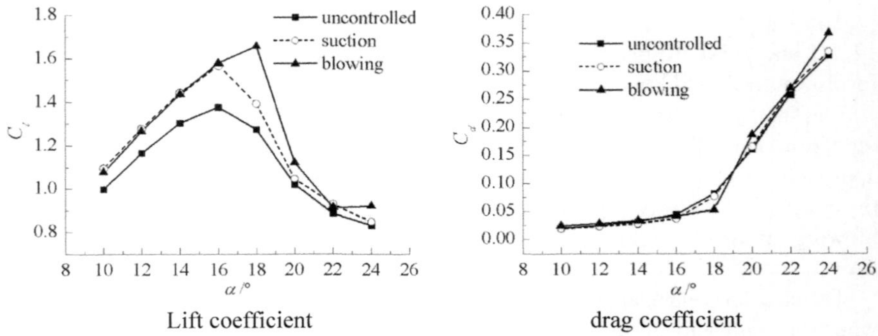

Fig. 17.7 Install different control location at 70%c

Blowing and suction methods meet an approximately equal impact of lift enhancement and drag reduction, when the control location is at 70%c. When AOA is less than 18°, both are efficient. However, either flow control method could be benefit when AOA is larger than 18° as the separation point is moving upstream far from the control location.

With the study and analysis of blowing/suction at different control locations, the variation tendency of separation flow control effect with flow control at 12%c is obtained in Figs. 17.8, 17.9 and 17.10 by the analysis of the streamlines near the control devices and airfoil surfaces, when the AOA is separately at 16°, 20° and 22°.

When AOA = 16°, the separation point is at the downstream of the airfoil surface, and the separation area is small. Suction at 12%c could attach the streamlines to the wall more closely, which results in a decreasing of the separation area; blowing approach pushes the streamlines away from the boundary layer, and the separation area grows larger. As the jet location is far from the separation point, blowing method has no influence in enhancing the fluid energy in separation area. On the contrary, normal velocity component weakens the ability of attaching to the boundary.

As AOA = 20°, the separation point moves upstream to the leading edge, and the separation area is larger. Suction at 12%c turns part of the recirculation region into attached-region and decreases the separation area. Blowing approach enhancing the fluid energy leads to the same result.

As AOA = 22°, the separation point is close to the leading edge, and the separation area is much larger. Suction at 12%c also could decrease the separation area. Although the blowing approach enhances the fluid energy, it just reduces a bit of the separation area.

17.4 Simulation of the Basic Flow Over a 3-D Airfoil

In order to distinguish the differences between 2-D and 3-D computations over a NACA0015 airfoil with DES k-epsilon turbulence model, three dimensional

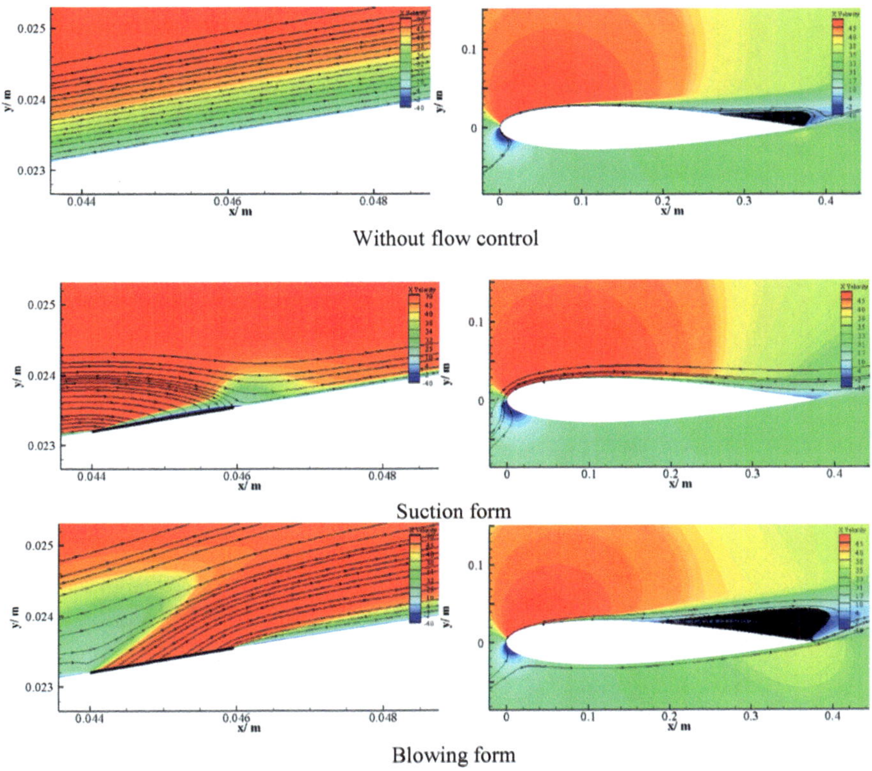

Fig. 17.8 Streamlines near the control device and airfoil surfaces (control location at 12%c, $\alpha = 16°$)

numerical simulation over a 3-D airfoil has been conducted as a reference for the further study of the effects of 3-D jet flow control model.

17.4.1 Computational Parameters and Grid Model

The computational grid is shown as Fig. 17.11. According to the experimental results from Siauw et al. [1, 2], the configuration of numerical simulation is designed as follow: the velocity inlet and pressure outlet conditions are separately imposed at the inlet and outlet boundaries, while the no slip wall condition is enforced on the airfoil surface, the periodic boundary condition is set in the spanwise direction; the inflow velocity is 35 m/s; AOA = 11°; the spanwise width is 0.1714c. The structured mesh used in present study contains 6,200,000 nodes in total, including 62,124 nodes in xy plane and 101 mesh points in spanwise direction.

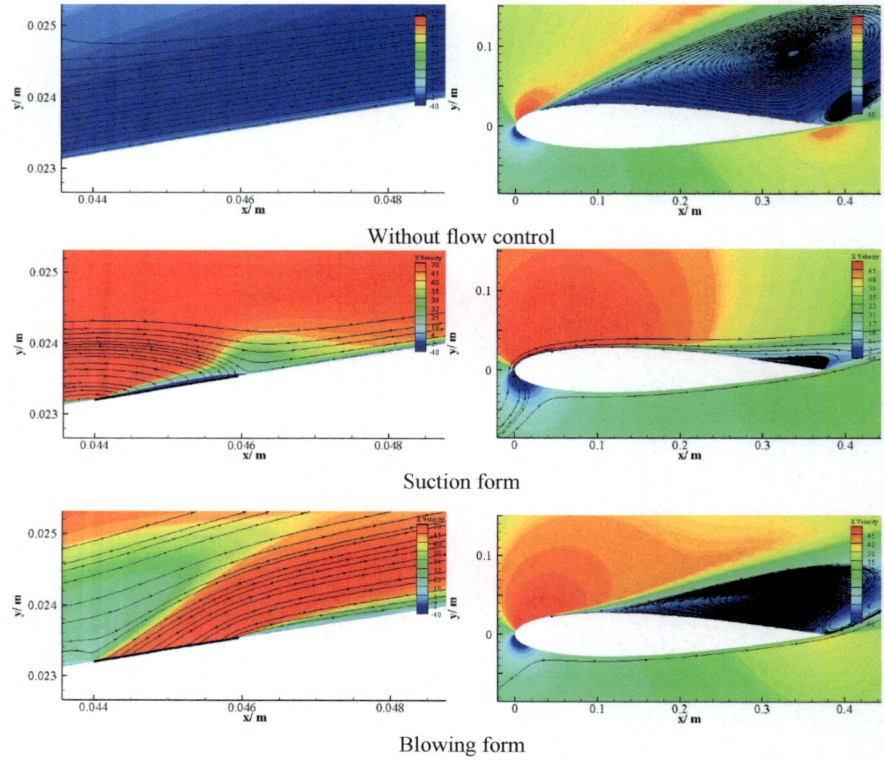

Fig. 17.9 Streamlines near the control device and airfoil surfaces (control location at 12%c, $\alpha = 20°$)

The special characteristics of the maximum lift coefficient and complex transition in separation flow at the critical AOA of 11° will be compared with the previous experimental data.

17.4.2 Simulation on the Basic Flow Over the NACA0015 Airfoil (Without Flow Control)

As shown in Fig. 17.12, the lift coefficient has been computed by using the DES k-epsilon turbulence model. It is shown that the variation of lift coefficient is periodic, and its averaged value of 0.915 is closer to the experimental value of 0.912 [1, 2].

The vortex structure obtained by the lambda-2 criterion is shown in Fig. 17.13. The figure shows that the vortexes are formed and shedding periodically in the downstream of the wing section. What's more, the separation structure of vortex is actually two-dimensional, but the vortex structures are not stable and the three-

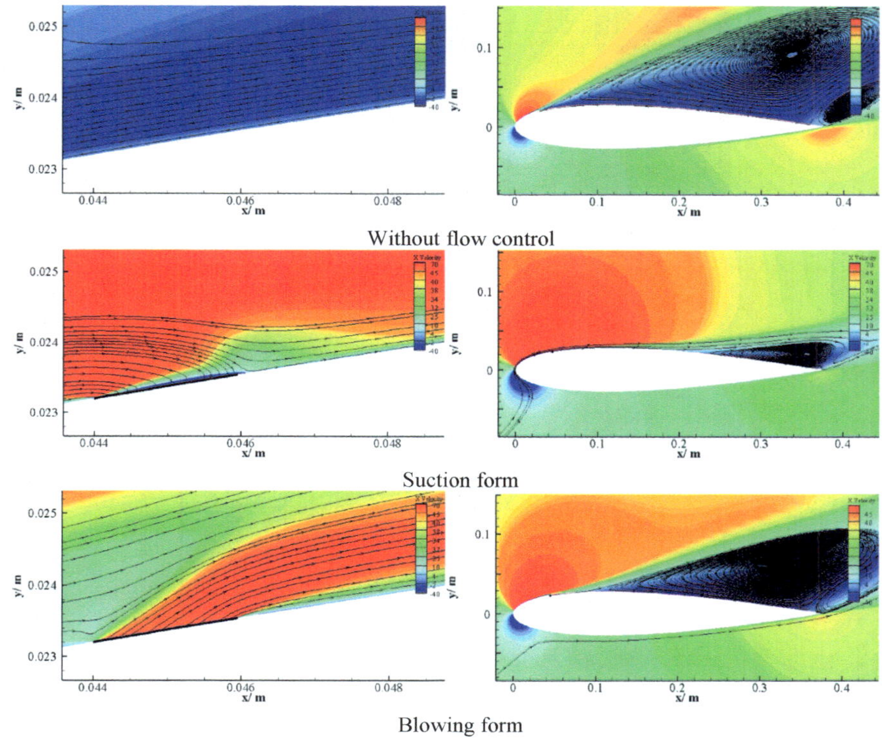

Fig. 17.10 Streamlines near the control device and airfoil surfaces (control location at 12%c, $\alpha = 22°$)

dimensional structures, such as hairpin vortex, are generated and further developed significantly in the downstream.

The flow field on the spanwise center plane of the wing section is shown in Fig. 17.14. The figure shows that the separation point is around 70% of the chord length, which is consistent to the previous two-dimensional results in the second section.

The Reynolds stress has 6 components of <u′u′>, <v′v′>, <w′w′>, <u′v′>, <u′w′> and <v′w′>, however, <u′u′>, <v′v′> and <u′v′> are the components being referred on account of the much smaller magnitudes of velocity fluctuation in spanwise than those in x and y directions. After the flow field reaching the status of vortex shedding periodically, the variables of the flow are statistically averaged and the components <u′u′> and <v′v′> of the Reynolds stress are obtained, as shown in Fig. 17.15. Due to the structures of the shedding vortexes, the components <u′u′> and <v′v′> in the downstream of the wing section are comparatively larger than <u′v′>.

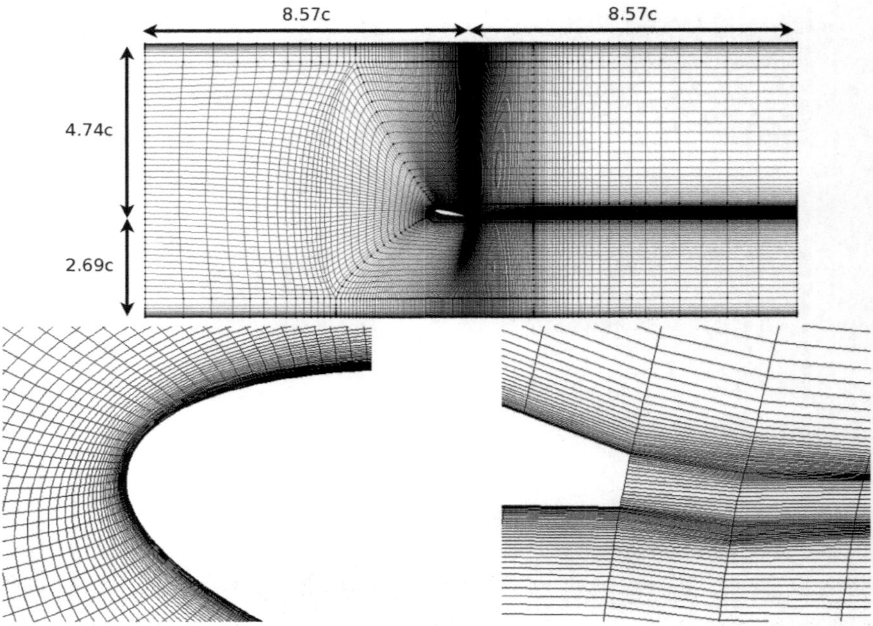

Fig. 17.11 NACA0015 computational grid

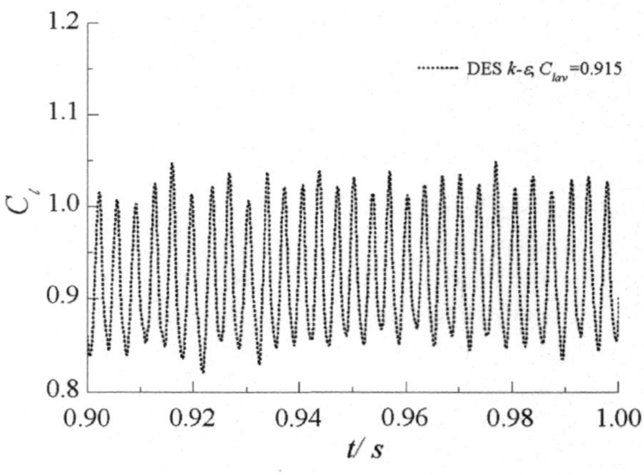

Fig. 17.12 Lift coefficient

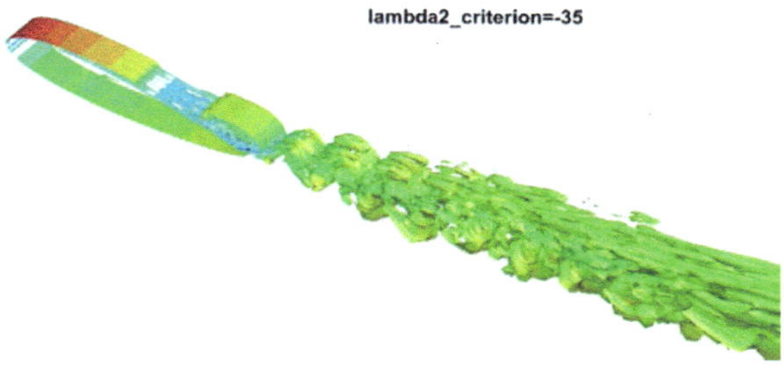

Fig. 17.13 Vortex structures of the flow past NACA0015 wing section

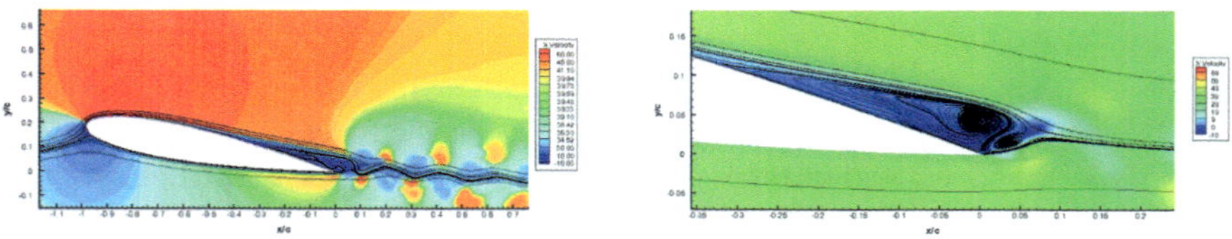

Instantaneous flow field

Average flow field

Fig. 17.14 Separation point of the central section on wing spanwise

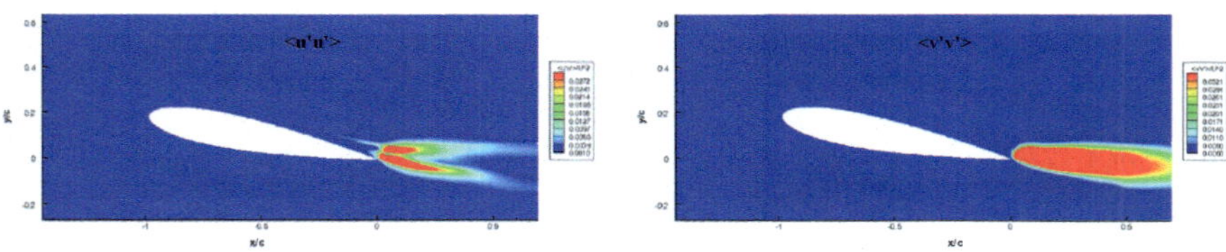

(a) <u'u'>

(b) <v'v'>

Fig. 17.15 Distribution of Reynolds stress surrounding the airfoil

17.5 Simulation of Blowing Control Over a 3-D Airfoil

17.5.1 Simulation of Slot Jet Control Device Over a NACA0015 Airfoil

According to the results of 2-D computations, two typical cases of 3-D blowing flow through a slot jet devise over a NACA0015 airfoil had been computed to check the consistency of flow control effect between 2-D and 3-D. The computational model is set up with the spanwise width of 0.1714c, the jet angle of 25° and the slot jet width of 0.53%c, which are same as the installing of 2-D model. Contours of vorticity magnitude with streamlines show the different control effects in Fig. 17.16 under the

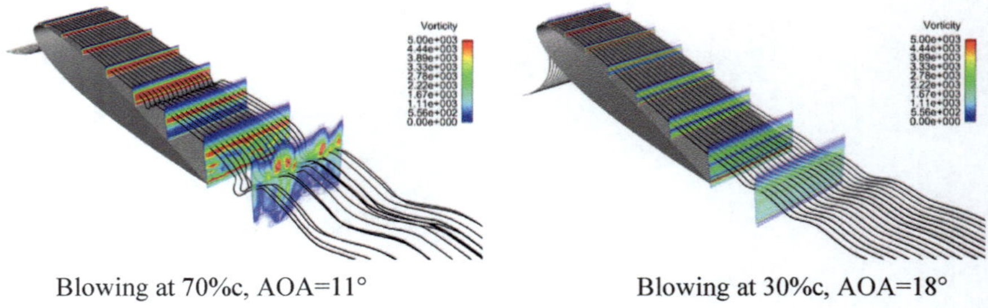

Fig. 17.16 Contours of vorticity magnitude with streamlines under two different blowing conditions

two different blowing conditions: blowing at 70%c with the AOA of 11° and blowing at 30%c with the AOA of 18°, respectively.

The results match those in Sect. 17.3.2 very well. The blowing control with locations close to the separation point does well in flow control. In addition, it seems that blowing at 30%c does an outstanding contribution to separation control when AOA = 18°, referring to maintaining stability of the vortex structures in downstream.

17.5.2 Simulation of Pulsed Jet Control Device Over a NACA0015 Airfoil

In order to control the flow separation well, such as reduction of separation area, 4 blowing holes had been installed at 30%c, and other conditions are given as: the distance between two adjacent jets is 15 mm; the maximum velocity is 200 m/s, and jet direction is shown in Fig. 17.17. Based on the standard grid offered in Fig. 17.11, finer mesh has been applied near the jet positions shown in Fig. 17.18.

The iso-surface of vorticity magnitude with flow control located at 30% near the airfoil surface and blowing holes of NACA0015 are shown in Fig. 17.19. The

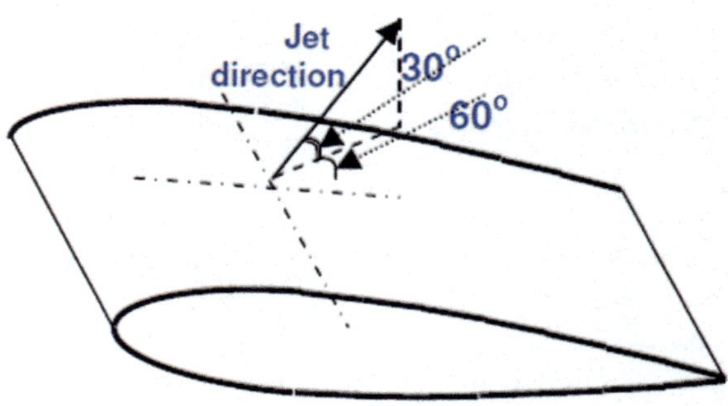

Fig. 17.17 Jet direction

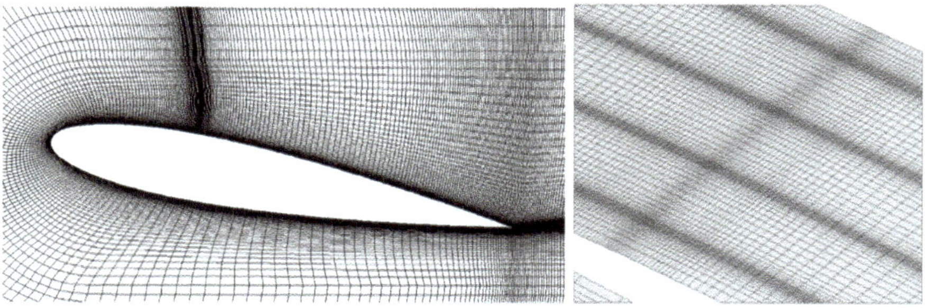

Fig. 17.18 Computational grid of jet flow control

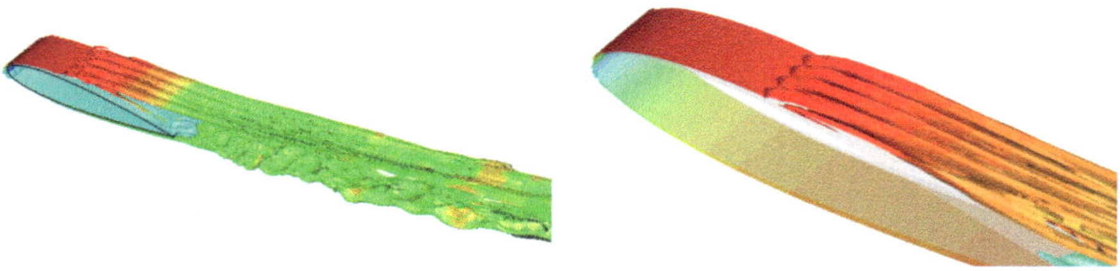

Flow field surrounding the airfoil Flow field near control position

Fig. 17.19 Iso-surface of vorticity magnitude near airfoil surface and control position

averaged flow and vorticity magnitude are obtained, as shown in Figs. 17.20 and 17.21. It can be found that the separation point is located at 70%c without control, however, it will move downstream to 80%c when the pulsed jet flow control was applied. In the meanwhile, the separation area is diminished.

After the flow field becomes periodic, the components $\langle u'u' \rangle$ and $\langle v'v' \rangle$ of Reynolds stress both in spanwise and streamwise directions are obtained by statisti-

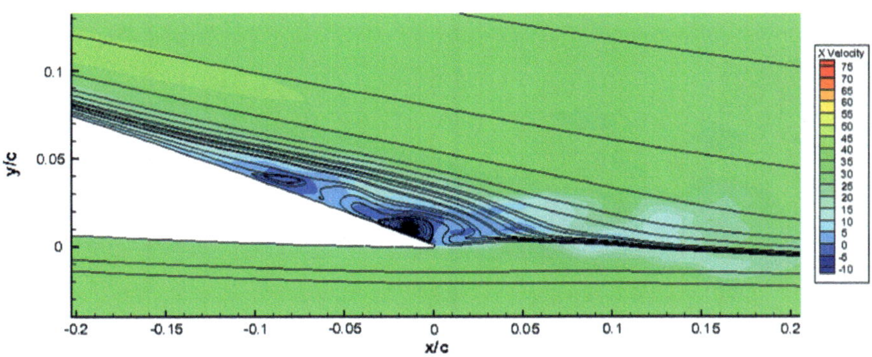

Fig. 17.20 Averaged flow distribution

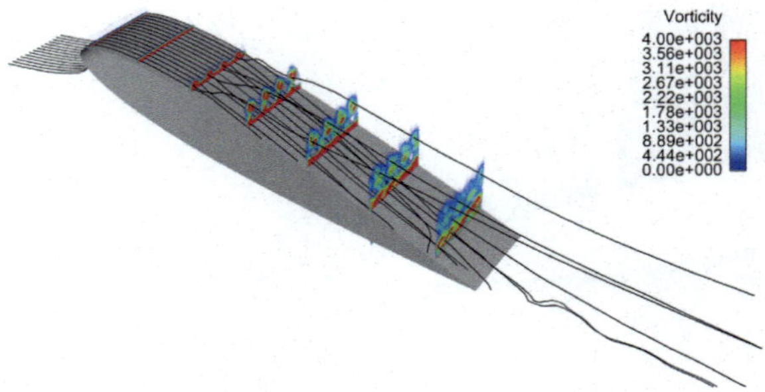

Fig. 17.21 Contours of vorticity magnitude with streamlines with Pulsed Jet

cally averaging the flow field, see Figs. 17.22 and 17.23. Obviously, both components present a remarkable change with blowing flow control surrounding the airfoil surface as follow: the components of <u′u′> and <v′v′> tend to be comparatively smaller than those in Sect. 17.4.2, respectively.

As a result, it clarified that pulsed jet flow control increases the fluctuation velocity u′ in the x direction, so that the pulsation component <u′u′> is increased and the pulsation component <v′v′> is reduced. Therefore, by increasing the pulsation

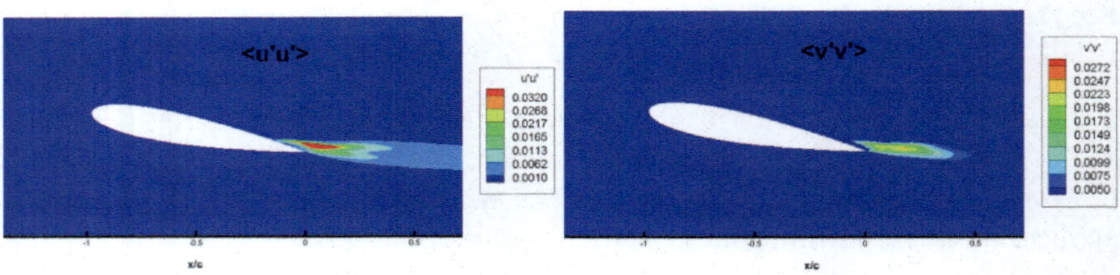

(a) <u'u'> (b) <v'v'>

Fig. 17.22 Distribution of Reynolds stress surrounding the airfoil in spanwise

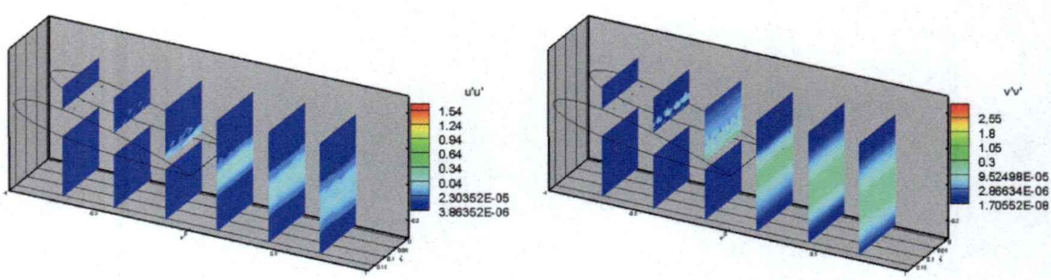

(a) <u'u'> (b) <v'v'>

Fig. 17.23 Distribution of Reynolds stress surrounding the airfoil in streamwise component $\langle u'u' \rangle$ or reducing the component $\langle v'v' \rangle$ on the airfoil upper surface, it may be possible to control the flow separation.

17.6 Conclusions

The numerical simulations of NACA0015 airfoil model have been carried out in the present study. Preliminary numerical computations of the flow around the airfoil without the flow control devices are performed with different turbulence models, and the DES k-epsilon is selected to solve the NACA0015 cases in present paper, because with this turbulence model, the unsteady separation of the boundary layer on the upper wall is reproduced well.

The modelling of the blowing/suction and pulsed jets as flow control devices over a NACA0015 airfoil has been performed. The numerical results demonstrate a great agreement with experimental results.

According to experimental results, some kinds of flow control models are established. For NACA0015 airfoil models, both of the flow control devices can make great contributions to the reduction of the separation region, as well as to the alteration of the Reynolds stress components. The differences of flow control effects between blowing and suction methods have been investigated numerically, and several conclusions are obtained as follow:

(1) The blowing control locations should be close to the separation point, by enhancing the kinetic energy on the boundary layer near the separation point, and strengthening the ability of resistance separation of fluid in the boundary layer. When the blowing position is located to the separation point far away, normal velocity component weakens the fluid ability of attaching to the wall, which leads to results of reducing lift, increasing drag and larger separation area.
(2) Blowing has a less efficient influence than suction, when the flow control devices are placed in 12 and 30%c. When the control device is set on 70%c, the effects of lift enhancement and drag reduction are approximately the same by blowing and suction flow control: when the AOA is less than 18°, both methods could reach the goal of lift enhancement and drag reduction; when the AOA is larger than 18°, neither blowing nor suction can be efficient.
(3) DES k-epsilon turbulence model is suitable to simulate the basic flow and flow control over a NACA0015 airfoil. The numerical simulation shows the flow characteristics of separation control over a 3-D airfoil agrees with 2-D numerical results on some aspects. The vortex structures reveal a strong 3-D property in the 3-D cases.
(4) Blowing at 30%c can reduce the separation area significantly and alter the distribution of Reynolds stress at AOA = 11°.

(5) Pulsed jet flow control device contributes a great effort in reduction of separation area, as well as increasing the fluctuation velocity u′ in the x direction, in order to increase the pulsation component <u′u′> and reduce the pulsation component <v′v′>.

Acknowledgements Research reported in this paper has been investigated in the MARS project entitled "Manipulation of Reynolds Stress for Separation Control and Drag Reduction", and supported by PRC-MIIT and EC-DG RTD Chinese and European governmental institutions.

References

1. Siauw W, Bonnet JP, Tensi J et al (2009) Physics of separated flow over a NACA 0015 airfoil and detection of flow separation. In: 47th AIAA aerospace sciences meeting, AIAA paper 2009-144
2. Siauw WL, Bonnet JP, Tensi J et al (2010) Transient dynamics of the flow around a NACA 0015 airfoil using fluidic vortex generators. Int J Heat Fluid Flow 31(3):450–459
3. Geissler W (2010) A family of CFD boundary conditions to simulate separation control. Aerosp Sci Technol 14(7):494–504
4. Hwang D (2004) Review of research into the concept of the microblowing technique for turbulent skin friction reduction. Prog Aerosp Sci 40(8):559–575
5. Hwang DP (2002) Experimental study of characteristics of micro-hole porous skins for turbulent skin friction reduction. In: Proceedings of the 23rd congress of international council of the aeronautical sciences, Toronto, Canada, 2002. Optimage Ltd, pp 2101.1–2101.7
6. Li J, Lee CH, Jia L et al (2009) Numerical study on flow control by micro-blowing. In: Proceedings of 47th AIAA aerospace sciences meeting and the new horizons forum and aerospace exposition, Orlando, Florida, USA, 2009, pp 8928–8946
7. Li X, Ren Y (2004) Detached-eddy simulation k-ε turbulence model and its application in fire simulations (in Chinese). J Tsinghua Univ (Sci Technol) 44(8):1126–1129
8. Spalart PR, Jou WH, Strelets M et al (1997) Comments on the feasibility of LES for wings and on a hybrid RANS/LES approach. In: Liu C, Liu Z (eds) Proceedings of 1st AFOSR international conference on DNS/LES, Louisiana, USA, Aug 1997. Advances in DNS/LES. Greyden Press, Columbus, USA, pp 137–147
9. Menter FR (1994) Two-equation eddy-viscosity turbulence models for engineering applications. AIAA J 32(8):1598–1605
10. Travin A, Shur M, Strelets M et al (2000) Detached-eddy simulations past a circular cylinder. Flow Turbul Combust 63(1):293–313
11. Gilarranz JL, Traub LW, Rediniotis OK (2005) A new class of synthetic jet actuators: part II: application to flow separation control. J Fluids Eng 127(2):377–387
12. Liu F, Zou J, Zheng Y (2013) Effect of synthetic jets physical parameters on flow separation control over airfoil (in Chinese). J Zhejiang Univ (Eng Sci) 47(1):146–153

Chapter 18
Industrial Assessment on Flow Control Technology Developed in MARS

Yi Ju Deng, Rui Fei Xu, Xiao Juan Guo and Fa Liang Wang

Abstract The purpose and application of flow control technology for aircraft design are analyzed, the requirements of flow control from the industrial viewpoint are also proposed, based on the research progress on flow control simulation, experiments, optimization in MARS (Manipulation of Reynolds Stress for Separation Control and Drag Reduction) project. Some flow control devices are assessed from the perspective of industrial applications. Feasibility of the application on real planes, effectiveness and reliability of theoretical and experimental research method, tradeoff between the payoff and the performance, as well as industrial requirements on flow control are especially focused on.

Keywords Flow control · Industrial assessment · MARS

18.1 Introduction

Flow control is one of the most potential research fields of current aerodynamics. The purpose of flow control involves in delaying/accelerating (in suppressed !) transition, inhibiting/strengthening turbulence, preventing separation, reducing drag, increasing the lift, strengthening mixing and heat conduction, as well as suppressing the flow caused noise. Modern aircraft design often involves multi-object design points, each point has its own design requirements, such as High-speed cruise design point, Climb design point, low-speed takeoff and landing design point. Single point design may lead to conflicting results, while taking all points into account will sacrifice some key performance and cause mediocrity. If the flow control devices can be used (the suppressed) more effectively, this problem will be solved. Whereas, micro flow control on boundary layer is hard to be applied widely, due to its complex mechanism and small scale, as well as lack of computational and experimental analysis method. Taking transport aircraft as an example, the areas which are prone to flow separation are wing-body fairings, wing tip, trailing edge of the wing, and after-body, if the

---

Y. J. Deng (✉) · R. F. Xu · X. J. Guo · F. L. Wang
AVIC the First Aircraft Institute, Xi'an, Shaanxi, China
e-mail: dengyeeju@sina.com

N. Qin et al. (eds.), *Advances in Effective Flow Separation Control for Aircraft Drag Reduction*, Computational Methods in Applied Sciences 52,
https://doi.org/10.1007/978-3-030-29688-9_18

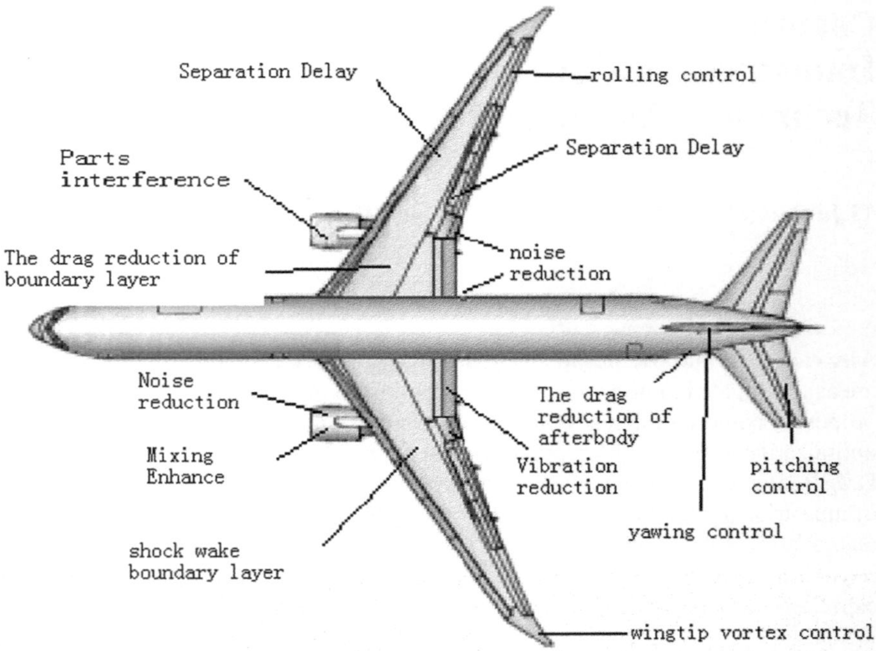

Fig. 18.1 Application of flow control technology on the transport aircraft

separation can be reduced, flight safety will be improved in addition to economy improvement. The wet area of wing and body surface brings the main frictional resistance, thus the control of this resistance will be significant to reduce drag. Steps on the plane may be brought in by control surfaces, flaps, and hatches, which may cause additional resistance and separation. The application field of flow control for the transport aircraft is shown in Fig. 18.1.

18.2 Industrial Requirement

In MARS project, active flow control technique is used to change the Reynolds stress to control basic flow. Generally, because the object is related to complex flow phenomena such as boundary layer flow, mixing flow, turbulence, shock wave, along with fluid-solid coupling, and multiphase flow, it is difficult to establish the exact model for the controlled object of active flow control technique; meanwhile, the flow is unsteady, so the flow phenomenon to be controlled may appear on an uncertain time, the flow structure is small in scale, short in life cycle, and sensitive to external disturbances. To be feasible in industry, the control system must satisfy (following requirements): 1. Sensor and actuator should be high resolution on time and space;

2. The control unit must be set in the whole flow area which is needed to be control, therefore, a highly integrated actuator is needed with low cost and light weight; 3. The installation of sensor and actuator should not change the aero-surface for reducing the disturbance to the flow; 4. Synchronization of control units request that the control system should not only get the good control effect, but also have low energy consumption; 5. It should have the good control effect under the high speed, real aircraft and high Reynolds number condition.

In general, applying flow control techniques in industry requires: 1. Structure is simple, volume is small, destruction to original structure should be as tiny as possible; 2. If additional energy provision is in need, additional energy requirement should be as low as possible, extra influence brought by the energy should be small; 3. Working frequency of the flow control system should be away from the natural frequency of the structure, response rate of the control system should be agile.

Requirements for flow control analysis and test methods are: 1. Simulation of the original flow control scheme should be as real as possible; 2. Precision and accuracy should be good; 3. Separation vortex in flow field and complex flow features can be captured; 4. Calculated consumption should be small; 5. Computation time should be short.

For flow control optimizing methods and tools, the optimization design loop should be as small as possible; the optimization design tools should be multi-point, Multidisciplinary, and global optimization.

18.3 Industrial Assessment on Flow Control Research in MARS

Flow control devices were studied in MARS, the actuators for different flow control were designed, the mechanism of flow control was investigated, the control parameters were optimized, and these flow control methods were applied to the BFS and NACA0015 airfoil. In MARS, Separation Control and Drag Reduction with the change of Reynolds Stress were focused on. In this paper the research in MARS on synthetic jet, plasma and moving surface are summarized and analysed, the computational methods, experimental, optimization research progress are also briefly investigated. Researches (the suppressed) provide useful reference and support for development on the flow control technique and its industrial application.

18.3.1 Synthetic Jet

According to the arrangement of MARS, the work for synthetic jet was accomplished mainly by NPU, BUAA, and DLR. Related researches on the BFS and NACA0015 airfoil were investigated by numerical simulation and wind tunnel for controlling Reynolds stress.

The test applying the synthetic jet to the BFS was carried out by the BUAA and NPU. Figure 18.2 shows the wind tunnel equipment and the flow control model used by BUAA research group. Figure 18.3 shows the flow control model of NPU. For the NPU research group, Piezoelectric and Electromagnetic synthetic jet were both applied to the BFS for the mechanism research on control of Reynolds stress in turbulent boundary layers. Experimental investigation indicates that by affecting Reynolds Stress the properties of turbulent boundary layers and shear layers can be altered and therefore the separation characteristic of boundary layers as well as the friction drag on the body surface affected.

The researches on the synthetic jet were also studied by the DLR, Figs. 18.4 and 18.5 show the results of the flow control on the BFS. The advanced measurement was adopted to obtain the details of flow field, and the mechanism research on control of Reynolds stress was also conducted. The numerical simulation for the synthetic jet was completed by ZJU. Figure 18.6 shows the change of lift and drag coefficient with and without flow control. It shows that the jet location should be close or

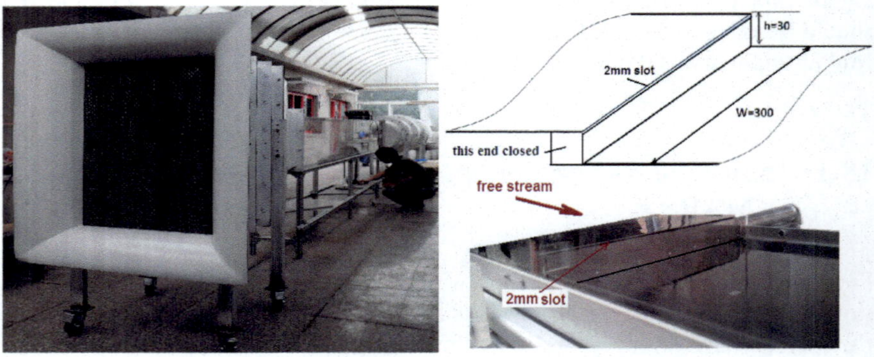

Fig. 18.2 Wind tunnel and flow control model for the backward facing step (BUAA)

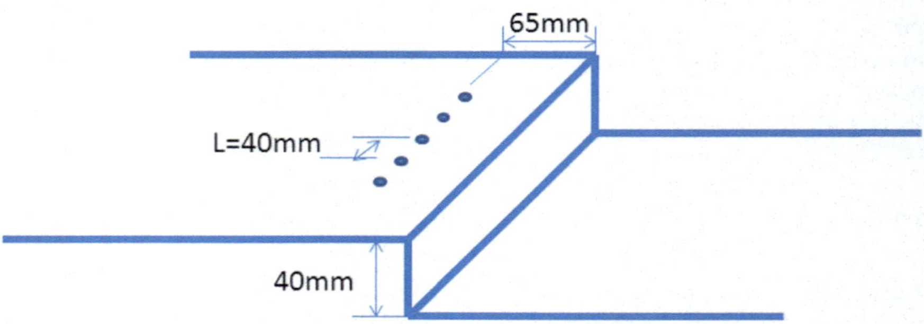

Fig. 18.3 Flow control model for the BFS (NPU)

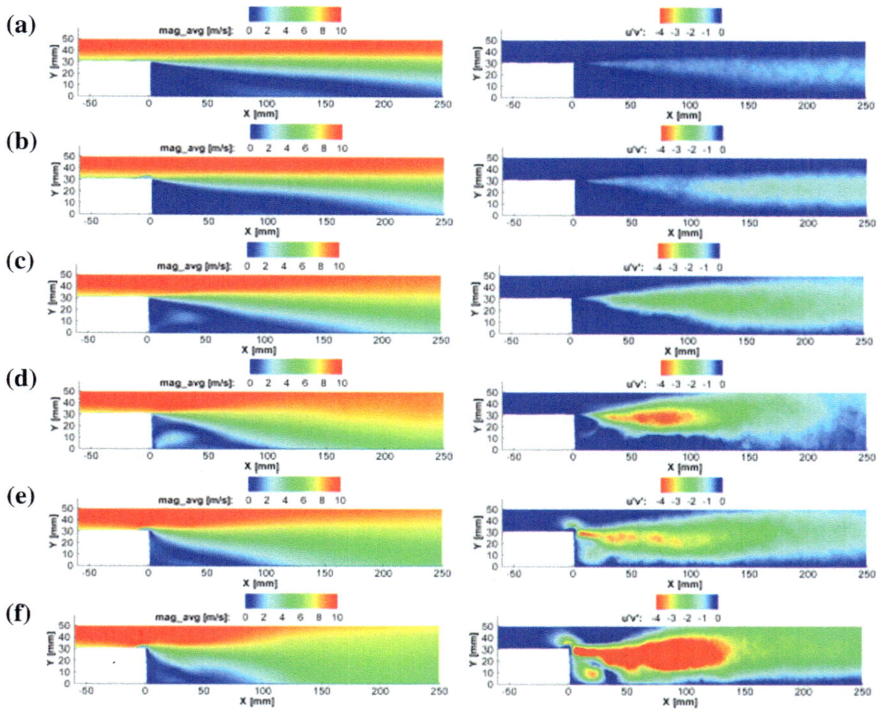

Fig. 18.4 (EU-DLR): Average velocity magnitude (left) and Reynolds stress u′v′ (right) of the BFS flow

prior to the separation point, only in this way the purpose of flow control can be well achieved. The performance of multi-location synthetic jet can be improved by choosing appropriate phase angle of jets.

Control mechanism, control parameters and control effects of Synthetic jet were carried out in the MARS project. But from the industrial application point of view, the Reynolds effect and scale effect should be considered at a large scale. In addition, the design level and work performance of jet actuator determine the application field and effect of Synthetic jet, therefore the design of actuator is a core problem for the development of Synthetic jet. Now, Piezoelectric actuator is one of the most widely used actuator for its advantage of simple structure, wide work frequency, quick response, good repeatability and easy miniaturization. The disadvantage is that the jet energy is not powerful enough. Electric discharge machining actuator makes it possible to control supersonic flow. It can form a supersonic jet, and the maximum speed can reach 500 m/s, but the change of cavity temperature and pressure is large, working voltage is high, and it has the high requirement for component materials. Therefore, in order to gain the good control effect, the actuator of synthetic jet must produce high momentum jet besides miniaturizing the actuator, high operating

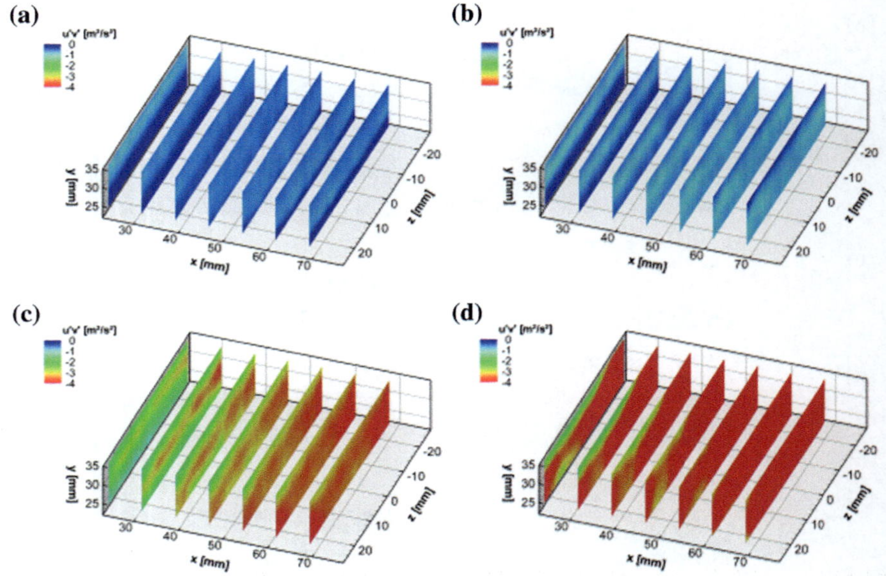

Fig. 18.5 (EU-DLR): Flow in the BFS shear layer depicted as selected planes color-coded with the Reynolds stress u′v′; without actuator (**a**), with plasma actuator (**b**) and with the synthetic jet actuator at 14 Hz in lower (**c**) and upper (**d**) position

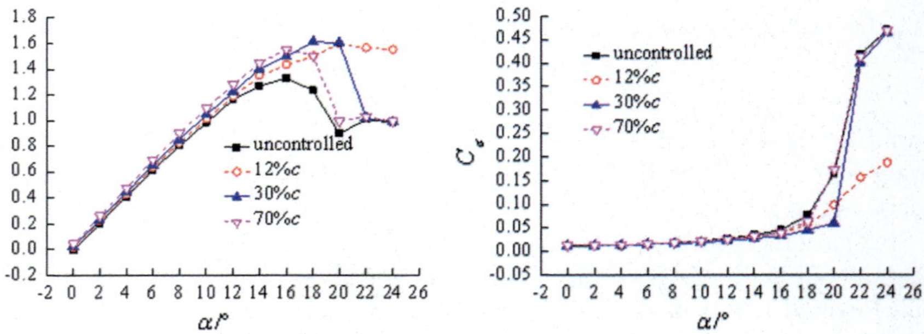

Fig. 18.6 The change of lift and drag coefficient with and without flow control (ZJU)

frequency, and wide frequency range, meanwhile, the actuator must be able to work effectively at high speed, high temperature, and high pressure.

18.3.2 Plasma

According to the arrangement of the MARS project, the work on Plasma was accomplished by POITIERS, UNIMAN, and BUAA. The researches on the BFS and NACA0015 airfoil were investigated by using the CFD and wind tunnel test for controlling Reynolds stress.

Numerous experiments on plasma have been conducted by EU-POITIERS in the MARS project. A large parametric study has been carried out as a first stage work, in order to define the most promising actuation in terms of reattachment reduction by finding out the influence of actuator location and electrical parameters. Figure 18.7 shows Reattachment mean position for each DBD location versus voltage and frequency.

Researches (the is suppressed) on plasma were also done by UNIMAN. Figure 18.8 shows the Schematic arrangement of SDBD plasma actuator. Figure 18.9 shows the comparison of the lift coefficient and pressure distribution with and with-

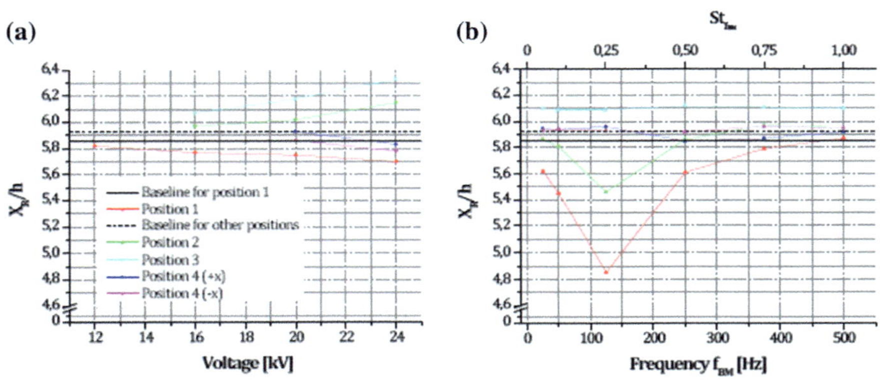

Fig. 18.7 (EU-POITIERS): Reattachment mean position for each DBD location versus **a** the applied voltage and **b** the frequency f_{BM}

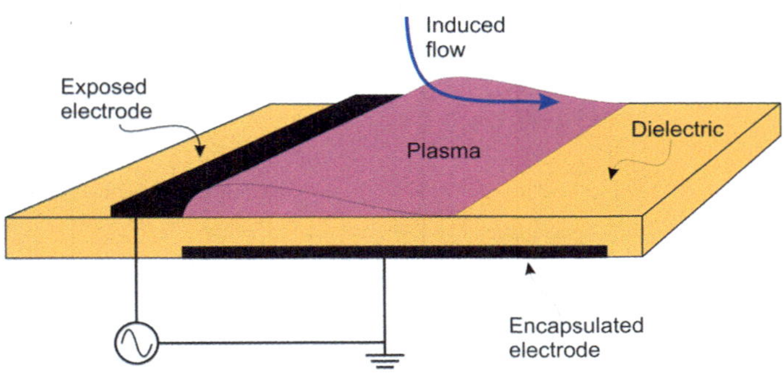

Fig. 18.8 (EU-UNIMAN): Schematic arrangement of SDBD plasma actuator

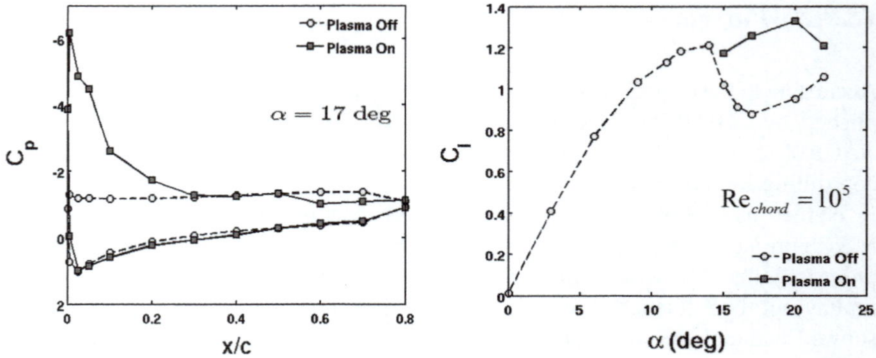

Fig. 18.9 The comparison of lift coefficient and pressure distribution with and without plasma (EU-UNIMAN)

out plasma on for the NACA0015 airfoil. It shows that the flow separation on the surface of the airfoil disappears, the lift coefficient and the stall angle of attack is obviously improved with plasma on.

The effects of dielectric barrier discharge (DBD) plasma-based body force on Reynolds stress and flow structures are studied numerically through three typical flows, which are channel flow, NACA0015 airfoil flow and BFS flow by the BUAA research group. Two effects of steady body force are identified, one is acceleration of flow velocity adjacent to solid surface, the other is the decrease of Reynolds stress. Figures 18.10, 18.11, 18.12 and 18.13 show some results of the BUAA group. The increase in kinetic energy of mean flow induced by plasma, is favourable to resist the adverse pressure gradient, however, the large velocity gradients at wall cause local

Fig. 18.10 The mean streamwise velocity profiles (BUAA)

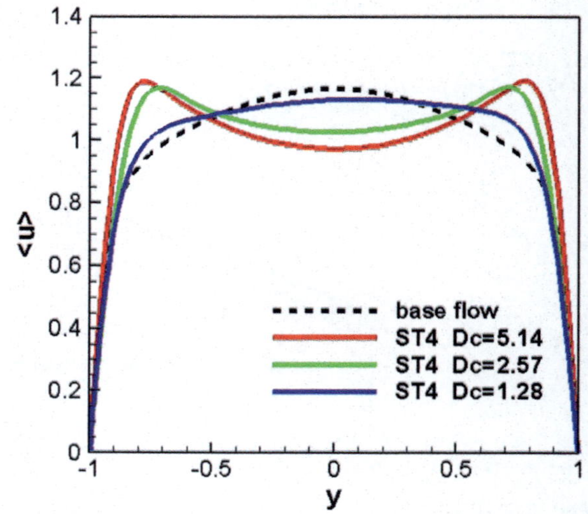

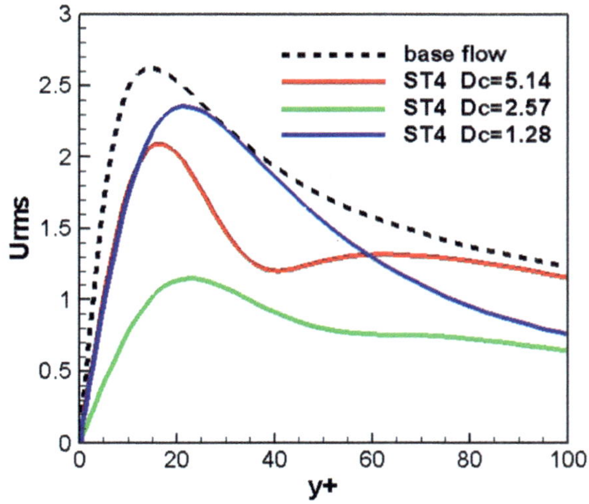

Fig. 18.11 The streamwise velocity fluctuation (BUAA)

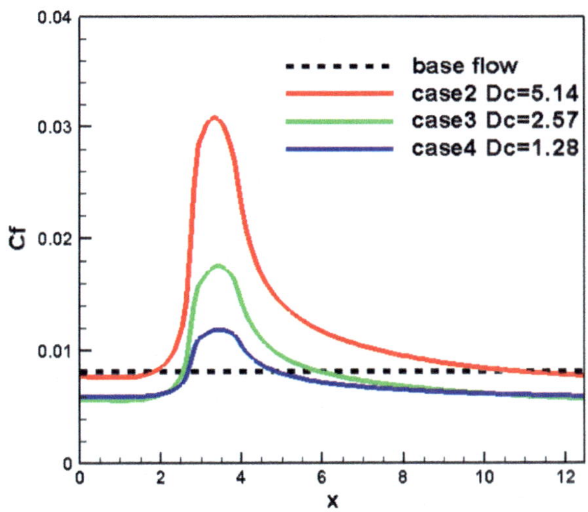

Fig. 18.12 Wall friction distribution (BUAA)

wall friction increase. The reduction of mean friction drag may be achieved with appropriate strength and gap of plasma actuator.

The plasma control mechanism was researched, and only low Reynolds number research for the plasma was done in MARS project. But in the view of industrial applications, Reynolds number has great relations with the result of flow control, so the research at high Reynolds numbers must be considered. As the same with synthetic jet, the application and effect are directly determined by the design level and working performance of the plasma actuator. The actuator should be high voltage, large current and small size. The miniaturization of plasmas excitation power is one

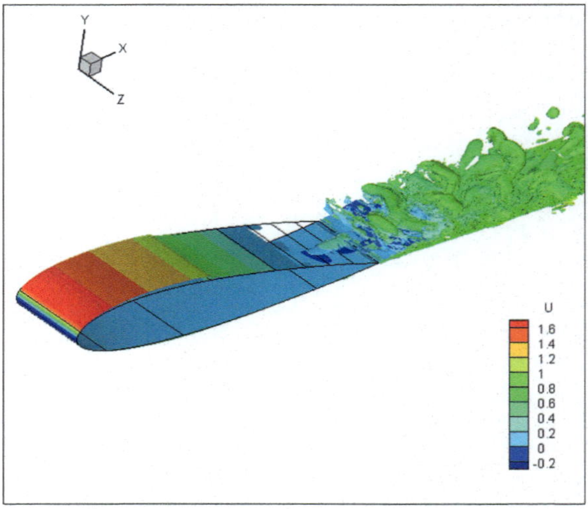

Fig. 18.13 The instantaneous vortex structure for NACA0015 airfoil (BUAA)

of the trends. At high speed, for plasma flow control, the most significant problem is the stability control of plasma excitation, such as electromagnetic interference. Therefore, it is necessary to develop a small electromagnet from the perspective of future application of plasma flow control technology for aircraft. In order to apply the plasma flow control technology on the future aircraft, it is necessary to improve the intensity of excitation for meeting the control requirements of complex flow.

18.3.3 Moving Surface

The moving surface can minimize the relative motion between wall and free stream, inject energy to the existing boundary layer, so it can prevent the boundary layer development and delay the flow separation. The moving surface control technology on the NACA0015 airfoil was analysed by the computations and experiments in the MARS project, and the cylinders location and ratio of cylinder linear speed were studied. The results include:

1. The leading edge enhanced the boundary layer energy and leaded to delay of the flow separation. The middle rotating cylinder had little effect on the flow separation at large angle.
2. The better control effect is obtained only when the speed of cylinder linear is higher than the free stream speed, in which situation, the airfoil deformation induced by cylinder application can be counteracted, as showed in Fig. 18.14.

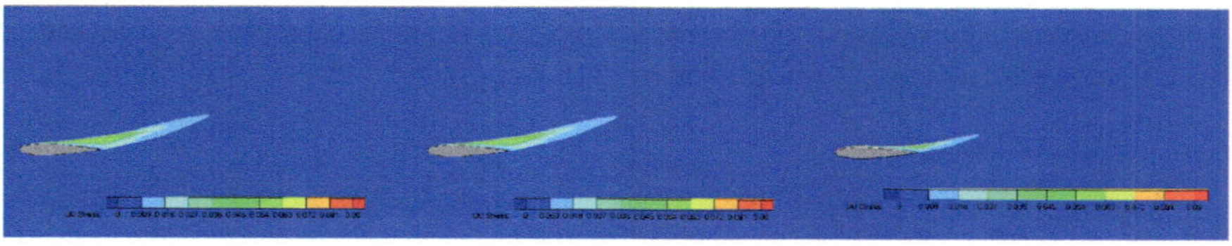

(a) Cylinder speed 0m/s (b) Cylinder speed 20m/s (c) Cylinder speed 60m/s

Fig. 18.14 Attack angle 18°, Reynolds stress of leading cylinder (ARI)

3. The drastic changes of turbulent velocity fluctuation vary from far field to the near wall region. The turbulence/separation character of outer flow decreases, while near wall flow, increases, as shown in Fig. 18.15.

The research on moving surface demonstrates that moving surface flow control technique has a good effect in increasing lift and reducing drag for the airfoil. Besides, one of the attractive characteristics is that under the condition of neglecting moment of inertia by using middle-vacuum cylinder, the active input energy only overcomes the friction force on axletree and aerodynamic drag force of the rotating cylinder, Therefore, the energy consumption is fairly less and even could be ignored. Take the North America ROCWELL YOV-10A airplane for instance, the gross weight is 4500 kg and wing area is about 27 m^2, equipped with two 715 HorsePower engines. According to the estimation, only 2% of the engine power is used on driving cylinder. The experimental data presents that the stall angel could be delayed from 10° to 48°, the maximum lift coefficient can be as high as 2.35. Nevertheless, it is found that the current calculation and experiment still stay in the principle research. Most of them are on the condition of low wind speed and low Reynolds numbers. For the real flight, the Reynolds number is much larger than the experimental Reynolds number. Thus, the influence of Reynolds number should be taken into consideration. The research also indicates that the high rotating speed ration of the moving surface could achieve remarkable effect on lift increase and drag reduction. However, in order to match the high cruise speed of airplane, the cylinder rotating speed must be higher than the

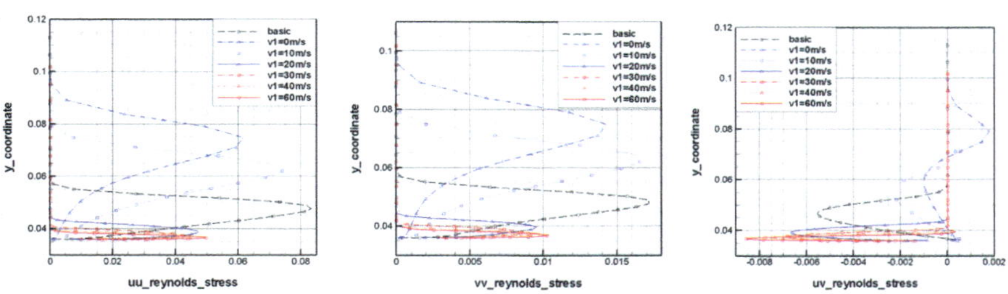

Fig. 18.15 Variation of Reynolds stress along the normal-to-wall direction with ratio increasing at x/c = 0.2 (ARI)

flow air speed, which will bring some new difficult problems of the design, control, abrasion and maintenance of high speed rotating device.

18.3.4 Computation and Experiment

Total pressure tube and micro pressure gauge measurement, hot wire measurement, based on PIV measurement technology were used in MARS project. Measurement technology based on total pressure tube and micro pressure gauge assisted by high precision space mobile test equipment has the advantage of high precision, but can only measure time evenly profile speed type. Measurement technology based on hot wire assisted by high precision space mobile test equipment has the advantage of fast response, unsteady measurement and can obtain comprehensive turbulence field information, and can measure time uniform velocity distribution, turbulence, turbulent kinetic energy and Reynolds stress. Whereas, hot wire system is complicated, inexperienced hot wire probe calibration, debugging and data processing will bring larger error. Time uniform velocity distribution in a certain area can be obtained based on the PIV measurement technology, thus it is suitable for understanding local flow pattern. Furthermore, PIV has its advantage on the recirculation zone of BFS velocity profile measurement. Whereas, PIV system is complicated, tracer particles are difficult to enter focused area such as the boundary layer. Surface reflection affects local velocity measurement, and sampling frequency is limited, so the accuracy for turbulence measurement is not good enough, and needs to be improved in the future.

Theoretically, the calculation accuracy and computational ability are increasing step by step, from RANS to LES, but there's still a long way to improve. From the specific situation of MARS project, each method has its inherent defection. For instance, the results of RANS method depend on calculation experience and turbulence model selected, while the results of LES method depend on calculation experience and sub-grid-stress model selected. At present, Hybrid RANS-LES is an important method worth paying attention, and it is playing a more and more important role in flow mechanism research. For the selected cases, RANS is lacunary in capturing the trailing edge separation, it can get results corresponding to IDDES in the concerned wake regions, based on proper turbulence model. So analysis on airfoil flow and control flow analysis by RANS can still be used as a relatively reliable tool and reference in practical research, however, application experience and combined with appropriate turbulence model are very important.

18.3.5 Optimization Method Study

For aerodynamic shape optimization design, the optimization process is totally dependent on CFD results. The final design results are directly related to the accuracy of CFD method, this may cause the serious conflict problem of optimization design quality and efficiency. Meanwhile, modern aerodynamic shape design problem usually has the characteristics of being sensitive to shape change, having large number of design variables, having multiple peaks, etc., which may easily lead the aerodynamic shape design to local optimization.

For actual aerodynamic optimization, a variety of design requirements and constraints such as high and low speed flight state often need to be considered, usually includes single objective and multi-objective problems, single point and multipoint optimization design problems. For different problems, adapted optimization model should be proposed and adapted optimization algorithm should be chosen. In traditional design optimization model, the design variables, objective function and constraints are deterministic, the optimal solution can be obtained by optimization algorithm. However, this optimal solution is usually hard to meet the need of actual engineering application.

Due to atmospheric turbulence, unstable flight status, and inaccurate manipulation, it is unable to know the exact value of input parameters, so the design state should be viewed as fluctuation near reference point, and it is more practical to consider working state in a range rather than at a single point, because the working state is uncertain. Meanwhile, the optimization design of aerodynamic shape at ideal state is difficult to reflect to the actual application, due to the manufacturing error and pollution on the surface of the geometry, aerodynamic geometric shape is uncertain either. The optimal solution obtained by traditional design optimization is sensitive to "uncertainty" factors, thus tiny fluctuations of input parameters may cause the whole design scheme to be sharp variation or unavailable. Therefore, how to consider the influence of uncertain factors during the design process under the condition of ensuring aerodynamic shape performance not to decrease, has become a very important task.

18.4 Conclusion

Calculation, testing and optimization of major flow control were deeply studied in MARS by both EU and Chinese research groups, based on two platforms of NACA0015 and BFS (backward-facing step). The research method, the parameters and the scheme design formed in the study can be used for reference in the aircraft design department. From the perspective of industrial applications, although good results were obtained by different flow control methods, Reynolds effect and scale effect should be investigated in the future. Moreover, the application, maintenance and reliability problem of the devices and the actuator should be considered for practical use.

References

1. Dong J, Sun J. Proceeding of the MARS M36 meeting from the Chinese delegation
2. Dong J, Sun J. Proceeding of the MARS M42 meeting from the Chinese delegation
3. Modi VJ (2000) On the moving surface boundary-layer control. In: Fluids 2000 conference and exhibit, Denver, CO, 19–22 June 2000
4. Bewley TR (2001) Flow control: new challenges for a new renaissance. Prog Aerosp Sci 37:21–58
5. Van Ness II DK, Corke TC, Morris SC. Turbine tip clearance flow control using plasma actuators. In: AIAA 2006-21
6. Laadhari F, Skandaji L, Morel R. Turbulence reduction in a boundary layer by a local spanwise oscillating surface. Laboratoire de Mécanique des Fluides et d'Acoustique, Ecole Centrale de Lyon-Université Claude Bernard-URA CNRS 263, BP 163, 69131 Ecu@, France
7. Xia X, Krishnan G, Mohseni K. An experimental investigation of synthetic jet in a coflow wake. In: 40th fluid dynamics conference and exhibit, Chicago, Illinois, 28 June–1 July 2010
8. Huu PN. Plasma-assisted high lift systems. In: 27th AIAA applied aerodynamics conference, San Antonio, Texas, 22–25 June 2009

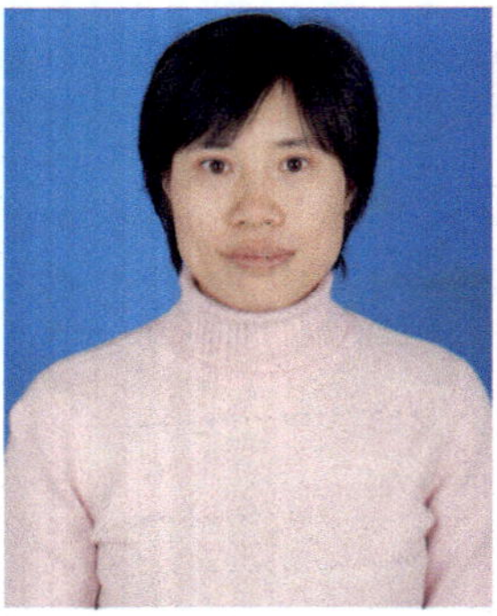

Yi Ju Deng works as a vice head of Aerodynamic Design Division of FAI, mainly on configuration and aerodynamic design.

Rui Fei Xu a senior engineer of First Aircraft Institution of AVIC. His major topic is aerodynamic design and computation.

Xiao Juan Guo a senior engineer of First Aircraft Institution of AVIC. Her major topic is aerodynamic design and computation.

Fa Liang Wang a senior engineer of First Aircraft Institution of AVIC. His major topic is aerodynamic design and computation.

Chapter 19
Hybrid RANS-LES Simulations of Incipient Airfoil Trailing Edge Separation

Spiridon Siouris, Wei Wang and Ning Qin

Abstract Hybrid RANS-LES, such as, Detached Eddy Simulation (DES) based models offer significant advantages over LES or RANS for massively separated flows. However, in cases of mild separation, their performance is not well understood, and for this reason this paper presents a comparative study of URANS, DES, DDES and IDDES simulations on a NACA0015 wing section at 11° incidence, and an airfoil chord based Reynolds number of 1 million. This poses a severe challenge for numerical simulations as the skin friction remains very close to zero after the separation point in the separated zone. The prediction of such flow was found to be very sensitive to the different modelling strategies. This test case with mild trailing edge separation from a smooth surface illustrates significant differences in the numerical simulations among the four approaches. The results showed that the IDDES formulation produced the more reliable results when compared to the experimental data in relation to surface pressure, velocity profiles and Reynolds stresses in the wake. The URANS, DES, and DDES methodologies varied significantly in results with URANS and DDES indicating fully attached flow, while DES predicting a much delayed separation point.

Keywords Incipient separation · Turbulence modelling · Hybrid RANS-LES · DES · NACA0015

19.1 Introduction

Detached Eddy Simulation (DES) [1] is one of the most popular hybrid RANS-LES methods and it was developed to overcome the large computational demands required for Large Eddy Simulation (LES) [2, 3] for practical applications. This is achieved by introducing the Reynolds Averaged Navier-Stokes (RANS) modelling in the flow region close to the wall, and LES in the separated flow region. The DES formulation has been successfully applied to complex turbulent flows such as aircraft

S. Siouris (✉) · W. Wang · N. Qin
The University of Sheffield, Mappin Street, Sheffield S1 3JD, UK
e-mail: s.siouris@sheffield.ac.uk

N. Qin et al. (eds.), *Advances in Effective Flow Separation Control for Aircraft Drag Reduction*, Computational Methods in Applied Sciences 52,
https://doi.org/10.1007/978-3-030-29688-9_19

at high angles of incidence [4], and massively separated flows such as in wings in deep stall [5]. In these cases, the sudden flow separation from the wall does not pose a numerical challenge in relation to the prediction of the separation point, and the benefits of DES can be clearly seen in terms of reduced computational demands compared to LES.

In DES, the only factor that drives the switching between RANS and LES regions is the grid spacing close to the boundary wall, and any grid modifications at this region will alter the switching between RANS and LES in relation to the boundary layer. In some cases, a large cell concentration normal to the wall may engage LES modelling within the boundary layer, but may not be sufficiently fine enough to resolve the shear stresses for LES [6]. Because of this drawback, Delayed DES (DDES) was developed in Spalart et al. [7] that delays the switching from RANS to LES outside of the boundary layer by taking into account the eddy viscosity in conjunction with grid spacing. Cases, such as a wing-body junction at Reynolds number of 7.5 million at 12.5° [8], and a fighter aircraft with chinned fuselage at Reynolds number of 2.68 million at 30° [9], demonstrated that DDES can produce more reliable results as validated by available experimental data in comparison to DES and URANS. A similarity among these cases is that turbulent eddies are induced from a leading edge with a high local curvature. In mild separation cases from a low curvature surface such as the suction side of an airfoil, the DDES formulation is overly dissipative and dampens the unsteadiness of the flow [10]. This is because DDES inherits the problems of DES in the log-layer mismatch between the RANS and LES zones.

The Improved DDES (IDDES) [11] was developed in an attempt to resolve the log-layer mismatch between the RANS and LES fields. The IDDES formulation has been applied to flows such as 2D periodic hills where the flow separates and reattaches [12], in a 3D pebble bed reactor [13], as well as in the original IDDES work for the case of smooth trailing edge separation from a hydrofoil and a backward facing step [11]. The study presented here focuses on the application of the different hybrid RANS-LES methodologies for the case of mild trailing separation in flow over a NACA0015 airfoil at an incidence at 11° and a Reynolds number of 1 million. This is carried out in order to provide a detailed insight into the behaviour of the DES based hybrid RANS-LES formulations in the case of mild separation, while comparing these with experimental data from Siauw [14] and Siauw et al. [15]. It has to be noted that in this case the boundary layer is tripped to turbulent near the leading edge according to the experimental conditions.

19.2 Numerical Method

The results presented in this work have been produced using an in-house Computational Fluid Dynamics (CFD) solver, called Dynamic Grid-Detached Eddy Simulation (DG-DES) [10, 16]. This is a finite volume density-based code that solves the unsteady compressible Navier-Stokes equations using a cell-centred approach on unstructured grids. The Spalart-Allmaras (S-A) turbulence model [17] is used as a basis for the hybrid RANS/LES models implemented in the code including DES [1], DDES [7] and IDDES [11]. The METIS partitioning library [18] is used to partition the computational domain for parallel computation using the Message Passing Interface (MPI). The following sections describe the governing equations, the S-A model, and the hybrid RANS/LES formulations. In this work, the naming of some variables has been changed from the original references in order to maintain consistency between all variants of the DES methodology.

19.2.1 Governing Equations

DG-DES uses a density-based dual time stepping approach based on the work in Weis and Smith [19] to enable the simulation of unsteady flows. For a control volume V and surface boundary A, the system of equations to be solved are of the form:

$$\frac{\partial}{\partial t}\iiint W dV + \Gamma\frac{\partial}{\partial \tau}\iiint Q dV + \iint (F - G) n dA = 0 \quad (19.1)$$

The vector W (physical time term) contains the conservative variables, Q (pseudo time term) that contains the primitive variables, and Γ is the conjugate $\partial W/\partial Q$. F and G are the inviscid and viscous fluxes respectively. For each physical time step t, there are a number of steps in pseudo time τ in which when convergence is achieved this term becomes zero and the solution for the flow field variables depends on the physical time. The pseudo time term is discretized using a third order Runge-Kutta method, while the physical time is discretized using the second order backward Euler scheme. The inviscid and viscous fluxes are discretized using Roe's scheme [20] and central difference respectively.

19.2.2 S-A Model

Hybrid RANS/LES schemes require the use of a turbulence model to provide closure to the RANS equations, and to the unresolved sub grid scales of the LES formulation. The DES based schemes have been originally developed based on the S-A model [17], however, other higher order turbulence models have also been used such as SST-DES [21]. In the S-A model, the dynamic eddy viscosity is defined as:

$$\mu_t = \rho\tilde{v} f_{v1} \quad (19.2)$$

The S-A model develops a single transport equation for the eddy viscosity $\tilde{v}$, defined as:

$$\frac{\partial\tilde{\upsilon}}{\partial t}+u_j\frac{\partial\tilde{\upsilon}}{\partial x_j}=C_{b1}\tilde{S}\tilde{\upsilon}+\frac{1}{\sigma}\left[\frac{\partial}{\partial x_j}\left((\upsilon+\tilde{\upsilon})\frac{\partial\tilde{\upsilon}}{\partial x_i}\right)+C_{b2}\left(\frac{\partial\tilde{\upsilon}}{\partial x_i}\right)^2\right]$$
$$-C_{w1}f_w\left(\frac{\tilde{\upsilon}}{d}\right)^2 \tag{19.3}$$

where

$$\tilde{S}\equiv\sqrt{2\Omega_{ij}\Omega_{ij}}+\frac{\tilde{\upsilon}}{\kappa^2d^2}f_{\upsilon 2},\ \ \Omega\equiv\frac{1}{2}\left(\frac{\partial u_i}{\partial x_j}-\frac{\partial u_j}{\partial x_i}\right),\ \text{and}$$
$$f_{\upsilon 2}\equiv 1-\frac{X}{1+Xf_{\upsilon 1}} \tag{19.4}$$

and

$$f_w(g)\equiv g\left(\frac{1+C_{w3}^6}{g^6+C_{w3}^6}\right)^{1/6},\ \text{where } g\equiv r+C_{w2}\left(r^6-r\right),\ \text{and}$$
$$r\equiv\frac{\tilde{\upsilon}}{\tilde{S}\kappa^2d^2} \tag{19.5}$$

The variable d is the minimum distance to the wall, and σ, C_{b1}, C_{b2}, C_{w1}, C_{w2}, C_{w3} and κ are constants derived from experimental observations, and their values are listed in Spalart and Allmaras [17]. The parameter $f_{\upsilon 1}$, defined in Eq. (19.6), is a damping function and depends on the kinematic eddy viscosity $\tilde{\upsilon}$ and molecular kinematic viscosity $\upsilon=\mu/\rho$, and a parameter $C_{\upsilon 1}$. This function tends to zero for high Reynolds numbers, and zero close to the wall.

$$f_{\upsilon 1}\equiv\frac{X^3}{X^3+C_{\upsilon 1}^3},\quad \text{where } X\equiv\frac{\tilde{\upsilon}}{\upsilon} \tag{19.6}$$

19.2.3 DES Formulation

One of the factors that the destruction term for eddy viscosity of the S-A model depends on is the distance closest to the wall, *d*. In the DES formulation, *d* is replaced by the following term:

$$d_{DES}\equiv\min(d,C_{DES}\Delta) \tag{19.7}$$

where C_{DES} is a constant, and Δ is the local grid spacing which can be defined as the largest spacing in all three directions. For the application of unstructured grids, the local grid spacing is calculated as the maximum distance between the cell's centroid and the neighbouring cell's centroid. When $d_{DES} \leq C_{DES}\Delta$ then RANS modelling is active in that region, and when $d_{DES} > C_{DES}\Delta$ then LES is active and the S-A model behaves like a sub-grid scale (SGS) LES model.

19.2.4 DDES Formulation

In DES, the switching between RANS and LES is grid dependent, and any grid refinement or coarsening will affect the switching between RANS and LES. For example, when the grid spacing close to the wall is refined, then LES mode can get activated within the boundary layer, and this can result in numerical relaminarisation due to modelled stress depletion if the grid is not fine enough for LES to resolve the stresses [6]. In order to overcome this, the Delayed DES was developed to take into account the state of flow (as well as the grid spacing) and to delay the transition from RANS to LES. The modified formulation for the DDES length scale becomes:

$$d_{DDES} \equiv d - f_d \max(0, d - C_{DES}\Delta) \tag{19.8}$$

where

$$f_d \equiv 1 - \tanh\left((8r_d)^3\right), \quad \text{and} \quad r_d \equiv \frac{\upsilon_t + \upsilon}{\sqrt{U_{ij}U_{ij}\kappa^2 d^2}} \tag{19.9}$$

The new variable introduced here is a weighting factor f_d, and with this, the definition of length scale now depends on the eddy viscosity as well as the grid spacing, therefore making it time dependent. This modification to the length scale $\tilde{d}$ has the effect of overriding the DES limiter and maintaining the RANS formulation over a larger area to capture the entire boundary layer.

19.2.5 IDDES Formulation

A significant drawback of the DES and DDES formulations is in the mismatch of log-layers arising from the RANS (inner log-layer) and LES mode (outer log-layer) resulting in under-prediction of skin friction [11]. For this reason, IDDES was developed to address this issue within a single formulation that can automatically treat the different flow features experienced within a flow over a complex geometry. The IDDES formulation is based on DDES and wall modelling LES (WMLES), and also on empirical functions to control the behaviour of the two branches as well as their coupling in order to resolve the log-layer mismatch of DES and DDES. The following section provides a brief outline for the construction of this model, according to the original work in Shur et al. [11].

Firstly, the sub-grid length-scale Δ is redefined so that the wall distance is included in regions close to the wall, in addition to the cell sizes. In the very close vicinity of the wall, the wall parallel steps are taken into account instead of the wall normal in order to avoid drastic decrease of Δ in this region. The definition of Δ is as follows:

$$\Delta \equiv \min\left(\max\left[C_w d_w, C_w \max\left(h_x, h_y, h_z\right), h_{wn}\right], \max\left[h_x, h_y, h_z\right]\right) \quad (19.10)$$

where h_{wn} is the cell size in the wall-normal direction, and C_w is an empirical constant.

The DDES branch for the length scale is based on Eq. (19.8) with the addition of a correction factor Ψ to compensate for the activation of low-Reynolds number terms in the LES mode. The modified DDES length-scale is:

$$d_{DDES} \equiv d - f_d \max(0, d - C_{DDES}\Psi\Delta) \quad (19.11)$$

with f_d and r_d are as defined in Eq. (19.9), and Ψ is given by:

$$\Psi^2 \equiv \min\left[10^2, \frac{1 - \frac{c_{b1}}{0.424 c_{w1}\kappa^2}\left[f_{t2} + (1 - f_{t2}) f_{v2}\right]}{f_{v1}\max\left(10^{-10}, 1 - f_{t2}\right)}\right] \quad (19.12)$$

and its parameters follow the ones in the S-A model [17].

The WMLES branch of IDDES defines the length scale as:

$$d_{WMLES} \equiv f_B(1 + f_e)d + (1 - f_B)C_{DES}\Psi\Delta \quad (19.13)$$

with f_B varying between [0, 1] to provide rapid switching between RANS and LES, and f_e is an empirical function that prevents the excessive reduction of the RANS Reynolds stresses to address the log-layer mismatch [11]. The combined IDDES length scale based on the DDES and WMLES branches becomes:

$$d_{IDDES} \equiv \tilde{f}_d(1 + f_e)d + \left(1 - \tilde{f}_d\right)C_{DES}\Psi\Delta \quad (19.14)$$

where $\tilde{f}_d = \max[(1 - f_{dt}), f_B]$ is a blending function and f_{dt} defined as in Eq. (19.9).

19.3 Computational Set-Up

The simulations carried out in this work follow the experimental setup from Siauw [14] and Siauw et al. [15] in which a NACA0015 airfoil with a chord length of $c = 0.35$ m and span of 2.4 m is placed in a wind tunnel of 2.4 m wide, 2.6 m high, and 6 m long. The angle of incidence is set at 11°, and an inlet velocity of 40 m/s results

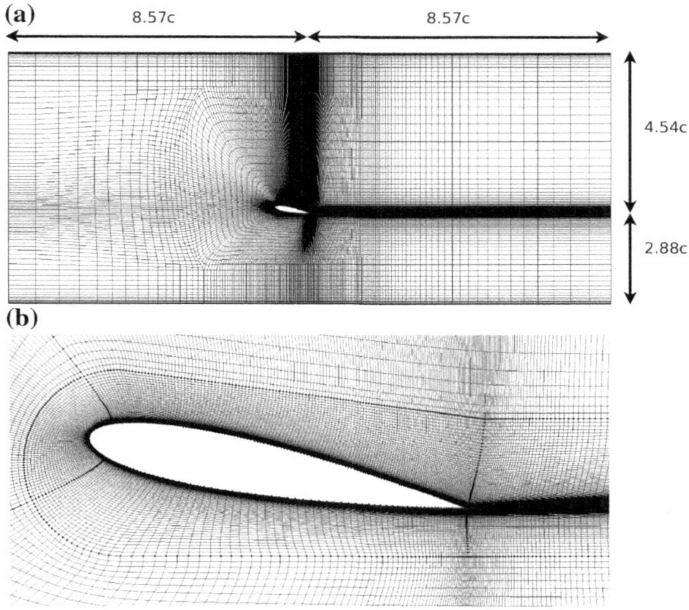

Fig. 19.1 **a** The computational domain and positional information for the NACA0015 airfoil, and **b** a close up view of the mesh around the airfoil

in a Reynolds number of 1 million based on the chord length. Figure 19.1a and b illustrate the position of the airfoil from its TE to the tunnel's height and length, and the mesh region close to the airfoil and its wake respectively.

The fully structured mesh has a size of 4.4 million cells, with a span length of 0.20c, and is discretized with 50 uniformly distributed cells along the spanwise direction. The first wall normal grid resolution y_1+ was between 0 and 0.8 throughout the whole surface area of the airfoil to ensure suitable boundary layer resolution. A velocity inlet was used for the incoming flow, and a pressure outlet for the outflow. The top and bottom walls of the wind tunnel were modelled as no-slip walls, and the periodic boundary condition was applied on the spanwise sides of the computational domain.

19.4 Results and Discussion

URANS and DES based simulations were run with a time step of 2.25×10^{-3} L/U for a total of 20,000 steps. Statistical averages have been calculated based on the last 10,000 steps, and the resulting flow fields have also been spanwise averaged. The coefficient of pressure, C_p for all methodologies can be seen in Fig. 19.2 and it is compared with experimental data from Siauw [14] in which the estimated maximum

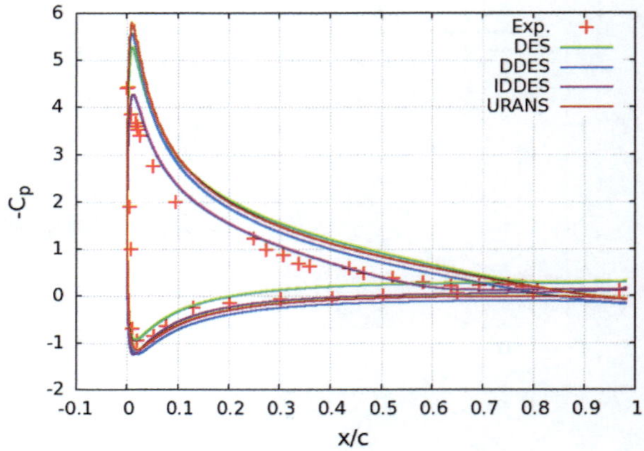

Fig. 19.2 Distribution of coefficient of pressure from experimental data, DES, DDES, IDDES and URANS simulations

uncertainty is $4 \times 10^{-4}\%$. At the pressure side, the URANS, DDES and IDDES results agree very well with the experimental data, whereas the DES results under predict C_p. When considering the airfoil's suction side, such good agreement with experimental data is not observed for the C_p wall distribution due to the large pressure gradients experienced along the airfoil's surface. The URANS, DES and DDES simulations clearly under predict the C_p, but the IDDES formulation shows very good agreement with the experimental data throughout the whole length of the airfoil's chord length. The agreement between experimental and numerical data in terms of C_p wall distribution is related to how well the resolved Reynolds stresses are estimated according to the DES methodologies, and this is dealt with in the next figure.

Experimental data for U/U_{inf} and resolved Reynolds stresses $u'^{*}u'/U^2_{inf}$, $v'^{*}v'/U^2_{inf}$ and $u'^{*}v'/U^2_{inf}$, are plotted against numerical results in Fig. 19.3 at a distance of x = 1.98c from the airfoil's LE. Firstly, the U velocity profiles in Fig. 19.3a show that the velocity distribution at the wake is under predicted in all simulations. The experimental data suggest a minimum velocity of U/U_{inf} 0.86, whereas the DES and DDES simulations predict a minimum U/U_{inf} of 0.77 and 0.75 respectively. The URANS results indicate a slightly less deccelerated wake than the DES and DDES, at a velocity ratio of 0.82. The IDDES simulation produced the least amount of U velocity ratio at 0.84. The URANS, DES and DDES simulations show a narrower extent of the wake, whereas the IDDES model predicts a slightly wider wake region, providing the closest agreement with experimental data.

The maximum normal Reynolds stresses $u'^{*}u'/U^2_{inf}$ in Fig. 19.3b are significantly under predicted for the cases of DES and DDES, but there is good agreement for the case of IDDES. Although the IDDES model has over predicted these normal stresses, it has to be noted that the shape of the stress distribution is in agreement with experimental data. For the case of the normal $v'^{*}v'/U^2_{inf}$ stresses, the IDDES simulation overpredicts the experimental results approximately by a factor of 1.6. The DES simulations significantly under predict these stresses, by a factor of 6, and the DDES does not suggest any normal stresses. It is evident here that the

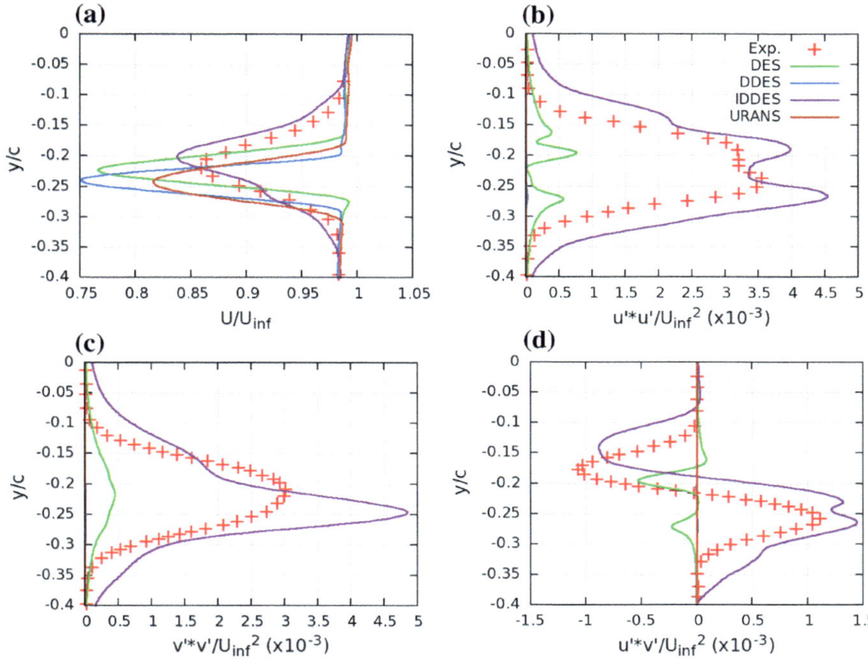

Fig. 19.3 Comparison between experimental data, and DES, DDES and IDDES simulations for **a** U/U_{inf}, **b** $u'^{*}u'/U_{inf}^{2}$, **c** $v'^{*}v'/U_{inf}^{2}$, and **d** $u'^{*}v'/U_{inf}^{2}$

normal $v'^{*}v'/U_{inf}^{2}$ stresses are very challenging to predict. A better comparison with experimental data for the IDDES case is for the shear stresses $u'^{*}v'/U_{inf}^{2}$ where the shape and amount of the stresses are in very good agreement. The DES results are under predicting the shear stresses, and the DDES data, as with the normal $v'^{*}v'/U_{inf}^{2}$ stresses, fail to produce any shear stresses.

From Figs. 19.2 and 19.3, it can be stated that the IDDES simulation produced the closest results to the experimental data, compared to the URANS, DES and DDES simulations. For this reason, it can be assumed that for theses flow conditions over the NACA0015 airfoil, the IDDES methodology is the most reliable amongst the URANS and DES methodologies. The difference in predicted flow conditions can also be seen through the study of wall skin friction, that can directly indicate the location of separation at the airfoil, and this is presented in Fig. 19.4. The URANS and DDES results do not indicate any separation to be present, contradicting with the experimental observations in Siauw [14] and Siauw et al. [15] of separation at approximately 0.7c. The DES results suggest a separation point at 0.75c from the airfoil's LE, and the IDDES simulation predict a much earlier separation point at 0.6c. The experimental observations indicate a case of TE incipient separation, and this is reflected in this figure where the skin friction approaches zero downstream from the separation point for both DES and IDDES cases.

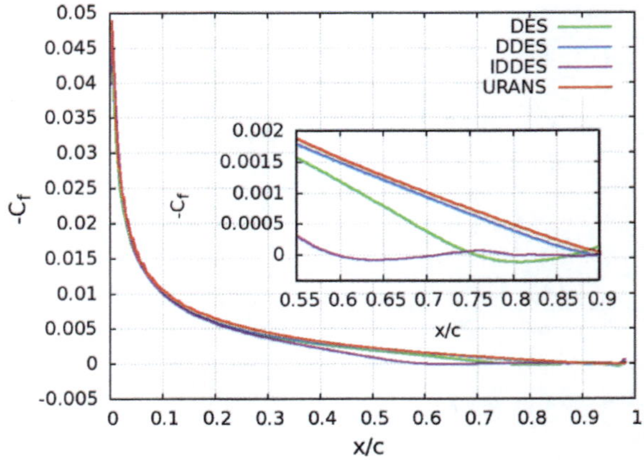

Fig. 19.4 Skin friction predicted by URANS, DES, DDES and IDDES simulations

The flow field over the NACA airfoil can be visualised by plotting the streamwise velocity contours, as shown in Fig. 19.5. The URANS results in Fig. 19.5a indicate that this formulation is unable to capture separated flow under these experimental conditions because the flow remains attached throughout the whole airfoil's chord. The DES model in Fig. 19.5b predicts separation at approximately 0.75c, as suggested by the skin friction plots in Fig. 19.4. After 0.2c from the airfoil's TE, the streamlines from the upper and lower surfaces fully combine and no turbulent regions can be seen further downstream. The DDES simulation (Fig. 19.5c) indicates a fully attached flow with no separation from the airfoil, similar to the URANS results, suggesting

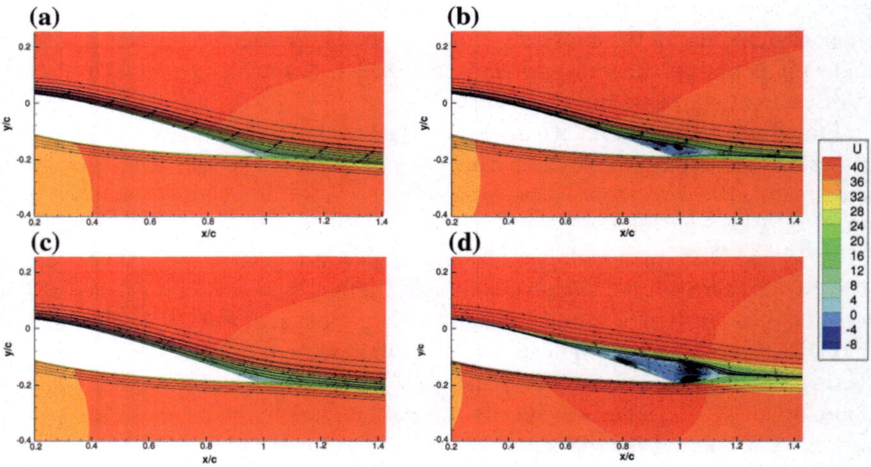

Fig. 19.5 U (m/s) contours and streamlines for **a** URANS, **b** DES, **c** DDES, and **d** IDDES simulations

that the RANS mode of DDES extends further out from the boundary layer region, subsequently resulting in an attached flow due to the associated numerical challenges with adverse pressure gradients and damping of the unsteadiness that can lead to separation.

The IDDES simulation in Fig. 19.5d shows a much different flow field compared to the other models, where the flow is shown to separate at approximately 0.6c from the TE. Even though a long time average has been applied for the statistical average, there are still notable regions of circulation that vary in extend. At the suction side of the airfoil, the primary circulation zone starts from 0.8c and finishes at the wake, at a distance of 1.12c from the TE. The secondary circulation zone originates from the airfoil's pressure side as it meets the flow from the separated region at the suction side. The extent of this circulation zone is approximately 0.15c. Here, the free shear layer initiates from 0.6c along the chord and between this and the point of separation (0.7c), there is a significant deceleration of the streamwise velocity, leading to reversed flow. The streamlines from the two largest circulation zones combine at 1.12c in the wake and beyond.

The switching between the LES and RANS regions can be clearly seen from the contour plots in Fig. 19.6. For the DES case (Fig. 19.6a), it can be seen that the LES region is very close to the wall, resulting in erroneous relaminarisation due to the reduction of turbulent viscosity arising by the DES switching without sufficient resolution of the Reynolds stresses to restore the balance [6]. The effect of this is a delay in separation (at around 0.75c) from the expected 0.7c point along the chord. The DDES limiter seen in Fig. 19.6b shows that the RANS region covers a

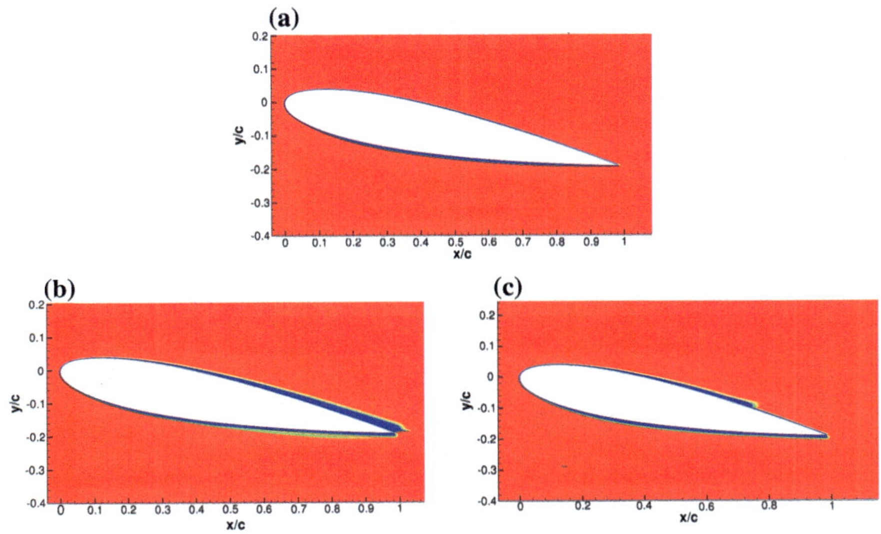

Fig. 19.6 Regions where RANS (in blue, close to the wall) and LES (in red, away from the wall) are active for **a** DES, **b** DDES, and **c** IDDES simulations

significantly wider area around the airfoil, and this delay follows the intentions of the DDES formulation. In most cases, DDES has shown an improvement over DES for suitable grid resolution close to the wall, but in this case the results deviate even more from the experimental observations, suggesting that the flow remains fully attached. Figure 19.6c (IDDES simulations) shows a different behaviour in the RANS/LES switching. As with the DDES model, an increase of boundary layer thickness from the airfoil's LE up to 0.8c is matched with the RANS region. However, at the point of 0.8c, there is an abrupt switch to LES whereas the DDES model carries on extending the RANS region. This switching to LES allows the adverse pressure gradients driving the separated flow to be resolved more appropriately, and subsequently the unsteady turbulent features within the separated region are appropriately simulated compared to the RANS mode. Inside the separated region, the LES mode is closer to the wall than prior to separation, as expected for the resolution of the turbulent structures.

According to the comparisons in Figs. 19.2 and 19.3, the IDDES formulation results in the most reliable simulations that can closely match experimental data better than the DES and DDES methodologies. This difference is due the varying amounts of modelled turbulent viscosity which can subsequently affect the turbulent fluctuations and ultimately how and where the flow separates. This is evident from Fig. 19.8 where it can be seen that high turbulent viscosity for the DES case, at the airfoil wall, is between 0.4c and 0.8c. The DDES formulation produces higher turbulent viscosity (Fig. 19.8b) compared to DES, and this is present from 0.4c up to the airfoil's TE, ultimately resulting in the damping of the turbulent fluctuations. For the case of IDDES in Fig. 19.8c, a significantly lower production of turbulent viscosity can be seen, and this in effect can allow more turbulent content to be present in the flow field (Fig. 19.7).

The extent of the turbulent structures can also be seen from the iso-surfaces of the Q-criteria, as shown in Fig. 19.8. The DES simulation (Fig. 19.8a) predicts a delayed

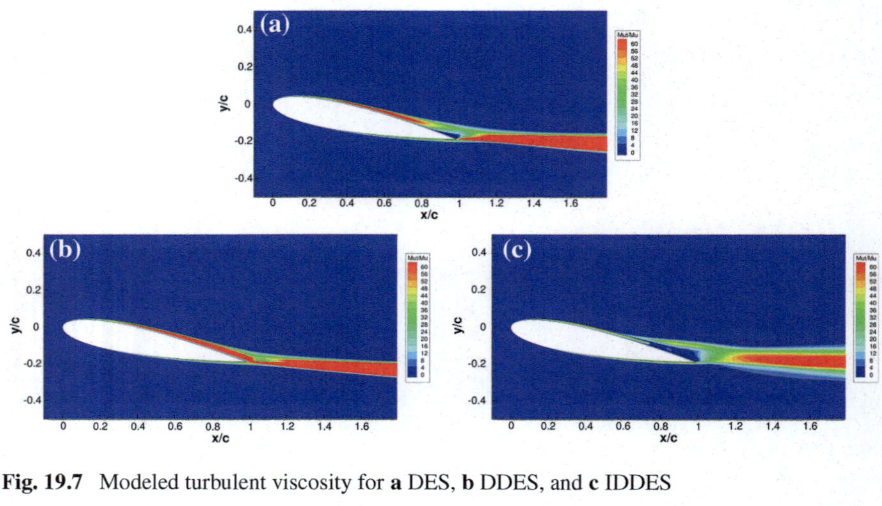

Fig. 19.7 Modeled turbulent viscosity for **a** DES, **b** DDES, and **c** IDDES

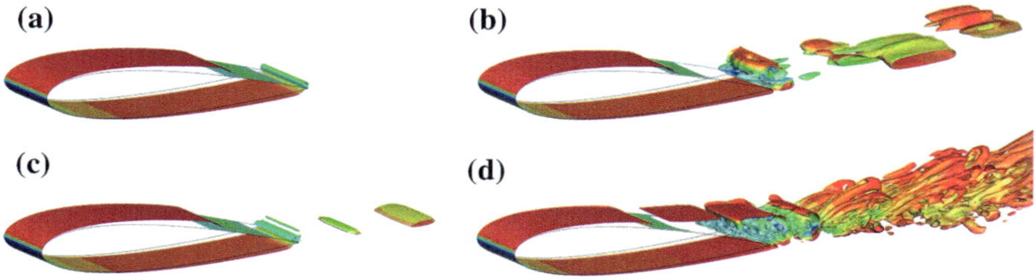

Fig. 19.8 Iso-surface of Q-criteria at 10,000 coloured by streamwise velocity for **a** URANS, **b** DES, **c** DDES, and **d** IDDES simulations

separation compared to experimental results, and therefore there is little turbulence generation and flow structures. The DDES formulation in Fig. 19.8b fails to produce any characteristics that are typical of a turbulent flow due to the issues highlighted previously. The more complex definition of wall distance in the IDDES formulation does provide an improvement from the DDES and DES simulations by accurately predicting the coefficient of pressure and Reynolds stresses. A range of turbulent structures can be seem here with the smaller scales being close to the separated region of the flow. After approximately 0.2c from the airfoil's TE, these small scale structures merge to form larger eddies between 0.2c and 1c, before experiencing vortex shedding.

19.5 Conclusions

The work in this paper dealt with the study of the performance of URANS, DES, DDES, and IDDES methodologies for incipient airfoil trailing edge separation from a NACA0015 airfoil at 11° incidence and Reynolds number of 1 million. It was shown that the DES simulations, although grid dependent, did allow separation to occur from a smooth surface, but at a point further downstream from the experimental observations. The LES region close to the wall, and within the boundary layer, results in modelled stress depletion and therefore delay on flow separation. The DDES formulation seeks to delay the transition from RANS to LES, but it leads to an extension of the RANS region beyond the boundary layer. The challenge of resolving adverse pressure gradients with RANS becomes significant, leading to a fully attached flow with much lower turbulence generation as expected according to experiments due to the larger content of modelled turbulent viscosity. The formulation of IDDES that combines DDES and wall modelling in LES results in flow features that differ considerably from the other approaches, and reproduces the turbulent phenomena as expected according to the experimental observations. This is due to the reduced amounts of turbulent viscosity that can allow for more unsteady turbulent features that can drive separation. Based on this study, the IDDES formulation offers the best potential amongst the current DES methodologies to capture correctly mild separated flow from a smooth surface.

Acknowledgements The authors are grateful for the support of this work from the EC FP7 MARS project. We also like to thank Prof Jean-Paul Bonnet for providing the experimental data for this paper.

References

1. Travin A, Shur M, Strelets M, Spalart P (1999) Detached-eddy simulations past a circular cylinder. Flow Turbul Combust 63:293–313
2. Ghosal S, Moin P (1995) The basic equations for the large eddy simulations of turbulent flows in complex geometry. J Comput Phys 118:24–37
3. Georgiadis NJ, Rizetta DP, Fureby C (2010) Large-eddy simulation: current capabilities, recommended practices, and future research. AIAA J 48(8):1772–1784
4. Forsythe JR, Squires KD, Wurtzler KE, Spalart PR (2004) Detached-eddy simulation of the F-15E at high alpha. J Aircr 41(2):193–200
5. Morton S (2009) Detached-eddy simulations of vortex breakdown over a 70-degree delta wing. J Aircr 46(3):746–755
6. Spalart PR (2009) Detached-eddy simulation. Annu Rev Fluid Mech 41:181–202
7. Spalart PR, Deck S, Shur ML, Squires KD, Strelets MK, Travin A (2006) A new version of detached-eddy simulation, resistant to ambiguous grid densities. Theor Comput Fluid Dyn 20:181–195
8. Fu S, Xiao Z, Chen H, Zhang Y, Huang J (2007) Simulation of wing-body junction flows with hybrid RANS/LES methods. Int J Heat Fluid Flow 28:1379–1390
9. Jeans TL, McDaniel DR, Cummings RM, Mason WH (2009) Aerodynamic analysis of a generic fighter using delayed detached-eddy simulation. J Aircr 46(4):1326–1339
10. Durrani N, Qin N (2011) Behavior of detached-eddy simulations for mild airfoil trailing-edge separation. J Aircr 48(1):193–202
11. Shur ML, Spalart PR, Strelets MK, Travin A (2008) A hybrid RANS-LES approach with delayed-DES and wall-modelled LES capabilities. Int J Heat Fluid Flow 29:1638–1649
12. Mockett C, Fuchs M, Thiele F (2009) Progress in DES for wall-modelled LES of complex internal flows. Comput Fluids 46:44–55
13. Shams A, Roelofs F, Komen EMJ, Baglietto E (2013) Numerical simulations of a pebble bed configuration using hybrid (RANS-LES) methods. Nucl Eng Des 261:201–211
14. Siauw WL (2008) Transient process of separation and attachment over a NACA 0015 airfoil controlled by fluidic vortex generators. Ph.D. thesis, University of Poitiers, France
15. Siauw WL, Bonnet J-P, Tensi J, Cordier L, Noack BR, Cattafesta L (2010) Transient dynamics of the flow around a NACA 0015 airfoil using fluidic vortex generators. Int J Heat Fluid Flow 31:450–459
16. Qin N, Xia H (2008) Detached eddy simulation of a synthetic jet for flow control. Proc Inst Mech Eng Part I J Syst Control Eng 222:373–380
17. Spalart PR, Allmaras SR (1992) A one equation turbulence model for aerodynamic flows. In: 30th aerospace sciences meeting & exhibit
18. Karypis G, Kumar V (1998) A fast and highly quality multilevel scheme for partitioning irregular graphs. SIAM J Sci Comput 20(1):359–392
19. Weis JM, Smith WA (1995) Preconditioning applied to variable and constant density flows. AIAA J 33(11):2050–2057

20. Roe PL (1981) Approximate Riemann solvers, parameter vectors, and difference schemes. J Comput Phys 43:357–372
21. Tucker PG (2006) Turbulence modelling of problem aerospace flows. Int J Numer Meth Fluids 51:261–283

Printed by Printforce, the Netherlands